GEOLOGY AND ECOSYSTEMS

GEOLOGY AND ECOSYSTEMS

International Union of Geological Sciences (IUGS)
Commission on Geological Sciences for
Environmental Planning (COGEOENVIRONMENT)
Commission on Geosciences for Environmental
Management (GEM)

Edited by

Igor S. Zektser
Russian Academy of Sciences

and

Brian Marker
Minerals and Waste Planning Division,UK
John Ridgway, *British Geological Survey*
Liliya Rogachevskaya, *Russian Academy of Sciences*
Genrikh Vartanyan
Russian Academy of Natural Sciences

 Springer

Library of Congress Cataloging-in-Publication Data

Geology and Ecosystems / edited by Igor S. Zektser

ISBN-10: 0-387-29292-6
ISBN-13: 978-0-387-29292-2   Printed on acid-free paper.

e-ISBN-10: 0-387-29293-4
e-ISBN-13: 978-0-387-29293-9

Library of Congress Control Number : 2005933139

Printed in the United States of America.

9 8 7 6 5 4 3 2 1   SPIN 11331193

springeronline.com

Contents

Contributing Authors

Alan Geoffrey Milnes, GEA Consulting, Chemin des Rochettes 58, 2012 Auvernier, Switzerland

Anatoly I. Krivtsov, Central Research Institute of Geological Prospecting for Base and Precious Metals, 117545, Varshavskoye sh., 129 "B" Moscow, Russia

Andrey Ye. Shapovalov, Water Problems Institute, Russian Academy of Science, 3 Gubkina st., 119991 Moscow, Russia

Anna P. Belousova, Water Problems Institute, Russian Academy of Science, 3 Gubkina st., 119991 Moscow, Russia

Arkady I.Sheko, Russian Research Institute for Hydrogeology & Engineering Geology, Zeleny village, 142452, Moscow region, Russia

Audrius Zvikas, Institute of Botany, Zaliuju ezeru 49, LT-08406, Vilnius,Lithuania

Brian Marker, Minerals and Waste Planning Division, Office of the Deputy Prime Minister, Head of Non-Energy Minerals and Environmental Geology Branch, Zone 4/A2, Eland House, Bressenden Place, SW1E 5DU, London, UK

Genrikh Vartanyan, Russian National Research Institute for Hydrogeology and Engineering Geology, Former-Director, Zeleny village, 142452, Moscow region, Russia

He Qingcheng, Department of Groundwater monitoring, China Institute for Geo-Environmental Monitoring, Director, Dahuisi 20, Haidian District, 100081, Beijing, China

Igor S. Zektser, Water Problems Institute, Russian Academy of Science, 3 Gubkina St., 119991 Moscow, Russia

J.H. van den Berg, Faculty of Geosciences, Utrecht University, The Netherlands, PO box 80115, 3508 TC Utrecht

John Ridgway, British Geological Survey, Kingsley Dunham Centre, Keyworth Nottingham, NG 12 5GG, United Kingdom

Jonas Satkunas, Geological Survey of Lithuania, Ministry of Environment, Deputy director, S. Konarskio 35, LT-2600 Vilnius, Lithuania

Jose A. Centeno, U. S. Armed Forces Institute of Pathology, Washington, DC 20306-6000 USA

Julius Taminskas, Institute of Geology and Geography, Sevcenkos 13, LT-03223, Vilnius, Lithuania

Leonid I. Elpiner, Water Problems Institute, Russian Academy of Science, 3 Gubkina St., 119991 Moscow, Russia

Liliya M. Rogachevskaya, Water Problems Institute, Russian Academy of Science, 3 Gubkina St., 119991 Moscow, Russia

Marek Graniczny, Polish Geological Institute, Centre of Geological Spatial Information, Head, 4 Rakowiecka Street 00-975, Warszawa, Poland

Naum G. Oberman, Mining and Geological Company "Mireko", Komi Territorial Centre for State Monitoring of Geological Environments, 75, Gromova St., Syktyvkar, 167983, Russia

Nina M. Novikova, Water Problems Institute, Russian Academy of Science, 3 Gubkina St., 119991 Moscow, Russia

Olga A. Aldyakova, Water Problems Institute, Russian Academy of Science, 3 Gubkina St., 119991 Moscow, Russia

Olle Selinus, Sweden Geological Survey, Head of Geochemical Division, Box 670 SE-751 28 Uppsala Sweden, Villavagen 18

Ricardas Paskauskas, Institute of Botany, Zaliuju ezeru 49, LT-08406, Vilnius, Lithuania

Richard E. Riefner, Jr., 5 Timbre, Rancho Santa Margarita, California 92688, USA

Roald G. Dzhamalov, Head of Lab., Water Problems Institute, Russian Academy of Science, 3 Gubkina St., 119991 Moscow, Russia

Robert B. Finkelman, U. S. Geological Survey, MS 956, Reston, VA, 20192 USA

Roy J. Shlemon, R.J.Shlemon & Associates Inc., Geological & Environmental Consultants, PO Box 3066, 92659-0620 Newport Beach, California, USA

S.Y. Treshkin, Institute of Bioecology, Uzbek Academy of Sciences, Berdacha pr. 41, 742000, Nukus, Uzbekistan

Tatyana Balyuk, Water Problems Institute, Russian Academy of Science, 3 Gubkina St., 119991 Moscow, Russia

Viktor T. Trofimov, Department of Engineering and Ecological Geology, Faculty of geology, Moscow State University, Leninskie gory, 119899, Moscow, Russia

Vladimir M. Shvets, Faculty of Hydrogeology, Moscow State Geology Academy, Mikluho-Maklya, 23, GSP-7, Moscow, Russia

Vladimir S.Krupoderov, Russian Research Institute for Hydrogeology & Engineering Geology, Zeleny Village, 142452, Moscow region, Russia

W. J. Langston, Marine Biological Association, Plymouth, UK

ZhannaV. Kuz'mina, Water Problems Institute, Russian Academy of Science, 3 Gubkina St., 119991 Moscow, Russia

Foreword

This book was prepared for publication by an International Working Group of experts under the auspices of COGEOENVIRONMENT - the Commission of the International Union of Geological Sciences (IUGS) on Geological Sciences for Environmental Planning and IUGS-GEM (Commission on Geosciences for Environmental Management).

The main aim of the Working Group "Geology and Ecosystems" was to develop an interdisciplinary approach to the study of the mechanisms and special features within the "living tissue - inert nature" system under different regional, geological, and anthropogenic conditions. This activity requires international contributions from many scientific fields. It requires efforts from scientists specializing in fields such as: environmental impacts of extractive industries, anthropogenic development and medical problems related to geology and ecosystem interaction, the prediction of the geoenvironmental evolution of ecosystems, etc.

The Working Group determined the goal and objectives of the book, developed the main content, discussed the parts and chapters, and formed the team of authors and the Editorial Board. The Meetings of the Working Group (Vilnius, Lithuania, 2002 and Warsaw-Kielniki, Poland, 2003) were dedicated to discussion and approval of the main content of all chapters in the Book.

Rational and sustainable development of society requires an ever-growing exploitation of natural resources, among which mineral and water resources play a dominant role. During recent decades, in many regions of the world, we have observed a high rate of human invasion into our geological structure, mainly through extraction of minerals, hydrocarbons, groundwater, etc. These impacts, together with day-to-day human activity, cause destruction to the Earth's surface and near-surface environment and upset the balance of sustainable development. In turn, due to increasing anthropogenic impact, the geological components, which serve as the basic substrate and foundation for all ecosystems, is changed from its natural original state and the relationship between the "living tissue and inert nature" (i.e. relationship between the biosphere and the geosphere) is significantly affected.

The book is devoted to a poorly developed, barely broached, problem related to different aspects of the relationship between biological communities of the Earth and objects of inorganic nature or the geological environment. In this respect, the experience of many geological schools of the world gives a quite new understanding of the real scales and the inherent link between independent components of our planet, such as lithosphere, vegetation and animal worlds and, at last, human society itself.

At the present, in spite of the abundance of publications devoted to environmental problems, our knowledge of the mechanisms and magnitude of natural interactions between living tissue (the biosphere) and inert geological materials (the geosphere) is still lacking. Available information on the influence of different geological phenomena on human society is often only schematic and requires more scientific substantiation. For example, it would be enough to mention poorly grounded opinions, currently available in the ecological-geological literature, on the inferred minimal harmful impacts of the injection of highly toxic and radioactive liquid wastes into geological structures (with unclear environmental consequences). Special attention is given to the subtle mechanisms that govern the influence of the atmosphere's composition on the evolution of ecosystems, and on permafrost development and degradation in relation to biological communities etc. These problems are the subject of much controversy in scientific communities and the general public.

This book includes an analysis of the relationship between the different geological, hydrochemical, hydrogeological, and engineering-geological processes and the processes within surface ecosystems. The analysis of specific interactions between the lithosphere and biosphere provides an integrated concept of the role of the geological environment in the evolution of the biosphere. The practical significance of the book is reflected by the analysis of modern engineering activity associated with the mining of minerals, excessive groundwater withdrawal, disposal of industrial and domestic liquid wastes (including radioactive wastes) and their impacts on all components of the environment.

The book includes a scientific approach to the complex monitoring of the environment under different natural and anthropogenic conditions, including the monitoring of permafrost regions. An important part of the book is the analysis of the "water factor" impact on ecosystems and sustainable development. In many regions, depletion of water resources and the impact of water quality on human health must also be taken into account. Influences of intensive groundwater extraction on river flow, vegetation and land subsidence are also considered in the book.

The unique outlook of our book is based upon a multi-aspect discussion of the most significant geoenvironmental factors that book exert an influence upon habitat conditions and stimulate or, on the contrary, hold back, human development. A great number of examples from different countries are given, illustrating a close link between the geological environment and the biota on our vulnerable Planet.

The group of authors includes a wide range of specialists: geologists with different specializations, ecologists, geographers, and specialists of neighboring sciences studying the interaction between different components of the environment. The complete list of authors and co-authors includes 34 authoritative and highly-qualified specialists from 11 countries. All chapters in the book were edited by the members of Editorial Board.

I feel compelled to emphasize the great goodwill and attention extended to each other among the authors and editors who promoted the successful completion of the book.

In this book, complicated and multi-aspect material is presented in a systematic and acceptable form, and can be a useful building block for the formation (or strengthening) of a fundamental nature conservation concept, which would serve as an impetus for sound and rational use of natural resources of the Earth.

The book covers a broad audience of readers. It is also intended as a professional update for environmental scientists, analysts, environmental health care professionals, geologists, ecologists, hydrologists, and other professionals with an interest in the Earth's environments and with environmental protection.

Acknowledgments

I would like to thank the leaders and members of the Commission of the International Union of Geological Sciences (IUGS) on Geological Sciences for Environmental Planning and Commission on Geosciences for Environmental Management for their attention, constant help and support of this work. I thank all authors and associate editors for their participation in the preparation of the book, for their benevolent and friendly attitude, which were constantly present at work and in a large degree promoted the mutual understanding that is key to a good working environment. I consider it my pleasant duty to express my sincere gratitude to Dr Brian Marker and Dr John Ridgway not only for their ceaseless efforts in the scientific editing of the book, but also as tireless English editors.

Igor S. Zektser
Editor-in-Chief

About the Editors

Igor S. Zektser is Head of the Hydrogeology Laboratory of the Water Problems Institute of the Russian Academy of Sciences. He is also a member of the Russian Academy of Natural Sciences, Russian Academy of Ecology and American Institute of Hydrology. Since 2003 he has been a member of the IGSP Scientific Board of UNESCO to whom he contributes his expertise in groundwater studies.

Prof. Zektser was a leader of several UNESCO International projects related to hydrogeological investigation. He is Editor-in-Chief of the International Monograph "Groundwater Resources of the World and their use", published by UNESCO in 2003, and an editor of "World Map of Hydrogeological Condition and Groundwater Flow" (2000). He is an expert contributor to the "Map of Groundwater Resources of the World in scale 1:25 000 000". And he is Chief of the Working Group of GEM "Geology and Ecosystems".

Professor Zektser conducts policy research in the field of groundwater and the environment. His research area includes groundwater flow, groundwater resources, quality and vulnerability to contamination, interaction between surface water, groundwater and seawater, groundwater contribution to water balance and general water resources. He is the author of 12 monographs and more than 220 papers, published in Russia, the USA and by UNESCO.

Professor Zektser has worked at the Water Problems Institute since 1968. He graduated from Moscow State University in 1959 and was awarded a PhD in Hydrogeology in 1965 and a Doctor of Sciences Degree in 1975. In 1982 he received the title "Professor of Hydrogeology".

Prof Zektser is a member of the International Association of Hydrogeologists and President of the Russian National Committee of the International Association of Hydrological Sciences (IAHS).

Brian Marker attended the University of London and was awarded a BSc in Geology in 1968 and a PhD in sedimentology in 1972. Following a research fellowship at the then City of London Polytechnic, he joined the then Department of the Environment in 1975 as an adviser on minerals planning.

He has remained in the same Division through allocations to various Government Departments (currently the Office of the Deputy Prime Minister). During that period he has provided advice, prepared guidance, and managed research on hazards associated with unstable and contaminated land; managed the Minerals, Land Instability and Waste Planning Research Program; headed the Waste Planning Branch; and currently heads the Minerals Planning Policy Branch. He has served on British National Committee of the International Geological Correlation Programme (1988-91); Natural Environment Research Council (NERC) Earth Science and Technology Board (1988-98); and the Board of the British Geological Survey (1990-present). Between 1975 and in 1997 he lectured adult education classes for the University of London and later the Workers' Education Association because of an interest in promoting public awareness of the geosciences.

He is Secretary of IAEG Commission No 1 Engineering Geology Maps (1992-present) and an officer of the IUGS Commission on Geosciences for Environmental Management (GEM) for which he chairs an Urban Geology Working Group (1998-present). He is a Chartered Geologist and a Fellow of both the Geological Society and the Royal Geographical Society. He has edited 4 books and authored over 35 papers on applied geological topics.

John Ridgway graduated from the University of Liverpool (UK) in 1966, specializing in geology and geophysics. He went on to teach structural geology, petrology and stratigraphy at the University of Keele for three years, whilst completing a PhD on the stratigraphy and petrology of Ordovician volcanic rocks in North Wales (UK). He then joined the British Geological Survey and was seconded to Zambia, working on Precambrian metamorphic rocks and managing the petrology, X-ray fluorescence and X-ray diffraction laboratories of the Zambian Geological Survey. This was followed by four years in Solomon Islands, South Pacific, managing a combined geological mapping and geochemical exploration project in the Western Solomons and running the geochemical laboratory. From 1980 to 1994, he ran research projects on a variety of geochemical exploration and environmental geochemistry topics involving soil, stream sediment, and floodplain sediment, rock, vegetation and gas geochemistry. This involved working in the South Pacific, Central and East Africa, the East Indies,

Central America and South America. During this period he set up and worked on stream sediment geochemical exploration projects in Kenya and Zimbabwe. In 1994 he turned his attention to using geochemistry as a tool to help solve problems of Holocene stratigraphy, began working on estuarine geochemistry with an emphasis on recognizing contamination and identifying its sources, and investigated ways in which geochemistry could help solve management issues related to contamination and international agreements. Now an Honorary Research Associate at the British Geological Survey and a member of the Management Board of the International Union of Geological Sciences' Geoindicators Initiative, he is active in the promotion of the geological sciences as a tool for environmental management. John Ridgway has published widely in the international scientific literature on topics ranging from Precambrian metamorphic rocks, through Neogene tectonics to modern sediment geochemistry.

Liliya M. Rogachevskaya, as a hydrogeologist, has a rich background in the study of geoecological problems in rational use of nature's resources, such as drinking water quality and sustainable development in conditions of their contamination. She has assessed the human impact on groundwater vulnerability, in particular, to radionuclide contamination. In addition, she has analyzed the radioecological state of natural waters, predicted possible changes of water quality under the conditions of intensive contamination, and prepared recommendations on safety of water supplies. She received the PhD degree in radioecology from the Water Problem Institute of the Russian Academy of Sciences. She has served as an expert scientific secretary of the UNDP project "Water Quality Evaluation and Prediction in the Areas Affected by the Chernobyl Accident". The scope of her specialization also includes sustainable development of groundwater systems under climate change. Dr Rogachevskaya has co-authored and co-edited over thirty articles on environmental assessment and ecological risk management.

Genrikh S. Vartanyan is a Russian geologist-hydrogeologist who holds a Professorship in Hydrogeology and Radiohydrogeology. His major achievements are: development of the Geo-environmental Ecological Research Principles; development of the Fluid-Physical Model of the Earth; discovery of "The Phenomenon of Globally Spread Short-living Pulsating Changes in the Hydrogeosphere" (known as the Vartanyan-Koulikov Hydrogeological Effect); development of a new approach to contemporary geological processes assessment, based on a deep-set ground water hydrodynamics investigation; formation of the "Regional Hydrodeformatics"; research Hydroinjection deposits; investigations of formation regularities of large provinces of carbon-dioxide and nitrogen-containing thermo mineral water; development of the

Hydrogeodeformation Monitoring System for the purposes of strong geodynamic events prediction (R-STEPS) and practical application of the system to geodynamic active regions.

Professor Vartanyan was for 20 years a director of the Russian National Research Institute for Hydrogeology and Engineering Geology. He is a member of the Russian Academy of Natural Sciences, New York Academy of Sciences, American Institute of Hydrology, International Association of Hydrogeologists, UN Expert Group on Land Use Planning in Asia and the Pacific (ESCAP), and the European Union Expert Group on Ecology.

Professor Vartanyan is an author of over 200 publications in scientific journals, proceedings of International Congresses and Conferences and numerous monographs and maps among which are:

Ecogeology of Russia. (Chief Editor). Publishing house "Geoinformmark", Moscow, 2000

Album of Geoenvironmental Maps of the Russian Federation 1:5 000 000. (Chief Editor), Moscow, 1997

Mining and the Geoenvironment. UNEP - UNESCO, (Chief Editor), Russian and English versions, Nairobi-Paris-Moscow, 1990

Erratum

The Publisher regrets that the above text was omitted during the printing of this title: *Geology and Ecosystems*, edited by Igor S. Zektser, Springer, ISBN 0-387-29292-6.

PART I: GEOLOGICAL ENVIRONMENT AS A BASEMENT FOR THE FORMATION OF ECOSYSTEMS

Chapter 1

CONTEMPORARY CONCEPTIONS OF THE GEOLOGICAL ENVIRONMENT: BASIC FEATURES, STRUCTURE AND SYSTEM OF LINKS

G. VARTANYAN
Russian National Research Institute for Hydrogeology and Engineering Geology, Moscow region, Russia

1. INTRODUCTION

Information on the geological environment, accumulated by different branches of the earth sciences, demonstrates the complexity of the system of cause-and-consequence links formed in the Earth during its geological evolution..

2. GENERAL FEATURES OF THE INTRA-PLANETARY EVOLUTION OF MATTER AND ENERGY

One of the most pronounced factors of the Earth's evolution is the intensive extrusion of matter and energy from the deeper parts of the Planet to its surface, and into outer space. In this connection, the role of the geological "stratum" appears to be very significant, having functional links between the physical state and dynamic development of geological strata.

In order to analyze the general evolutionary features of some of the most environmentally significant natural processes one should, obviously, proceed from the fact that the geological "cover" of the Earth consists of an inexhaustible quantity of mineral agglomerations discretely-continuously distributed in four-dimensional space. Each of these agglomerations constitutes a quasi-autonomous and, in principle, open thermodynamic system which, directly or indirectly, continuously enters into multi-functional matter-energy interactions with all the other constituent systems. Individual systems are aggregated into mezo- and macro-systems. In this case, the development of geological cover is controlled by self-regulating mechanisms.

These processes of evolution of geological matter are characterized by:
- potential energy-mass transfer and multiple field effects as universal mechanism of functioning;
- production of matter and energy fluxes and fields as a result of physico-chemical transformations occurring within an active system;
- transformation of qualitative and quantitative compositions of components (as a result of physico-chemical reactions), with an appropriate change per unit of the mass state and matter's initial volume where these reactions are proceeding;
- a tendency of the active system to disperse the matter- energy fluxes into outer space, and to equalize the matter- energy potential with outer space;
- completion of active physico-chemical processes inside the system turning it from an active into a passive form, thus subjecting it to matter-energy influences from other systems.

It therefore follows that the development of the geological cover proceeds with the participation of both internal and external factors that determine the property and state of a particular agglomeration, as well as the nature and intensity of its evolution.

Internal factors include the chemical (mineral) composition of individual substances participating in the building up of a given agglomeration.

External factors are pressure and temperature at a given level within the geological cover. These predetermine the commencement and rates of processes of matter transformation.

An important component of the Earth's geological cover fluids of various origins. Under given thermodynamic and physico-chemical conditions these migrate into potential fields (fields of temperatures, pressures, stresses), filling existing cavities in geological bodies and, thus, globally modify the geophysical and geochemical properties of the geological cover. Essentially, this fluid state provides the mechanism for mass transport of elements and compounds in the subsurface. The fluids are both the medium and reagents for various geochemical and physical processes in geological strata.

3. CONTEMPORARY GEOLOGICAL PROCESSES

The evolution of the Earth's geological cover is implemented by various physical and physico-chemical processes of varying scales, intensities, rates of development, and duration.

These can be subdivided into endogenic processes based on factors and mechanisms of abyssal origin, and exogenic ones generated by external factors (cosmic, climatic and even anthropogenic).

Endo- and exogenic processes have close links and mutual causality between each other.

Endogenic geological processes include numerous forms of evolution of geological matter under the action of abyssal factors, such as temporal variations of the geothermal field; extrusion of matter from the deep-seated parts of the Earth; and gravitational subsidence of extended areas (for example, above the zones of intensively developing regional metamorphism), etc.

The most significant to the environmental evolution are neotectonic motions (including eustatic fluctuations of the earth's surface), earthquakes, and volcanic activity.

Major natural geological processes generally have slow (in the physical scale of time) rates of development. At the same time, each of the developing processes usually becomes a starting point or starting step in the cause-and-consequence chain of transformations which, finally, lead to considerable changes in the geological situation at local, regional or even global scale.

Neotectonic processes are caused by many factors associated with changes in the mantle and lithosphere, and manifested through horizontal and vertical motions of the crustal blocks.

Vertical movements of the crust are, sometimes, very intensive. Thus, during the Tertiary-Quaternary period the Tian-Shan fold-mountain system was uplifted by 11 km with differential rising and falling of different blocks.

Modern horizontal tectonic movements have been insufficiently investigated, however displacements of lithospheric blocks for many tens and sometimes hundreds of kilometers are known (e.g. Parkfield Area of the San Andreas Fault, Pamirs and others). Investigations of tectonic structures, mass overthrusts, drag folds and features of lithospheric plate movements show that horizontal tectonic motions have had wide activity in the past and present. One of the pieces of evidence confirming the global scale of horizontal tectonic motions during the geological evolution of the Earth is the "mirror-image" similarity of the facing contours of present-day continents now set apart by many thousands of kilometers.

Modern geodynamic processes can be characterized with the aid of remote photogrammetry (i.e. satellite multi-zonal survey, altitudinal aero-location, etc.), as well as different adaptations of ground and air geophysical methods, and systematic monitoring of the regional HydroGeoDeformation (HGD) field of large areas (Vartanyan, 1990, 1997). The HGD-field of the Earth reflects the evolution of stress-strain state over huge areas, gives a picture of subglobal geodynamic evolution of the lithosphere and, through this, demonstrates either concordant or negative behaviour of very large geological structures located in different parts of the globe. This can be demonstrated by the following example.

Trans-regional reconstructions of the HGD-field, for the Caucasus and Middle Asia, made it possible to identify strong extension of the earth's crust during 1989-1990 in the mountainous areas of Pamir, Tian Shan, Turan and, the Arabian Plates. This may be related to the final preparatory phase of the catastrophic earthquake in Iran on 21 June 1990. The event involved areas totaling not less than 3 300 000 square kilometers.

By analyzing this process for the Kopet-Dag and Caucasian fold systems, that was established since the end of 1989, the Kopet-Dag fold system had undergone considerable extension, which reached its peak during January-March 1990. Following the end of March, these short-lived extension structures began to diminish and were gradually replaced by compression structures. The principal compressions were registered after mid October 1990.

During the same period in the Caucasus, short-lived compression structures predominated. After the end of April 1990 these were intensively replaced by extension structures with a peak reached at the beginning of September 1990. After that, extension structures rapidly diminished to be replaced once again by compression..

The picture is one of rather complicated processes in which periods of maximum extension within the Kopet-Dag area coincided with periods of greatest compression in the Caucasus and vice versa .

Similar deformation curves have been obtained for the Caucasus and California; for Japan (Suruga polygon) and Kazakhstan, and elsewhere.

The comparison of the data on HGD-field evolution, obtained from various regions of the globe, demonstrates that regional endogeodynamic processes usually involve tremendous masses of rocks far more extensive than mega-structures. Moreover, it also shows that deformation processes in the lithosphere have a pronounced cyclic character with periods of predominant extension (relaxation) alternating with short intervals of compression lasting a few months.

Analysis of specific features in the geodynamic evolution of large regions provides evidence that, in real time, particular, sometimes closely located, geological blocks are developing geodynamically in contrasting ways. Geodynamic processes have an oscillatory-pulsating quasi-wavy nature whether within adjacent or closely located geological structures, or more remote geological bodies. Processes of contrasting type and intensity can take place at the same time.

Such tectonic movements cause the development (in real time) of stresses and deformations in large lithospheric blocks and, therefore, numerous earthquakes of various magnitudes. Evolution of the stress-strain state in rock massifs causes an accumulation of excessive stresses up to the strength limit of the geological strata. Further growth of stresses-strains leads to the strength limit being exceeded and, as a result, to the formation of cracks and faults.

This has the following consequences:

- formation of a new geologico-structural situation in a region;
- a change in the permeability of fluid-containing strata;
- the appearance of a geodynamic filtration regime with formation of short-living drainage windows (so-called "shifting sieves");
- active drainage of aquifers and aquifer systems through the formed rupture dislocations;
- general changes to the hydrodynamic and hydrogeochemical regime in a region subjected to the above-mentioned geodynamic loads, etc.

Much slower, but having more significant global consequences, are processes of sedimentation that alter rock composition and relief (i.e. water intake areas, river beds, valleys) over considerable areas with resulting changes to geochemical, geophysical, hydrodynamic, geothermal and other conditions. Similarly, more "rapidly" proceeding eustatic fluctuations lead to changes in sedimentation, and alteration in the drainage conditions of underground-surface waters. The latter are accompanied by a fall in water level, changes in sediment transport and sedimentation. Depending on the nature of these processes, aquifers may have sharp water level decreases, that can sometimes lead to partial or complete reduction in drainage and, hence, aridization of the territory.

Another group of the natural geological processes that reflect internal changes to the planet's geological cover consists of catastrophic phenomena such as volcanic eruptions, earthquakes, landslides, etc. The processes can lead directly to significant changes from the preceding geological (hydrogeological, geochemical and other) situation, leading, sometimes, to transformation   of   landscapes,   hydrologic-hydrogeological, hydrogeochemical and other conditions in large areas and regions.

Exogenic geological processes include a wide spectrum of irreversible discrete changes in the composition, structure and state of geological strata (or in particular their less stable elements) due to natural energy- and mass-exchange in the contact zone of the litho-, atmo- and hydrospheres, as well as due to man's activity.

Exogenic geological processes vary in mechanisms, nature and intensity at the earth's surface and can sometimes give rise to threats to life and property.

Thus, according to the data of R.Shuster, direct and indirect damage from landslides and mudflows in the USA are estimated as high as over 1 billion dollars per year. Comparable (by size) damage has been reported for such countries as Russia, Japan, Italy. In some cases, the catastrophic activizations of landslides and mudflows have resulted in thousands of human deaths (China, 1920; Peru, 1970; Colombia, 1980, etc.).

In general, the degree of a hazard of exogenic geological processes is a function of the genetic features, mechanisms, intensity, and multi-year and annual frequency.

Most hazardous among the exogenic geological processes are landslides because of the variety of the sources activating them and the high levels of damage. The metric volume of some landslides can reach billions of cubic meters. Thus, for example, the strong earthquake of 1911 in the Pamirs caused a landslide with a volume of 2.2 km^3 . This buried Usoi village and blocked the Murgab river. As a result of this event, Sarezskoye Lake formed with a volume of 18 km^3, at a height of over 3.2 km, presenting a major threat for the entire river valley downstream.

Exogenic geological processes are especially dangerous in mountainous areas due to: a high energy of the relief, seismicity, a cumulative influence of many factors causing instability of such widely spread geological elements as loosely clastic deposits, large mudflow-catchment areas, presence of constant free-water sources (glaciers), etc.

Development or activization of a single process may not present a direct threat, but indirect consequences of it can cause catastrophic results.

The state of geological massifs is significantly impacted by human activities, the influence and action of which upon the geological objects can be similar to or even exceed natural processes.

The sharp growth of mineral resources consumption in the 19th and 20th centuries increased the impact of man on the lithospheric cover and, as a result, affected the state of many natural systems, causing their degradation and considerable deterioration of human habitat conditions in some areas or, sometimes, in large regions. These tendencies led to revision of conceptual positions in relation to traditional use of natural resources, including a more profound understanding of the place and role of the lithospheric strata in the natural systems combined under the term "environment".

As a result the new terms of "geoecology" and "ecogeology" appeared and underlined the importance of geological formations as a basis for existence of all biological life on the Earth. And, finally, introduction of the term "geological environment" has completed the "ecologization" of geological sciences and given impetus to new directions of research and practical activities under the terms "ecological geology" or "ecogeology".

By "geological environment", one should understand the upper lithosphere including genetically varied rocks and the related (closely interacting with them) intra-rock fluids, gases and microorganisms which appear to be subjected to a direct or indirect anthropogenic impact and, in turn, affect the habitat conditions of biological communities.

The most important results of material-energy interactions between all of the components in the geological environment are the generation and functioning of various geophysical and geochemical fields which, in turn, exert a strong influence upon the rates and direction of evolutionary processes in the lithosphere.

Various large-scale subsurface processes that are affected by anthropogenic factors are commonly difficult to observe but these give rise to many ecologo-geological events which, although having a comparatively small impact on the whole geological environment, do demonstrate its ecological "fragility".

Through intensive mass-energy fluxes, the geological environment demonstrates two basic properties – adaptability and self-development – and constantly, in real time, changes its physico-chemical and thermodynamic status.

The term "adaptability" means the ability of geological systems to adjust the properties and state of rocks (fluids, fields) to new external conditions formed under the influence of external factors. These properties serve to some extent as a protective mechanism, allowing the system to adjust to new conditions without radical changes from the initial state.

At the same time, the ability of the geological environment to adapt to new conditions without notable damage to its current state are not unlimited. With appearance of critical or "beyond-limit" (for a given geological set of factors – material and mineralogical composition, structure, stress-strain state, etc.) conditions, the protective mechanism becomes unable to adjust the disturbed mass and energy balance to the new situation and, hence, restore the pre-existing situation.

Typical scaled adverse consequences of human impacts on the geological environment are sharp changes in moisture- and redox potential (because of some engineering measures) within the zones of areal groundwater drainage.

In particular, wide-ranging work on the so-called deep reclamation of lands has led to long term lowering of groundwater level and, in some countries, to considerable changes in landscapes and hydrogeological conditions, soil degradation, and so on.

Similarly, construction of tunnels, linear structures, motor-car and railway lines cause, in some cases, either drainage of aquifers or, swamping of large areas (for example, while filling railway beds).

Another serious consequence resulting from the disturbed equilibrium of the geological environment is the undercutting of unstable slopes during construction of roads and other linear structures inducing long-lasting landslide-forming processes. Once being set in action, the mechanism of landslide-formation proceeds over a long period of geological time. Sliding of tremendous rock masses is accompanied by other adverse phenomena: drainage of aquifers; lowering of groundwater levels, and sometimes depletion of groundwater reserves; changes to groundwater composition and temperature regime; etc. Anthropogenically induced changes in groundwater composition and state can persist in a layer or aquifer for a time period comparable to several human generations.

A special emphasis should be placed on the impact of hard and liquid (mainly, hydrocarbon) mineral mining on the geological environment, resulting in numerous consequences, some of which are impossible to remedy, such as spoil heaps, quarries, and subsidence above mined areas and oil fields. There are known cases of considerable (to 10 m) subsidence of the earth's surface due to withdrawal of large amounts of water (e.g. Tokyo, Mexico, California and others).

Episodes are also observed where anthropogenic or exogenic processes turned out to be a direct cause (a "triggering" mechanism) of intensification of catastrophic endogenic geological processes in endogenically stressed regions. Thus, a strong earthquake on the Turan Plate (Middle Asia) was caused by intensive development of gas deposits (Gazli), related decrease in the stratum pressure, and stress redistribution in rocks. Seismologists suppose also that a similar mechanism caused the anthropogenically-induced seismic catastrophe in Neftegorsk in 1995 where 60 % of population were killed.

Chapter 2

THE GEOLOGICAL ENVIRONMENT AND ECOSYSTEMS

G. VARTANYAN
Russian National Research Institute for Hydrogeology and Engineering Geology, Moscow region, Russia

Judging from the absolute age of the most ancient rocks, the geological cover of the Earth has been developing for about 4 billion years. The processes of origination and evolution of all forms of life on the planet are inseparably connected with two important media – geological and aquatic (oceanic) ones. First, about 1 billion years ago, life in the form of protozoa originated in an aquatic medium and then, with sophistication of the forms and accretion of populations, moved to the land. The geological matrix served always as a substrate for it.

The history of the development of the lithogenic basement and life presents itself as a chain of complicated processes of a tight, multiform and continuous interaction and interconnection between biotic and abiotic matter. After being died, bio-organisms have been and are forming now caustobioliths, organogenic limestones and some of siliceous rocks. The bio-organisms through the physiological activity of some of their biological communities participated in formation of many types of other rocks, accumulations of iron, manganese, copper, and many useful mineral compounds. Simultaneously with this, the biological forms were undertaking an ever growing complexity and improvement; they were becoming more diversified, and more adjustable to the conditions of staying in various thermodynamic zones of the lithosphere, and in all geographical points of the Planet. It was established, for example, that some forms of bacteria and microorganisms are able to exist in the subsurface (in particular, in the conditions of oil- and gas accumulation) at temperature up to +80°C; germinative spores and plant pollens can survive at negative temperatures in permafrost for millions of years.

According to V.I.Vernadsky's evaluations, the total biotic-matter mass of the Earth remained constant during the entire Phanerozoic period, with an ever growing role for biotic organisms in the lithosphere evolution at the same time. It may be supposed that this process was accompanied by the formation of an abundance of various ecosystems, making its own "contribution" to evolutionary processes in the Earth's geological cover.

So study an ecosystem one should take a functionally unified and mutually dependent totality of vegetative and live organisms, the reproduction and destruction of which is regulated by the internal equilibrium laws of biological communities participating in the building of the this totality. It follows from this that the ecosystem is a basic functional unit of the planetary cycle of life. In essence, the concept "ecosystem" is close (though wider in interpretation) to another concept, " biocenosis", used in the biological literature. In this book we will also mean by the concept "biogeocenosis" a community of functionally interconnected vegetative and animal organisms and their habitats (landscapes, top soils, surface and ground waters, rocks, geophysical and geochemical fields), homogenous by topographical, soil, hydrologic, micro-climatic and other conditions.

The biotic organisms (from microscopic to large forms) through interacting with the basis of their existence (rock minerals, groundwater, gases etc), perform a big geochemical "job" on transformation, accumulation and dispersion of the earth crust's matter, leading to a change in the initial rock properties.

In turn, the evolutionary processes of the geological environment cause changes in the conditions of ecosystems' functioning, helping, thus, their development-adjustment to new P-T and Ph-Eh conditions or, in case of sharp changes in the life activity's conditions, destructively affecting the steady forms of the biological communities' existence. A direct sequence of large changes in the geoenvironmental state may be a deterioration or aggravation (up to the complete decay) in the functioning conditions of the earlier formed ecosystems.

It follows from the above-stated that the litho-mineral matrix of rocks, being at the same time a basis for habitats' formation, "nutrient" material for many biological forms, and, very often, a carrier of "field atmosphere" where biotic organisms are functioning, serves to be the most important factor which determines a direction and intensity of developing life activity's processes.

An impact of geological processes upon the ecosystems is especially acute while catastrophic scenarios of the geoenvironment develop. First of all, most notable is this impact on man and the ecosystems' elements related to him.

In particular, in the long chain of destructive geological processes, the most dangerous (by rapidity, areas of affectedness and scales of human and material losses) role belongs to earthquakes and tsunami accompanied actually by instant and terribly devastative acts.

According to the data of the Scientific Board of Japan (1989), only in the 20th century were the highest share of catastrophic events of the above-mentioned type of natural disasters and amounted to 50.9 % of their total number.

The presently available statistical information does not give sufficiently unambiguous estimates of the impact of seismic catastrophes upon ecosystems and human society. However, it follows from comparing the estimates reported (World Conference on Mitigation of Natural Disasters Hazard, 1994; Despande, 1987; Zschau, 1996) that the human victims of all the earthquakes that happened in the world in the 20th century amounted to about 3.5 mln.persons.

It is enough to mention only some of the strongest earthquakes of the 20th century, which turned large areas into fully ruined zones and killed thousands of people: Tan-Shan, 1923 (200,000 persons): Ashkhabad, 1948 (100,000); Spitak, 1988 (25,000); Rudbar-Tarom, 1990 (45,000). The scales of material losses and total economic damage from some particular seismic events are depressive. Thus, the economic losses from the earthquakes in Managua (23.12.1972) were twice higher than the value of the national gross output in this country; the earthquakes in Honduras (04.02.1976) and Salvador (10.10.1986) have brought economic damages in these countries equal to 32 % and 27 % of their national gross outputs, respectively (Energy and mineral potential of the Central American-Caribean region, 1989).

Environmentally severe consequences accompany various catastrophic phenomena of volcanic origin.

Not speaking about such historically known catastrophes of the above type as Pompeya and Herculanum, one should mention relatively recent events (August,1986) on the crater Lake Nyos located within the so-called "Cameroun volcanic zone" when as the result of a sudden outburst of huge amounts of carbon dioxide, 1746 persons and many wild and domestic animals died from asphyxia. A similar catastrophe but with a lower level of victimization (37 persons) happened near another crater, lake Monoun in 1984 in Cameroun (G.W.Kling, M.Kusakabe et al., 1990; Sigurdsson et al, 1987).

It is quite obvious that such extreme manifestations of the geoenvironmental evolutions are followed also by a rather considerable change in the ecosystem's balance in a region subjected to a natural cataclysm.

While not so notable no such dramatic events, nevertheless, mass impacts on the ecosystems are exerted by the phenomena that follow from the normal activity of geophysical and geochemical processes in the geological environment.

In particular, with geodynamic processes' activization, seismic events are usually preceded by anomalous changes or an increase in the concentrations of different chemical compounds penetrating from the subsurface in liquid or gaseous state. Besides, prior to seismic events, one observes considerable variations of different geophysical fields exerting an adverse influence both on some individual persons and whole groups of population. The joint impact of all these factors on society can cause psycho-biochemical (biophysical) pre-conditions for formation of a collective reaction of aggression, apathy, fear, etc., leading, in turn, to destructive social acts.

The most dangerous types of such natural "poisoning" of man are those among the naturally existing impacts which, though not manifesting themselves openly, at the same time through a long micro-influence on the human organism, lead to impalpable, but socially hazardous modifications in the behaviour of large groups of people.

That is, in some cases, phases of the social agitation (or non-standard behaviour) can turn out to be derivative elements of the geobiophysical anomalies being periodically generated in nature (Vartanyan, 1997).

Taking into consideration the above-listed statements, Caesar Voute, after having fulfilled the multi-year historical-archeological and ethnographic investigations in the area of the religious construction Borobadur (Central Java) spoke his opinion on a possible influence of geophysical (atmo-geochemical) anomalies upon society during the preparatory periods of the catastrophic earthquakes of 925-1006 (Society and Culture of Southeast Asia. Continuities and Changes, 2000).

It was established by researchers that during the period prior to a series of destructive earthquakes on Java Island, sudden and logically unexplainable (by all other reasons) exodus of the political and religious circles of the Hindu Buddhist Java State happened from the highly habitable and comfortable part of Central Java. The circumstances, established by the investigations, gave grounds to suppose that the motivation of such mass departure could be initiated and governed by tiny mechanisms of psycho-geophysical-geochemical links between "abiotic" (geological) and biotic matter, that have a new impulse due to activization of geodynamic life in the Earth's interior.

It follows from the above-stated that the investigations of fine mechanisms and intensity of an impact upon society from different deformation-geodynamic, physico-chemical anomalies and the related geophysical field effects is becoming an important problem.

Changes in the geological environment exert the most influence upon the biological component during an intensive anthropogenic impact on rocks and fluids contained in them. In particular, hydrotechnical construction, connected, as a rule, with a change in the groundwater natural regime, causes either draining and actually complete dewatering of earlier fertile lands with degradation of soils and vegetation or, vice versa, a rise of water level to the earth surface and swamping of large territories.

Thus, for example, the works on regional draining of the swamped territories in Belorussia (including withdrawal of large groundwater amounts), where has existed a stable ecosystem, containing sphagnum moss and the related vegetative and biological communities, led to destruction of a considerable number of swamps and their biological components, degradation of vegetation, top soil. As a result, it brought and harm to natatorial birds' breeding sites, beavers' colonies, and forced other larger animals to migrate out of those areas. That is, the above mentioned engineering actions affecting the geological objects have destroyed the natural chain in the ecosystem formed there.

In turn, the disturbance in the thermal and moisture balance of the atmosphere over these rock massifs has led to appearance of an earlier unknown (in these regions) phenomenon of "black storms" – a factor deteriorating considerably the human habitat conditions.

The cascade of water storage reservoirs constructed on the Don and Dnepr rivers and the system of navigable canals connecting them have sharply changed the basis of groundwater draining within a large territory and led to a water level rise to the earth surface. Due to this, basement spaces in the buildings, outdoor washrooms and other sanitary engineering constructions turned out to be flooded actually over the entire southern part of Russia and in Ukraine (namely, in Rostov and Dnepropetrovsk Regions). In many cases the basement spaces had high concentrations of dangerous gases, including radon.

It is important to note that the above-mentioned adverse effect has manifested itself 20-25 years later after completion of the water storage reservoirs construction. This period was needed for the system of held-up aquifers to get adjusted to the anthropogenically formed draining conditions and therewith induced rise of the groundwater level everywhere in the region. In the practice of large-scaled constructive works accompanied by creation of mine working-outs, there are known cases of a disturbance in the natural regime of groundwater aquifers supplying large water intakes or municipal and resort establishments.

Thus, in Tbilisi – the Capital of Georgia – which is famous for its therms (in Georgian language, "tbili" means warm; "isi" – a lake), the thermal hydrogen-sulphide groundwaters, since long served as a hydro-mineral basis first for public baths in the city and since the 1930s – for resort-curative purposes), were drained for construction of the subway.

The consequent hydrogeological works, carried out to restore the curative thermal water resources in the city, have solved the problem only partially: the yields and especially qualitative characteristics of the groundwater (i.e. temperature, H_2S content) became lower than the earlier ones. At the same time, some of the subway stations have, in spite of intensive ventilation, a persistent smell of hydrogen sulphide, which points to serious, though not yet reaching toxic levels, H_2S concentrations in the air. It is quite obvious that one should speak in the given case about the uncomfortable conditions for passengers and, especially, for the subway employees who have to be in the hydrogen sulphide-polluted atmosphere for a long time.

The above-listed examples are far from covering all the possible variants of an interaction between biotic and abiotic matter (i.e. ecosystems and the geological environment) and are discussed in this chapter to demonstrate a rather wide variety of forms and scales of reaction of live matter to an action of the surrounding medium: from mass psycho-physical consequences in a scale of the socium to discomfortable habitat conditions in concrete geometric spaces; from destruction of the earlier-formed ecosystems to their modifications and so on.

It should be outlined as a result of the above short discussion of the problem that the world practice of the recent 40-50 years enabled us to collect a great amount of information on the high dependence of biological communities on the processes occurring in the Earth's interior. At the same time, this information needs a thorough analysis in order to assess the role of particular geological factors and a degree of their influence upon the life cycles by obtaining reliable qualitative characteristics and the related subsequent ecologo-geological conclusions.

Due to the above-stated, the necessity appears to create a special apparatus of scientific research using both the methods and technology of modern geological science, physics, chemistry, and the techniques of solving the problems of the medical-biological cycle.

It is obvious that in this case it will be required to form teams consisting of specialists of different disciplines, but working on the same scientific problem "Regularities of the interaction between ecosystems and the geological environment – mechanisms of modifications and evolution, intensities, rates".

Chapter 3

INFLUENCE OF MODERN GEOLOGICAL PROCESSES ON EVOLUTION OF ECOSYSTEMS

G.VARTANYAN
Russian National Research Institute for Hydrogeology and Engineering Geology, Moscow region, Russia

The evolution of the geological environment proceeds through numerous physico-chemical and geodynamic processes and these directly influence the functioning of associated ecosystems. Modern geological processes and phenomena thus exert direct or indirect effects on biota and the overall environmental situation over large areas.

Modern processes are defined here as all known processes, related to evolution of the Earth's geological cover, occurring at present or during recorded history. Thus, for example, the eruption of Mount Vesuvius that led to the destruction of Herculanium and Pompeii is modern on the scales of geological time or of the total duration of the development of human society.

In this book we discuss two large groups of natural geological processes that considerably affect different forms of biological life:

- endogenic geological processes (EnGP);
- exogenic geological processes (ExGP).

These two groups of processes interact closely but are also partly mutually independent. Abyssal factors that serve as a trigger for the endogenic processes can, at the same time, act directly on exogenic geological processes. Thus, in seismically active regions, geodynamic motions induced by earthquakes can activate processes of landslide formation, as has been shown in investigations by the Geological Surveys of USA, Russia, Japan, China, etc.

In turn, external (exogenic) factors can influence the velocity and character of abyssal processes. In particular, there are many cases when human-induced or exogenic processes directly caused the intensification of endogenic processes. For example, when an increase in seismic shaking is observed, the general groundwater level may rise relative to the pre-existing level. This often happens as a result of activated exogenic processes such as groundwater level changes due to sliding of large rock masses so as to dam river valleys, or engineering works.

Endogenic processes involve multiple forms of the geological changes produced by actions of abyssal factors such as heat field fluctuation, migration of products of metamorphism products and of volcanic material, and ground surface subsidence as a result of the formation of defects in rock masses at depth, and so on.

From the environmental point of view, of the greatest interest are various forms of the neotectonic movements as well as earthquakes and volcanic activity.

Neotectonic movements can cover both very extensive areas and groups of geological blocks. It is noteworthy that if at a given time some blocks undergo compression or uplift, the adjacent blocks can be, at the same time, in a state of extension or sinking.

Differential movements of lithospheric strata have both geological and environmental consequences especially in areas between geodynamically active blocks which may become geomechanically weakened and likely to develop a high surface water content (lakes, swamps, rivers) with correspondingly diverse fauna and flora communities.

It is obvious that abrupt changes of the geoenvironmental state in such regions lead to a structural change of the area such as increase or decrease of permeability (depending on the sign of geodynamic stresses), and to a change in the distribution and behaviour of surface water bodies leading to consequent faunal migration to (or from) that region. For example, channels of large rivers can be found distant from the present channel location.

Vertical motions of the earth's crust are observed everywhere, but are most intensive and provide the greatest contrast in mountainous areas. In some cases, high-accuracy levelling data has detected parts of some fold-mountain systems that are being uplifted with a velocity of 10 to 15 mm per year.

Although at present the environmental consequences of such displacements may not be apparent, it can be supposed that during geologically long time periods directional movements of large blocks of the earth's crust should have led to considerable changes in the habitat conditions.

Modern tectonic motions also give rise to stresses and deformations in geological formations. If the long-term strength limit of the rocks is exceeded, new dislocations form and the release of accumulated elastic energy is accompanied by an earthquake.

Each large seismic event in a densely populated area ends with a catastrophe.

Even earthquakes of moderate intensity in areas where seismic-proofing measures have not been incorporated into buildings and potentially hazardous constructions (e.g. chemical industries or power-producing plants) can be catastrophic.

Consequences of destructive earthquakes can last for tens of years. Expenses for dealing with these can amount to a considerable part of the national budget (see Chapter 2).

Although the highest seismicity is typically within geodynamically active regions of the Planet (the so-called Pacific Seismic Belt, Mediterranean Belt and others), strong earthquakes occur also within platform-type formations and shields (e.g. Scandinavian, Canadian, Anabarian, etc.). Though these earthquakes are infrequent, nevertheless they bring a large shock to people.

There is medical and biological evidence for the influence of geophysical processes that occur prior to earthquakes and associated phenomena upon people and animals. Rather significant in this respect is, in particular, the reactions of different (cold-blooded and warm-blooded) animals to abnormal changes or increased concentrations of different chemical compounds (in the liquid or gaseous state), or variations of physical fields prior to a seismic energy release.

It can be supposed that the influences of such natural factors upon humans are similar to those on other organisms. However, the developmental level and social structure of human society, in combination with the political and economic situation within a region, also determine the characteristics of collective responses to natural biophysical variations. Thus, if a similar situation develops against the background of activization or slowdown of geobiophysical processes, those processes exert a constant but continuous influence upon the psychosomatic state of individuals, and, together, can reduce or raise the level of reaction of an entire society.

Proceeding from this, it becomes possible in some cases to consider phases of social alarm as derivative elements of periodic narural geobiophysical anomalies. Such phases can form a certain succession of natural and social phenomena that culminate in a limiting event – a geodynamic catastrophe that ends one phase and initiates the subsequent chain of development of the biotic and inert matter in a concrete area (Vartanyan, 1997; Vartanyan, 1993; Vartanyan, 1994).

Medical specialists have noticed a direct dependence of the frequency of cardio-vascular and nervous diseases, as well as clinical characteristics of diseases, on the strength and frequency of seismic fluctuations.

Negative emotions caused by an earthquake and its aftershocks affect the endocrine system, leading to deep reactions. One observed functional affect on the nervous and cardio-vascular system, has been called by doctors "the disease of the earthquake".

The geophysical phenomenon under consideration and its medico-biological and psychosomatic consequences represent one of the most acute ecogeological problems, the solution of which requires involvement of joint efforts of geological and medical specialists.

One of the most vivid and powerful manifestations of modern geological life – volcanic processes – are accompanied by eruption of burning hot lava, ashes, volcanic gases, steam, geyser outbursts, and thermal springs.

Most modern volcanism is located in the Pacific and Mediterranean geosynclinal belts, as well as within the insular arcs and middle-oceanic ridges.

Environmental consequences of eruptions of terrestrial volcanoes are various and often tragic. Lava fluxes with a temperature of 1000-1200°C burn all living organisms in their path and volcanic ash and bombs destroy constructions and buildings. During eruptions, ash can be thrown into high layers of the atmosphere and troposphere (45-50 km) and distributed over many thousands of kilometers. In some cases, volcanic eruptions are followed by strong earthquakes with a partial disturbance to the volcanic structure, as with the Mount Saint Helens eruption when the blast wave circuited the globe several times.

Relatively short volcanic eruptions are followed by long periods (sometimes for a few hundreds of years) of postvolcanic, relatively "calm life" during which potentially hazardous areas are settled by people. In such cases, labile ecosystems, in spite of attractiveness of the area under development (fertile soils, microclimatic conditions, etc.), are in a dangerous zone, exposed to a high risk of a sudden natural disaster.

Exogenic geological processes (EGP) exert a significant influence on the habitat conditions of living organisms and, above all, people.

Within the term exogenic are included a complex of processes of physical and physico-chemical changes in the composition, structure, properties and state of geoenvironmental objects. These occur due to an intensive energy- and mass-exchange in the zone where three media are adjoined (litho-, atmo- and hydrospheres). In a number of cases, human activity can activate these natural processes and sometimes even strengthen them, giving rise to catastrophic forms on scales that impact the ecosystems.

Exogenic geological processes, varying in mechanism of development, type and intensity at the earth's surface, sometimes create situations that are incompatible with the minimal requirements of surrounding habitats.

The EGP mechanism and genetic features determine basic parameters of a hazard, including such parameters as size, area and velocity of manifestations, and distance of influence, etc.

Thus, for example, landslides and mudflows – one of the most hazardous groups of the exogenic geological processes – are able to move huge rock masses comprising hundreds of thousands to hundreds of millions cubic metres (Sheko & Krupoderov, 1994).

The volume of some landslides can reach billions of cubic meters. For instance, the Usoisky landslide, 2.2 km^3 in volume, occurred in 1911 in the Pamirs and buried the settlement of the same name with all the people living in it.

Intensities of impact upon the geological environment and ecosystems of exogenic geological processes can be subdivided into the following sequence (in decreasing order of influence): landslides, mudflows, avalanches, karst, abrasion, channel erosion, subsidences, ravine erosion, underflooding, swamping, etc.

The greatest hazard to habitats and the continued functioning of ecosystems are the exogenic geological processes that occur in mountainous areas.

Chapter 4

GEOLOGICAL AND GEOCHEMICAL INFLUENCES ON ESTUARINE ECOSYSTEMS

W. LANGSTON[1], & J. RIDGWAY[2]
[1]*Marine Biological Association, Plymouth, UK,* [2] *British Geological Survey, Nottingham, UK*

1. INTRODUCTION

The physical structure of an estuary is governed by geological circumstance and shaped by a combination of river flows, tidal characteristics, current speeds and wave action. An over-riding constraint on estuarine biota is the nature of the variable salinity regime, since the capacity for ionic and osmotic regulation varies greatly between species and sets the limits for their distribution. Of equal importance, if the organism is to settle and survive, are the properties of deposits. Superimposed on these primary drivers are numerous other factors that influence estuarine biota, either directly or indirectly. These include light attenuation and oxygenation patterns (natural characteristics), together with an assortment of anthropogenic impacts. The current chapter focuses on the ways in which geological and geochemical features (substrate properties) impinge on estuarine ecosystems, including modifications made as a result of contaminant bioavailability and toxicity. We also consider ways in which biological activity can mobility in estuaries through processes such as bioturbation and biodeposition.

In simple terms an estuary is a semi-enclosed coastal body of water, with free connection to the sea, in which salt water is diluted by terrestrial fresh water flow. Most estuaries are the product of the inundation of river valleys during Holocene sea-level rise following the end of the last major glaciation.

At first sight estuaries would appear to present biota with severe tests for survival due to their dynamic tidal nature and accompanying changes in salinity. Nevertheless, estuaries are among the most productive of aquatic habitats. The biotas, which make up present day estuarine assemblages, have their origins in three different systems: the sea, fresh water and the land. Species of marine origin tend to dominate, however within each of these categories there are sub-components that reflect varying degrees of success in penetrating estuaries. As a result of this abundance and diversity, estuaries are of major importance to fisheries, acting as nurseries to many forms of aquatic life. Intertidal areas in particular are also significant feeding grounds that attract a variety of bird life, often being globally important sites for migratory species.

Positioned between marine and terrestrial environments, estuaries are zones of sediment transfer between fluvial and marine systems and often form sinks for sediment from both their hinterland and adjacent coastal zone. Such sediment may vary greatly in grain size, mineralogy and chemistry, and there is also the possibility that parts of the estuary will be dominated by bedrock. These factors, together with salinity, tidal and turbidity regimes, all may play a part in determining the type of organism and hence community, capable of colonizing different parts of the estuary. In addition, estuaries have been the focal point for a variety of human activities, becoming sites of major port, industrial, urban and recreational development. Agriculture is often a common feature of the coastal lowlands along their shores. Estuaries may thus be affected by dissolved and particulate contaminants from recreational, farming, manufacturing and extractive industries, both on land and offshore, together with domestic inputs from sewage. Estuarine biota, therefore, have to adapt to a unique combination of natural and anthropogenic forcing features.

This examination of geological and geochemical influences is by no means exhaustive and is largely centered on a few well-studied examples from temperate estuaries and ecosystems, although mention will be made of some tropical and sub-tropical estuaries. The emphasis is placed on the benthic component of estuarine ecosystems since it is here where geological and geochemical influences are likely to be greatest.

2.   GEOLOGICAL INFLUENCES

The most obvious and direct geological influence on estuarine ecosystems is the nature and extent of various types of substrate. This pattern is largely pre-determined by the unique geological history of the area. Clearly, a rock substrate provides a firm base for the attachment of epibenthos, which contrasts significantly with the infaunal assemblages typically associated with mud and sand substrates. This diversity in form and function will, in turn, affect the biota that depends on bottom-living organisms for food or protection. A useful account of the relationships of estuarine organisms to varying substrates is given in Perkins (1974).

By virtue of their genesis through the drowning of pre-existing river valleys, estuaries inherit features from past landscapes, which in turn have been strongly influenced by the interplay between geology and climate. Thus, estuaries may reflect the inundation of relatively narrow, steep-sided valleys, as in typical arias, or of wide floodplains, leading to the classic, funnel-shaped development (for a classification of estuaries see Dyer, 1997). Some of the complexities of estuary formation under different geological and climatic conditions are described by Lampe (1996) from the shallow, polymict, tideless estuaries (boddens) of the Baltic coast of Germany, Woodruffe (1996) from macrotidal estuaries in northern Australia, and Healy *et al.* (1996) from a wide diversity of estuary types in New Zealand. Estuarine environments range from sub-tidal, through intertidal, landward to the limits of inundation by storms, and include beaches, dunes, eroding margins, stream and tidal channels, rocky platforms, saltmarshes, mangroves, sea grass meadows, algal beds and tidal mud or sand flats (Roman and Nordstrom, 1996). Substrates may be dominantly rocky, sandy or muddy, with any combination of the three possible.

The nature of the substrate at any particular site is governed by the complex interaction between tidal forces, freshwater flow and sediment supply, the latter being particularly dependent on the geology of both the hinterland and offshore areas. Shoreline exposure to waves is also of great importance in determining the type of environment that will be established and maintained at a specific location. For estuaries, fine-grained material is terrestrially derived, whilst coarser material is of marine origin, transported landward by tidally driven, bottom currents (Ridgway and Shimmield, 2002). However, in some systems fine-grained muddy material comes from both terrestrial and marine sources, as in the case of the Scarcies Estuary of Sierra Leone, West Africa (Anthony, 1996) and the Humber Estuary in the UK (Dyer et al., 2001). The distribution of mud and sand in an estuary will clearly be influenced by such factors.

Relationships between sediment type and supply and the hydrodynamics of the estuary affect the degree of turbidity and this in turn has a strong influence on the ecosystem, governing the amount of light penetrating through the water column to the bed. In highly turbid estuaries photosynthetic activity of phytoplankton can only take place in the surface (photic) layers and as a consequence productivity may be reduced to a minimum. In these conditions salt marsh-plants will dominate overall carbon production of the estuary.

Where tidal currents are particularly strong, sediment may be unable to accumulate and the bed of the estuary will be of rock or other well-indurated material, such as glacial till. The nature of the geology, from hard igneous and metamorphic rocks, through sandstones to softer limestones, chalks and very soft clays affects the surface texture on which organisms can live (e.g. very smooth, highly pitted) and also the larger physical features of the environment, from even rock platforms to highly fissured and folded terrain. Rocky substrates provide stable surfaces for the attachment of sessile and sedentary organisms, whereas the shifting, unstable environments of estuarine sediments are host to only infauna as permanent residents (Perkins, 1974). In turn, the variable properties of these soft bottom estuarine deposits (in particular grain size) may be of considerable importance in shaping estuarine communities.

The Joint Nature Conservation Committee of the UK (JNCC) has conducted a survey of Britain's marine habitats - the Marine Nature Conservation Review (MNCR) - and some examples of benthic habitats from this review (Hiscock, 1996) serve to illustrate the importance of the geological influence on ecosystems. Four major habitats (with numerous sub-divisions), all of which can occur in estuaries, are recognised in the MNCR: littoral rock; sublittoral rock; littoral sediment; and sublittoral sediment.

Littoral rock environments provide a range of structures (rock platforms, cliffs, overhangs, caves, pools, boulder fields etc.) that encourage species diversity. Zonation of species is common, due to tidal immersion and emersion and biological interactions such as competition for space, predation and grazing. Stable rocky substrates support a wider range of species than unstable hard substrates of boulders, cobbles and pebbles. However, stable bedrock-type communities can develop on shingle and cobbles in areas of estuaries and sea lochs sheltered from wave action. Rock type is important: rich algal communities can develop if the rock retains water during tidal emersion and soft rocks allow animals to bore into them to provide security from predators; on harder rocks, crevices allow distinctive faunas to develop. In low or variable salinity zones of estuaries, rocky substrates are characterized by a low number of species that also occur in full salinity conditions. Sublittoral rock can support a richer diversity of species than littoral rock because it is always submerged, placing less stress on organisms, but the turbid conditions of many estuaries may limit biodiversity.

In littoral sediment environments, particle size, the mixture of sediment grades and the stability of the sediment are important controlling factors in determining the types and numbers of species present. For example, fine sediments support more diatoms and bacteria than coarse sediments, because of the larger surface area for attachment and growth. In contrast, ciliate protozoa are generally unable to live in sediments of < 0.1 mm diameter. Moreover, those protozoa species that inhabit sands with a mean particle size between 0.4 and 1.0 mm are generally rather square in form, whereas those living in sand of 0.1 to 0.4 mm tend to be long and slender in form (Perkins, 1974). Particle size can also affect burrowing and feeding habits and is of considerable importance to filter-feeding organisms. Many suspension-feeders have an optimum range of particle sizes and abundance (in suspension) that can be retained by gills and other feeding structures. The predominance of a particle population within the favoured range, at any given site, will tend to promote colonisation and growth. Depending on the organism, preferred particle sizes vary over a spectrum from <1µm to 100µm – equivalent to the material carried in suspension in most estuaries. Surface chemistry of particles also has an important role in the selection of appropriate material for digestion. By comparison, deposit-feeders may be less selective in the choice of particles.

Capilliary lift and water retention are also a function of grain size and are generally greatest in fine sediments as opposed to coarse sands. This is of obvious importance to inter-tidal organisms, where water retention during low tide may be critical for survival. The phenomenon accounts, partly, for the zonation of species seen in estuaries, though, perhaps with the exception of saltmarsh plants, this is generally less striking than the more familiar zonation patterns seen on rocky coastlines. On the upper parts of a mud flat where there is a risk of drying out, some species have developed restistance to dessication, in the case of certain diatoms through the ability to secrete mucilagenous envelopes. Sediment porosity and permeability are related features of importance in littoral estuarine deposits, since they govern resistance to the burrowing activities of biota. Sediments with a high water content generally offer less resistance to penetration by organisms, however if they become too fluid the permanent burrow structures preferred by some species cannot be maintained and settlement is impossible. Colonisation by saltmarsh plant species can also be determined by the firmness of the substrate: *Spartina* (cord-grass) may take root in a variety of sediment types though *Salicornia* (samphire) is unable to gain a hold in very fine muds.

Sublittoral sediment environments provide extensive sedimentary habitats and associated communities. The most species-rich sediments are those that are stable over time and have a heterogeneous mixture of coarse and fine sediment grades. Sediment composition varies according to the strength of wave action, tidal streams and sediment supply. Strong tidal streams generally lead to a coarse sediment bed, but in the Severn Estuary

(UK) the suspended sediment load is so high that silt settles out and the bed may be muddy over large areas, despite the presence of very strong tidal currents (see below and Hiscock, 1996). Deposition in estuaries is sometimes interspersed with episodes of erosion (associated with high flow, extreme tidal conditions or variable river channels): clearly, these periodic alternations in depositional patterns will tend to wash out infauna and prevent settlement of established communities, especially those with life-spans greater then a few months.

An example of the distribution of biotopes in the Swale estuary in southeast England is shown in Figure 4-1. Other examples, from Germany and New Zealand, respectively, are given in Lampe (1996) and Healy *et al.* (1996).

2.1   Geological Influences: Case Study – The Severn Estuary (UK)

One of the best examples illustrating the importance of natural physical and geological forcing agents on estuarine communities is that of the Severn Estuary (Langston *et al.*, 2003a). The exceptional tidal range (in excess of 14.5 metres) and classic funnel shape (Figure 4-2) make the Severn Estuary unique in Britain and rare worldwide. Large tidal-currents are a dominating feature, reducing vertical stratification (compared with the rias typical of south west England) and providing a mechanism for transport of particles up to sand-size (moving as suspended solids or mobile bed-load). There is a continual exchange of material from areas of erosion to areas of deposition, through the turbidity maximum - which occupies the whole of the Estuary east of Bridgwater Bay. The associated variable frictional stresses result in variations in bed-types despite the ever-present turbidity of the water. In turn, the relatively sharp divisions between muddy, sandy and rocky areas dominate the distribution of benthic organisms. Composition will obviously differ, fundamentally, between hard- and soft-bottom communities, whilst among the latter, sediment grain-size is a dominant characteristic determining the abundance and type of estuarine infauna (Warwick *et al.*, 1989; Moore *et al.*, 1998).

A high proportion of the estuary is subtidal, hosting some rare communities. The intertidal flats and rock platforms support a range of

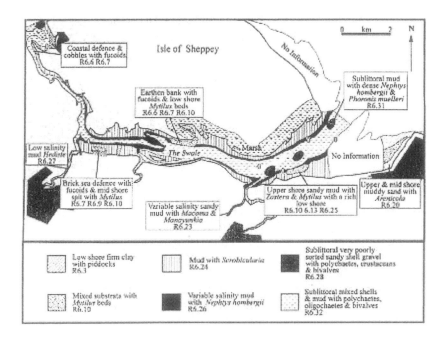

Figure 4-1. The distribution of biotopes in the Swale, Kent, Southeast England (reproduced from, Hiscock, 1996, with permission JNCC).

invertebrate species and the upper Severn Estuary includes an extensive area of mudflats and sandflats bordered by large fringes of saltmarsh. This variety of habitats has, in fact, led to its proposed designation as a Special Area of Conservation (SAC) under the EU Habitats directive. Despite this label, however, the Severn as a whole supports a relatively impoverished fauna and flora, characterized by low biodiversity when compared to other sites (Warwick *et al.*, 1989; Moore *et al.*, 1998).

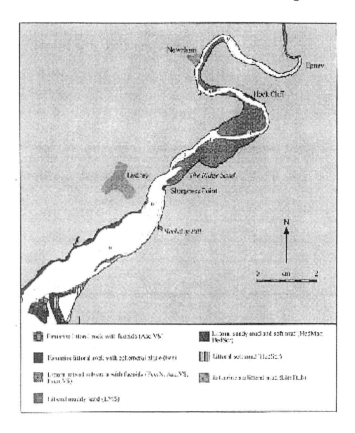

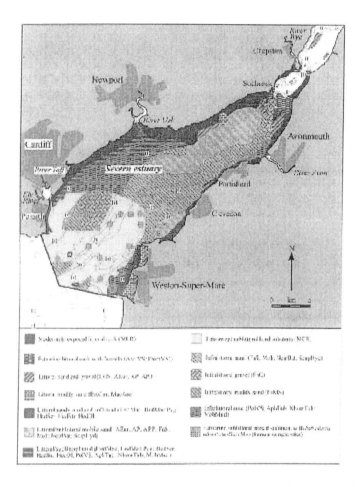

Figure 4-2. The distribution of biotopes in the Severn Estuary, Southwest England. Upper Severn (top), Lower Severn (bottom) (reproduced from Moore *et al.,* 1998, with permission JNCC).
Legend to the top figure: Estuarine littoral rock with fucoids, Estuarine littoral rock with ephemeral algae, Littoral mixed substrata with fucoids, Littoral muddy sand, Littoral sandy mud and soft mud, Littoral soft mud, Estuarine sublittoral mud.
Legend to the bottom figure: Moderately exposed littoral rock, Estuarine littoral rock with fucoids, Littoral sand and gravel, Littoral muddy sand, Littoral sandy mud and soft mud, Littoral/sublittoral mobile sand, Littoral/sublittoral mud, Tide-swept sublittoral hard substrata, Infralittoral sand, Infralittoral gravel, Infralittoral muddy sand, Infralittoral mud, Estuarine sublittoral mixed sediment.

High turbidity is largely responsible for this characteristic, impacting on biota in a number of ways. Generally, the high-suspended solids loading limits light penetration and hence algal productivity. Thus, a striking feature of much of the estuary is the absence of a subtidal zone of macroalgae, due to the effect of the high turbidity, which reduces available photosynthetic light, coupled with the scouring effect of the silt, which interferes with the settlement of algal spores. Similar effects may impact on eelgrass (*Zostera*) beds in mid-estuary. Here, additional deposition of sediment during the construction of the nearby second Severn crossing has coincided with a reduction in area and density of this important and sensitive biotope. Invertebrate populations associated with the *Zostera* bed may also have been affected by extreme episodes of erosion and deposition of sediment (allied to coffer dam construction).

Whilst limiting vegetation, high turbidity provides abundant particulate surface area for microbial processes. As a result, organic carbon may be enriched (and biological oxidation demand elevated) in fluid muds that can disperse at spring tides to produce dissolved oxygen sags in the upper estuary. In those areas frequently covered by turbid layers, colonisation is likely to be sparse. Much of the sub-tidal Severn mud is impoverished and even some sandy areas may be depleted because of the mobility of silts at spring tides. Extreme conditions (sediment instability, turbidity and scouring) also limit the range and abundance of 'expected' estuarine organisms to be found inter-tidally (see typical food-web associated with estuarine mud-flats in Figure 4-3, re-drawn from Green, 1968). Consequently, productive areas are restricted to the more stable, marginal regions. Throughout much of the Severn Estuary, the virtual absence of suspension-feeding bivalves including *Cerastoderma edule* and *Mya arenaria* (and other suspension-feeding invertebrates), can be attributed to the very high levels of turbidity. The deposit-feeding bivalve *Scrobicularia plana* is also uncommon, though probably because of sediment instability. For similar reasons the colonisation of rooted saltmarsh plants is limited to areas which are least affected by strong tidal flows.

Sedimentary conditions not only affect ability to colonize, but can also influence the longevity of those estuarine organisms that do settle. In comparison with other estuaries, populations of several invertebrate species in the Severn are dominated by small individuals, suggesting a shorter lifespan (Warwick *et al.,* 1989).

Suspended sediment concentrations in the Severn vary throughout the diurnal and spring/neap tidal cycles, and modify the distribution of benthos accordingly. Sediments in the upper reaches tend to reflect deposition of fine sediment during neaps and erosion of fines (together with deposition of coarser sediment) during springs. This pattern is in turn exacerbated by seasonal influences such as conditions of river flow and storm surges, leading to both short-and long-term cyclical influences on biota.

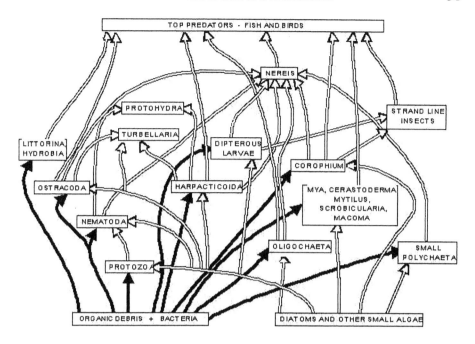

Figure 4-4. Generalised example of a food web from the mudflats of a UK estuary (adapted from Green, 1968)

The fate of the large quantity of sediment that is resuspended and recirculated on each tide will also play a major role in the mobility, bioavailability and impact of associated contaminants. Because of the energetic hydrodynamic regime in the Severn, and resultant high turbidity, there is considerable mixing and redistribution of fines, resulting in unusually homogenous distributions of contaminant loadings. Since physical conditions dominate the composition of biota in the Severn, and communities are already modified as a result, it may be more difficult to gauge impact due to contaminants than in typical estuarine systems, except, perhaps, in the immediate vicinity of major discharges.

3. GEOCHEMICAL INFLUENCES

Geochemically, one of the most important elements in the biology of estuaries is sulphur, which is involved in a complex interplay with other elements including oxygen. If oxygenation is low, sulphate in estuarine water is reduced to sulphide. Reduced circulation in pore waters, low O_2 levels and the presence of Fe and decaying organic matter will lead to the formation of ferrous sulphide. The depth and characteristics of this black reducing layer in sediments influences macrofaunal communities and the activities of a range of microorganisms. If Fe salts are limiting, excess hydrogen sulphide liberated by sulphate-reducing bacteria permeates into pore water and overlying water, compounding the problems for sensitive species caused by low O_2 levels. Generally, sulphur bacteria can tolerate a wider range of redox conditions and some are capable of oxidising H_2S, to produce free S, which in turn may be metabolised in various ways by other microorganisms, to complete the sulphur cycle.

Rather than dwelling on this well-reported sequence, the emphasis in this section is on how geochemical properties, including salinity, can influence biota in other ways - in particular, the relationships between organisms and contaminants in sediments (using metals as the primary example). This encompasses consideration of chemical speciation and bioavailability, together with an appraisal of assimilation pathways and finally ecological effects. Firstly, however, it is important to recognise that it is not just the activities of bacteria and other microorganisms that can shape sediment characteristics, other biological processes, including bioturbation and biodeposition, are significant in terms of sediment geochemistry and transport.

3.1   Bioturbation and biodeposition

Estuarine sediments can act as a store for buried contaminants, which may be isolated from organisms until remobilized by storms, dredging or changes in currents brought about by natural events, or construction works such as sea defenses or port installations. However organisms may themselves modify sediment properties and play a role in remobilising contaminants from sediment to overlying waters (and *vice versa*).

Bioturbation and its consequences can vary considerably in nature as a result of the temporal and spatial heterogeneity in abundance and composition of benthic communities, coupled with the extent and type of sediment disturbance. The review by Lee and Swartz (1980) categorises bioturbatory processes according to guilds (groups of organisms that function similarly in terms of sediment reworking) of which twelve dominant types are recognised. This scheme is primarily based on feeding type (suspension or deposit feeders; injection and ejection of sediments at differing depths), different degrees of mobility (sedentary to highly mobile) and burrowing activity (complexity and extent, horizontally and vertically; epifaunal or infaunal). Within each of these guilds particle reworking can be further categorised, depending on the relative importance of the different activities (burrowing, excavation, irrigation, feeding, tube production, locomotion at the surface and effects on topography). To date, evidence of the relative importance of these activities, particularly on contaminant movements, has largely been qualitative.

Individual reworking rates vary tremendously between species – values range from ~1 g d^{-1} up to >200g d^{-1} depending on category of bioturbation, organism size and seasonal parameters such as temperature. In terms of total reworking rate per unit area of sediment (dependent also on abundance of biota), values are equally variable. They extend over several orders of magnitude - from a few hundred grams m^{-2} y^{-1} for small polychaetes and bivalves, up to several hundred kg m^{-2} y^{-1}, and occasionally more, in large callianassids and holothurians. Care must be taken, however, in extrapolating to impacts on pollutant budgets – high rates of sediment reworking may not necessarily result in proportionate effects on sediment or contaminant transport; more quantitative data is needed on this aspect.

To add to this complexity, bioturbation can affect not only transport of particulate material, but also the geochemical composition of the sediment and pore waters - including adsorption/ desorption properties at the sediment-water interface. In attempting to quantify bioturbation the most common solution to date has been to assume that it can be described as a simple diffusive process. However, although there have been some attempts at modelling sediment redistribution, the signs are that, in reality, the calculated biodiffusion coefficient does not correlate well with faunal density or composition since it represents the summation of a complex mixture of different processes. There is still a need to relate animal activities with their mixing consequences.

Rhoads (1974) has indicated that biological transport of sediments in benthic ecosystems is likely to outweigh molecular diffusive fluxes for some, but not all, ions. Therefore, where sediments are relatively stable and well populated, it would seem that biological processes could dominate the distribution of sediment contaminants. In an early demonstration of this, measured diffusion coefficients derived from mixing curves for [106]Ru in Irish Sea sediments were found to be four-to-five orders of magnitude higher than predicted molecular diffusion coefficients and the disparity was attributed to biogenic influences (Duursma and Gross, 1971).

Sediment redistribution of pollutants by benthic biota may vary considerably, depending on conditions. For example, where biological reworking rates are low (or deposition rates are high), episodic pollutant inputs will be preserved after a time as a discrete layer buried below the mixing zone. Cores with this type of structure are most useful in tracing historical trends in contamination. Where biological reworking is high and sedimentation relatively low, an equivalent pollution episode will rapidly become dispersed throughout the reworked layer, though its eventual disappearance below the mixing zone may be prolonged due to continual reworking at depth. Bioturbation depths vary considerably among different types. Probably most activity takes place in the upper 10cm, though larger burrowing shrimps may be active at depths of over one metre. The potential importance of biological processes on pollutant distribution in sediments is evident.

Studies with burrowing shrimp *Callianassidae* indicate considerable variability in burrow structure and depth, and therefore bioturbating properties, which is related partly to the nature of the substrate. In soft muds there is a single shaft of 30-80cm –much deeper and less complex than the chambered systems (9-23cm) constructed in coarser substrate (Rowden and Jones, 1995). These organisms have also been used to demonstrate the possible geochemical consequences of sediment turnover, which generally relate to the introduction of oxidising conditions, and resultant changes in speciation and solubility, at depth in sediments. Effects are most pronounced for redox-sensitive elements such as Mn, Fe, Pu and As. Transport of suboxic sediments to the surface by other 'conveyor-belt' species is also likely to be an effective mechanism for introducing reduced metal sulphide to the interface; subsequent microbial and chemical oxidation will influence the speciation and mobilisation of a number of elements. Related mechanisms by which bioturbation can effect speciation and remobilisation include the pumping of large volumes of water through tubes and burrows to the surface (irrigation), and the movement and mixing actions of biota in surface layers: the burrowing activities of *Nereis diversicolor,* for example, increase fluxes of Zn by between three and sevenfold (Renfro, 1973).

Burrowing organisms can affect the geotechnical properties of estuarine sediments, which may also influence, indirectly, contaminant transport, speciation and distribution. The most important of these effects concerns sediment stability, which tends to decrease in the presence of abundant macrofaunal populations. This decrease in stability results from an increase in water content, degradation of organic binding and altered surface microtopography and is likely to increase erosion and resuspension at the sediment-water interface. In some cases there may be a direct relationship between both abundance and type of bioturbation and erodibility. Consequently, the effect on erodiblity may be seasonal: biological activity is usually highest in summer, implying that the bed may be less resistant to erosion, particularly during summer storms. Destabilisation, though probably only affecting the top few cm of sediment, can result in near-bottom turbidity zones even in relatively weak tidal currents.

Contaminants such as metals, oil and pesticides may, if present in sufficient concentrations, influence the rate of bioturbation by affecting processes such as filtration rates and re-burial times and, in extreme cases, species composition. Pollutant-induced community alterations could, in turn, influence sediment turnover rate and stability, for example where large sub-surface burrowers are replaced by less active or sedentary opportunists such as small polychaetes. This type of effect might be important in estuaries, particularly close to discharges.

Whilst most bioturbatory activities tend to lead to greater remobilisation, some biogenic processes associated with benthic ecosystems can help to decrease erosion. For example, high organic loadings originating from microorganisms and their exudates tend to increase cohesive properties, whilst dense tube mats of some polychaetes, algal mats and roots of sea grasses and salt marsh plants also help to stabilise the benthic sediments. An illustration of the importance of this occurred during the 1930s when *Zostera* was eliminated extensively around the UK by disease: the absence of stabilising rhizome systems resulted in substantial erosion and collapse of mud banks. Dense growths of some estuarine salt-marsh plants such as *Spartina* and *Salicornia* may encourage deposition and consolidation by reducing current speeds. The same applies within the tangled rhizophores (prop roots) of mangrove vegetation (the equivalent niche to salt-marsh in many tropical and sub-tropical regions).

High densities of suspension-feeding benthic organisms can remove significant quantities of suspended matter from the overlying water which, via faecal pellets or pseudofaeces, eventually becomes incorporated into bed material. Biodeposition by benthic organisms will therefore increase the rate at which particulate contaminants are sedimented and reduce dispersal through currents, correspondingly. Estimates of biodeposition rates in bivalves (up several kg m^{-2} yr^{-1}) and descriptions of the influence of factors such as temperature, suspended solids loadings and substrate type, are given in the review by Lee and Swartz (1980).

Processing by zooplankton can influence the characteristics and settlement of particles and associated pollutant loads, though deposit-feeders probably play the most important role in terms of amounts of material sorted. In some areas 100% of the surface sediment may be pelleted and in productive regions with low sedimentation rates it is estimated that the sea floor may be passed through benthos several times each year (Lee and Swartz, 1980). These pellets sometimes become enriched in pollutants compared with surrounding sediment as a result of modification during passage through the gut (Brown 1986). In terms of transport processes, however, faecal-pellet production by deposit-feeders may alter characteristics and microbial activity but does not add new material to benthic sediments and thus differs from the activity of suspension feeders: strictly speaking, only suspension feeders are involved in biodeposition.

Pellets are compacted and fuse with the sediment matrix as they are buried. This may be a slow process in relatively low energy environments (muds and silts) but elsewhere may take place within one day, aided by metabolic activities of microbes and disturbance from meio- and macro-fauna. By comparison, pseudofaeces tend to be less compact and are readily disaggregated. The effects of feeding on the size distribution of particles (sorting), rates of processing, rates of biodeposition, and properties of sediments have been reviewed by Rhoads (1974) and Lee and Swartz (1980). As an example of the potential scale of effects of pelletisation on deposition, it has been calculated that faecal pellets settle at a rate which is four orders of magnitude faster than particles of 2-3µm diameter, typical of the food of many suspension feeders (Haven and Morales-Alamo, 1972).

In general, therefore, suspension-feeding organisms are important agents of deposition in estuarine ecosystems due to their influence on increasing apparent grain size (through pellet production) and reducing settlement time. Once entrained in sediments, other biological processes may come in to play. Highly pelletised sediments have a further, indirect effect on the transport processes in that the increase in grain size increases porosity and water content and can lead to greater erodibility.

3.2   Salinity and chemical speciation

Salinity, along with substrate type, is a dominant environmental variable directly affecting the distribution of organisms in estuaries, although, as indicated, other factors may modify communities, significantly. Salinity at any one point varies with the tides, and will be dependent on stratification, flushing characteristics and conditions of river-flow, and gradients may be both horizontal and vertical. The osmoregulatory range of any organism will determine its ability to survive and colonise any given stretch or tidal level of the estuary. Significantly, however, it is not only the magnitude of the salinity regime, but also its rate of change, which affects survival: the more gradual the change the more likely it is that the organism will adapt. In this context it is important to recognise that reduced exchange between sediment pore-waters and the overlying water column can buffer infaunal organisms from rapid change and extremes of salinity.

Salinity not only exerts direct control on survival but can impact on ecosystem health, indirectly, via processes such as contaminant geochemistry and bioavailability. The negative effect of increasing salinity, on uptake and toxicity of divalent cations such as Cd^{2+}, Cu^{2+} and Zn^{2+}, has been established for a number of estuarine organisms. The assumption made is that increasing complexation of the free ion at high salinities (principally with chloride) reduces bioavailability (Engel *et al.*, 1981; Wright and Zamuda, 1987; Campbell, 1995). Competition from Ca at high salinities, together with osmotic and other physiological changes, may also contribute to these observations.

Most complexing agents would be expected to inhibit metal bioavailability (by reducing the free metal ion concentration or activity), although it appears that complexation sometimes enhances uptake. For example, cadmium bioaccumulation rates in mussels *Mytilus edulis* are doubled (when compared with the ionic form) by prior sequestration with pectin or humic and alginic acids, suggesting that these uncharged forms can be transported across membranes preferentially (George and Coombs 1977). Furthermore, binding of Cr^{3+} to proteins in tannery waste appears to increase bioavailability disproportionately, resulting in exceptional body burdens in mussels collected near to outfalls (Walsh and O'Halloran, 1997). Significantly, some of the most important pollution events to affect estuarine ecosystems have been caused by metal-organic moieties - including methyl mercury, alkyl lead and tributyl tin (TBT). The lipid solubility of these metals is greatly increased by the presence of the associated alkyl groups, facilitating entry across biological membranes. This is particularly notable for tin, which, in inorganic form in estuarine sediments, is relatively inert and seldom accumulated substantially, even in heavily contaminated estuaries receiving mine tailings containing cassiterite (SnO_2). In contrast, organotins such as TBT are bioconcentrated from sediments to a significant degree, especially by some infaunal bivalves (Langston *et al.,* 1990).

3.3   Bioavailability and assimilation pathways

Benthic plants and animals of both the sub-and inter-tidal zones are important components of estuarine food chains, which ultimately depend on organic debris, bacteria and diatoms associated with estuarine deposits as sources of nutrition, nutrients and minerals (Figure 4-3). Inevitably, bioaccumulation of sediment-bound contaminants may occur *via* a variety of pathways and results in their transfer to higher trophic levels, including fish, birds, mammals and, eventually, humans. Contaminants in sediments, and their bioavailability, thus have important and widespread implications throughout the estuarine ecosystem and beyond.

Direct assessments of bioavailability in estuaries usually involve field surveys with bioindicator fauna, such as bivalves, gastropods or worms. However, the choice of organism is crucial. Because different organisms have different ranges in estuaries, and accumulate metals from different sources, there is no 'universal bioindicator' present throughout the range of environments found in a typical estuarine-coastal zone. Filter-feeders such as mussels *Mytilus edulis* and cockles *Cerastoderma edule* may give some measure of the importance of contamination on suspended solids, whilst infaunal deposit-feeding organisms and detritivors (e.g. clams *Scrobicularia plana, Macoma balthica* and ragworm *Nereis diversicolor*) are often useful bioindicators of bioavailability in estuarine benthic deposits. In contrast, being primary producers, seaweeds such as *Fucus spp* are generally considered good indicators of dissolved (rather than particulate) contamination. Nevertheless, there may be situations where even *Fucus* may assimilate contaminants directly from sediments, for example where fronds lay in direct contact with highly contaminated muds during low water (Luoma *et al.*, 1982). It is important to keep in mind that geochemical and geological influences may vary among different components of the ecosystem, and in different conditions.

The determination of metal concentrations in bioindicators is of obvious relevance in ecotoxicological terms, since, in effect, direct measurements are made of bioavailable - and hence potentially deleterious - metal. Analysis of sediment (and water) is undertaken more routinely in statutory monitoring programmes and, though helpful in terms of defining comparative loadings in the environment, is usually restricted to measurement of 'total' metal (including refractory forms which may be of little biological significance). Metal 'speciation' techniques hold out more promise in terms of understanding and predicting bioavailability, though they have yet to be widely validated and adopted.

Surrogate chemical measurements of bioavailable fractions are often derived from sediment extraction schemes used by soil geochemists - incorporating for example, dilute mineral acids, hydroxylamine hydrochloride (reducible metals), EDTA and ammonium chloride (ion-exchangeable forms), or their biomimetic equivalents (e.g. enzymes and gastric fluids). Similar attempts have also been made to apply this approach to synthetic organic contaminants, such as pesticides, PCBs and TBT, using semi-permeable membrane devices filled with organic solvents to determine extractable sediment fractions. If used appropriately, such extraction schemes provide information on how contaminants are bound to sediment - essential in interpreting the geological and geochemical controls on bioavailability. Important modifying factors which may need to be considered, alongside the concentration of the contaminant in extracts, include: grain size, the role of complexing agents (such as iron and manganese oxides and organic matter), effects of early diagnosis, and the role of suspended particulate matter.

Some of the earliest demonstrations of the assimilation of particulate metals by estuarine organisms, and the importance of geochemical associations, were provided by Luoma and Jenne (1976, 1977). These included a study of sediment-Cd uptake in clams *Macoma balthica* which demonstrated the importance of organic matter in suppressing Cd bioavailability; in contrast, Ag, Co and Zn were assimilated readily from detrital organics (though Fe and Mn oxyhydroxides inhibited the uptake of the latter two metals). These and similar findings eventually led to the general hypothesis that bioavailability declines as the strength of binding to various sediment components increases (in line with principles derived from complexation studies with dissolved metals).

Speciation techniques for oxidised sediments are thus based, primarily, on operational methods of uncertain selectivity to extract specific (labile) metal forms. Nevertheless, despite drawbacks, these schemes provide some of the most meaningful and practical assessments of sediment-metal availability. By examining the goodness-of-fit between extracted metals and body burdens in ubiquitous benthic species, over a range of sites and conditions, it is often possible to quantify the influence of sediment contamination and to evaluate the strength of anthropogenic contributions from different sources. Scrutiny of outliers in the data can also help to highlight the more important geochemical parameters which modify bioavailability (Luoma and Bryan, 1982; Langston, Bebianno and Burt, 1998; Ying et al.,1992; Ridgway et al., 2003).

Particularly useful models for predicting metal bioavailability in estuarine surface sediments have arisen from field-based studies with infaunal clams and polychates. These confirm the frequent role of the major metal-binding components Fe and Mn oxyhydroxides, or organic matter, in mediating uptake. Thus, for many of the examples, metal burdens in biota $[M_b]$ are best described by the linear equation:

$$[M_b] \quad = m \frac{[Ma]x}{[x]} + c \qquad (1)$$

where $[Ma]x$ = the concentration of metal associated with sediment component x, and $[x]$ = the concentration of the metal binding component. Such normalising routines quantitatively account for the influence of these major geochemical parameters and indicate that metal impact is unlikely to be the same in all sediments. The success of extractable Fe or organic content as normalisers reflects their importance in the partitioning of adsorbed (non-detrital) fractions in oxidised, estuarine surface-sediments (which commonly form a major part of the diet of deposit- and suspension-feeders).

Notwithstanding the excellent correlations between tissue burdens and sediment fractions that are sometimes achieved, there is still much speculation as to the physiological mechanisms involved and the processes surrounding geochemical controls. The simplest explanations assume that assimilation of particulate metal takes place in the gut and that metal-binding sediment phases either compete with uptake sites in the digestive epithelium or render the metal less labile. Nevertheless, even in estuarine organisms, which are thought to derive most metals from ingested surface sediments, accumulation of some metal from interstitial and overlying water cannot be ruled out (Langston and Spence, 1995).

Provided that the system is at adsorptive equilibrium, it may not be necessary to separate uptake sources in order to model bioavailability. The amount of soluble metal [M] will be a function of that which is adsorbed [Ma] to a complexing sediment phase x (i.e. [Ma]x), together with the concentration of the solid phase [x] responsible for binding that metal. Thus, the ratio

$$\frac{[Ma]x}{[x]}$$

(used in equation 1 as a predictor of bioavailable sediment metal) would also represent a surrogate of dissolved metal. If more than one sediment phase is important in complexation, then additivity of this term, for each phase, is assumed (Tessier *et al.,* 1993).

This hypothesis receives some support from experimental studies with artificially manipulated particulates (e.g. Luoma and Jenne, 1977) in which sediments exhibiting the greatest rates of sediment to water desorption (lowest K_d values) were also those from which metal bioaccumulation was greatest. Where such labile sedimentary sinks occur in the field, bioavailability of particulate metals could be enhanced both through increased assimilation from ingested material and from the higher concentrations of desorbed metal. However, whilst this concept might be expected to work well with Cd and other metals whose particle reactivity is relatively low, the assumption of steady -state conditions would be less realistic for metals whose behaviour is dominated by strong and complex interactions with sediment.

Not surprisingly, attempts to predict metal bioavailability, based on equilibrium partitioning, are sometimes unsuccessful, even after normalisation with respect to Fe or organics. This applies particularly when sediments of widely differing characteristics are compared. Intuitively, other site-specific geochemical parameters such as sulphides would be expected to exert more control under anoxic conditions and might conceivably determine bioavailability of metals from reducing sediments. The validity of traditional extraction schemes is questionable in these circumstances, however, and the relevance of buried, reducing sediments as a metal source for many organisms is uncertain and perhaps highly variable.

Anomalously high levels of Cu have been observed in estuarine clams *S.plana* and *M. balthica* from relatively anoxic sites, even though contamination with Cu is not evident in the sediments themselves. An increase in bioavailable cupric ions during bouts of anoxia is one possible explanation; alternatively, enhanced Cu burdens in clams could result from immobilisation in tissues, as CuS, following an influx of hydrogen sulphide (Bryan and Langston, 1992). The presence of high levels of acid volatile sulphide (AVS) in sediments is reported to modify the impact of Cu and Ni in amphipod bioassays (Di Toro *et al*, 1992), though, in contrast to the clam scenario, high levels of sulphide in this case are thought to *reduce* accumulation and toxicity (where sulphur-to-metal ratios in sediment are >1, metal is assumed to precipitate as insoluble, unavailable, metal sulphide). A high total S content of sediments has also been shown to lower Pb availability to mussels *M.edulis,* close to a Pb/Zn smelting complex, though whether this is a competitive phenomenon, or the result of the anomalously sulphide-rich nature of ore-impacted sediment, is unknown (Bourgoin *et al.*, 1991).

In summary, concentrations of metals (and presumably other contaminants) in estuarine biota are significantly influenced by the major geochemical parameters. For aerobic surface sediments, metal availability is often determined, predictably, by adsorption/desorption characteristics on Fe/Mn oxyhydroxides and organic coatings. The bioavailability of metals in anoxic (usually sub-surface) sediments is more uncertain but appears to fall under the control of sulphide reactivity: metal ions whose solubility is less than FeS (e.g. Cd, Ni, Cu, Zn, Hg and Pb) may be precipitated from pore-water as insoluble metal sulphides which are relatively unavailable, at least in some organisms. In contrast, in other estuarine infaunal species anoxic conditions appear to enhance bioaccumulation.

In situations where adsorptive equilibria cannot be assumed, predictions of bioaccumulation may be confounded by the fact that estuarine organisms accumulate metals from a combination of water and diet (including sediments) in varying proportions. Development of kinetic models, where individual pathways are treated additively, may help to interpret uptake routes and efficiencies, and transfer through aquatic ecosystems. Metals in solution are usually considered to be more bioavailable than solid-phase metal; however, the higher concentrations in the latter often render dietary vectors more important. This may be further enhanced in filter feeders through the selection of certain particle types, relative to bulk sediments.

The importance of assimilation efficiencies and particle type in determining bioaccumulation of solid-phase metals has been illustrated in estuarine sediment-dwelling clams where, for example, highly efficient assimilation of Se (93% from epipelic diatoms) accounts for a predominantly particulate uptake route. Similarly, the presence of adherent bacteria and extra-cellular polymers on the sediment particles selected by *M. balthica* enhances the digestibility and bioavailability of Ag, Cd and Zn. Cr assimilation also appears to be much more efficient from bacteria and polymers than from diatoms or purely inorganic sediment fractions (Decho and Luoma 1991,1994, 1996; Harvey and Luoma, 1985; Luoma *et al.,* 1992). Further insights into assimilation pathways and efficiencies, from particulates of different type, would be useful.

3.4   Geochemical Influences: Case Studies - Ecological effects of contaminated estuarine sediments.

Tributyltin (TBT) is worth singling out as a 'model' pollutant which has been responsible for significant impairment to estuarine ecosystems (as result of leaching from antifouling paints on boat hulls), imposex in dogwhelks *Nucella lapillus* being a particularly notable effect (Bryan *et al.*, 1986; Minchin *et al.*, 1995). In the 1980s, recognition of damage to non-target organisms, particularly molluscs, instigated legislation to ban the use of TBT paints on boats <25m length (which encompasses the majority of the leisure market). TBT levels in waters of many marinas and small-boat harbours diminished considerably, following restrictions. However, a slow-down in the rate of disappearance of TBT has been described for some ports and estuaries. Delays to further improvements in water quality are partly related to sedimentary sinks of the compound, together with certain dockyard operations and the continuing presence on larger vessels (still entitled to use existing TBT antifouling until 2008, though no longer permitted to apply new coatings, according to edicts of the International Maritime Organisation). There is also a possibility that TBT paints are still being used on some craft, illegally.

TBT concentrations in sediments are highest, as expected, close to dockyards, marinas, and hull-cleaning facilities, whilst chronic contamination may be detected in deposits at considerable distance from TBT sources. Partitioning is reversible, allowing release back to the water column, but because of the relatively high affinity for particulate matter, residence times for TBT in sediment are usually prolonged in comparison with overlying water, especially in unperturbed, organic-rich fines. Paint particles containing TBT may also become entrained in sediments, increasing the persistence of the biocide (Thomas *et al.*, 2003). Degradation of TBT can occur in surficial sediments through the activity of microorganisms, and involves stepwise debutylation to inorganic tin. Nevertheless, TBT removal rates are relatively slow, with half times of the order of 1-5 years. In undisturbed anaerobic muds, and at sites where there is a continuing input of TBT from antifouling, temporal reductions in TBT sediment-loadings may be undetectable; half-lives under such conditions are likely to be of the order of decades (Langston and Pope, 1995).

Consequently, the retention of TBT in fine-grained muds is seen as one of the major reasons for the continuing threat from the compound in poorly flushed estuaries and harbours. The concerns are two-fold: firstly, slow release of TBT held in the sediment reservoir, to the water column, may extend exposure to concentrations above no-effects thresholds for pelagic species (including larval forms). Secondly, some burrowing species, particularly deposit-feeding clams such as *Scrobicularia plana,* derive potentially deleterious burdens of TBT directly from sediments and, as a result, have been in decline at TBT-contaminated sites. Where TBT persists in sediments, clam populations are unlikely to recover quickly. This scenario, where sediments increasingly become a major source, is probably representative for a broader range of contaminants, following the widespread introduction of measures to reduce discharges to estuarine waters in recent years.

Areas exhibiting metal contamination from mining or smelting operations often provide further examples of ecological impact and serve to illustrate both geochemical and geological influences on estuarine ecosystems. Reduced biodiversity is attributable to metals in such cases, though in fact this phenomenon is symptomatic of most forms of anthropogenic disturbance (Langston, 1990). The Fal Estuary in south-west England is a case in point of such a metal-impacted system (Figure 4-4).

Compared with most estuaries in south-west England, the Fal, as a whole, has a very low abundance of a number of sensitive taxa, notably benthic crustaceans and molluscs, whilst certain small annelid worms are more abundant (Rostron, 1985). Metal pollution is implicated as a major factor responsible for these differences (Langston *et al.*, 2003b). The conspicuous absence of bivalve and gastropod molluscs from highly metal-contaminated sites in the Fal (notably Restronguet Creek) is a consequence of the long history of metal mining in the region (Bryan *et al.*, 1987). Here, for example, high levels of Cu and Zn in sediment inhibit the settlement of juvenile bivalves, including *Scrobicularia plana, Cerastoderma edule* and *Mytilus edulis.* In contrast, enhanced metal tolerance is observed in some organisms (e.g. the polychaete *Nereis diversicolor*), and is effective in ameliorating impact. Different sensitivities to metal-laden sediments, between taxonomic groups, may thus translate into community-level effects and Restronguet Creek has developed a distinct macrofaunal community composition

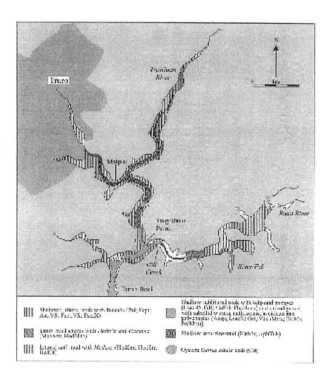

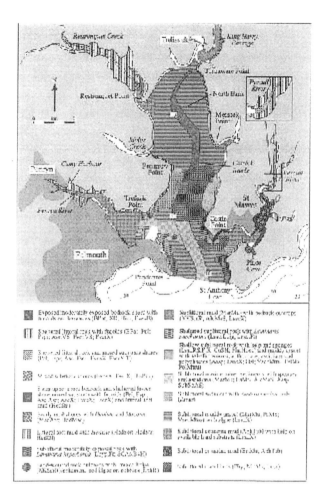

Figure 4-5. The distribution of biotopes in the Fal Estuary, Southwest England. Upper Fal (top), Lower Fal (bottom) (reproduced from Moore *et al.,* 1999, with permission JNCC). Legend to the top figure: Sheltered littoral rock with fucoids, Sandy mud shores with *Hediste* and *Macoma,* Littoral soft mud with *Hediste,* Shallow sublittoral rock with kelp and sponges and mixed gravel with sabellid worms, anthozoans, ascidians and polychaetes, Shallow estuarine mud, Oysters *Ostrea edulis* beds. Legend to the bottom figure: Exposed/moderately exposed bedrock shore with fucoids, Sheltered littoral rock with fucoids, Sheltered littoral rock and mixed substrata shores, Mixed substrata shores, Steep upper shore bedrock and sheltered lower shore mixed substrata with fucoids, Sandy mud shores with *Hediste* and *Macoma,* Littoral soft mud with *Hediste,* Sublittoral moderately exposed rock with L*aminaria hyperborean,* Sand-scoured rock outcrops with mixed kelps, Sublittoral mud with bedrock outcrops, Sheltered sublittoral rock with *Laminaria saccharina,* Sheltered sublittoral rock with kelp and sponges and muddy gravel with sabellid worms, anthozoans, ascidians and polychaetes, Sublittoral marine mixed sediments with sponges and ascidians, Sublittoral sediments with *Zostera marina* beds, Sublittoral muddy gravel, Sublittoral estuarine mud with kelp on available hard substrata, Sublittoral estuarine mud, Sublittoral beds.

compared to other, less-contaminated creeks in the Fal system (Warwick, *et al* 1998). Sediment copper concentrations in Restronguet Creek are in the region of 2,500 $\mu g\ g^{-1}$, whereas in other creeks in the Fal, Cu concentrations range from 100–1,200 $\mu g\ g^{-1}$.

There is a gradation in meiofaunal (nematode and copepod) community structure in different parts of the Fal, which, as with macrofauna, is consistent with increasing metal concentrations, particularly Cu (Warwick *et al.*, 1998). Large differences in the Cu tolerance of nematode communities from different creeks have also been shown to correlate with their previous history of exposure to Cu in sediment (Millward and Grant, 2000). A level of 200$\mu g\ g^{-1}$ sediment Cu (1M HCl extractable) has been suggested as the threshold above which 'pollution-induced community tolerance' is initiated. It is uncertain whether these thresholds would be of universal significance however, as many communities in the Fal may be partially adapted to survive high metal concentrations.

Similar ecological effects have been observed in polluted sediments from other locations. For example, studies in metal-contaminated Norwegian Fjords indicate considerable reductions in faunal diversity at sediment Cu concentrations above 200 $\mu g\ g^{-1}$, with sensitive species, including molluscs, lost, leaving a high proportion of (tolerant) polychaetes (Rygg, 1985). Large reductions in shellfish production have been linked to impact from metalliferous mine spoils in estuaries in Goa (Parulekar *et al.*, 1986) and in Australia, smelting wastes (hazardous because of their chemical and biological reactivity) have been shown to result in changes in seagrass communities (Ward *et al*, 1984).

Despite recent trends towards improved water quality, contamination in estuarine sediments often remains consistently high (as in Restronguet Creek, following closure of the mining industry). Benthic communities also appear to retain their modified characteristics for long periods. The persistence and behaviour of contaminants in sediments is clearly an important geochemical feature, which may govern the rate of 'recovery' of impacted estuarine ecosystems long after man's polluting influence has abated. There is still much to be learned concerning timescales and mechanisms involved in this process.

4.   CONCLUSIONS

In conclusion, estuarine ecosystems are significantly influenced in a variety of ways by geological and geochemical characteristics, particularly the nature of benthic deposits. The over-arching effects of varying tidal and salinity regimes introduce a further level of intricacy to an already complicated and dynamic environment. Understanding the key processes and interactions responsible for spatial and temporal variability in biological communities is an essential requirement for predicting future change in the functioning of these most complex of ecosystems, and for distinguishing between natural and anthropogenic causes.

Chapter 5

THE ROLE OF TECTONIC PROCESSES IN THE INTERACTION BETWEEN GEOLOGY AND ECOSYSTEMS

ROY J. SHLEMON[1], RICHARD E. RIEFNER, JR[2]
[1] P.O. Box 3066 Newport Beach, California, USA
[2] 5 Timbre Rancho Santa Margarita, USA

1. INTRODUCTION

Planet Earth is but one large complex ecosystem. Indeed, every centimeter of the planet is itself an ecosystem, and yet part of a larger, interacting community of organisms and environment that give rise to the natural world (Pickett and Cadenasso, 2002). Habitats, the places where organisms live, are primarily fashioned by geologic processes. These processes started and still control the distribution of the earth's major land biomes: tundra, desert, grassland, chaparral, forest and savanna. Of particular interest is chaparral, a shrub-dominated ecosystem found on several continents, including the Mediterranean, the African Cape region, central Chile, southwestern Australia, and California Though climate directly determines the distribution of these vegetation types and their related fauna, the climate itself is influenced by mountain building, and by the size, shape and location of landmasses. Accordingly, ecosystems from microscopic to greater than continental-size, generally depend on the magnitude and frequency of geologic processes affecting a given area. The importance of geologic and ecosystem interaction is reflected by increasing scientific recognition and new fields of university study now generally termed "geoecology," "geobiology," and "ecohydrology" (Bastian et al., 2002; Rodriquez-Iturbe, 2000; Safford, 2002).

A major geologic process is tectonism, a product of the earth's internal movements that give rise to the formation of continents, oceans, mountains, streams and the many other large- and small-scale geomorphic features that strongly influence, if not locally control, the distribution of life on the planet.

This paper first briefly summarizes typical effects of long-term (hundreds of millions of years) tectonic processes on regional ecosystems, · such as epeirogenic uplift, the related impact on climatic change, and the regional distribution of forest, grassland and soils. It then notes some medium-term (tens to hundreds of millions of years) impacts, such as those affecting fluvial systems, groundwater regimes and slope stability. Finally, it provides examples of relatively short-term, and often dramatic tectonic impacts on small-scale ecosystems, such as those produced by earthquake-induced ground rupture and related local hydrological change.

2.   LONG-TERM TECTONIC IMPACTS

Plate tectonics are the ultimate drivers of land uplift and mountain building. The earth's major and myriad of minor continental and oceanic plates move at variable rates of up to several cm/yr (Rowley, 2001). It is the spreading and collision of these plates that has given rise to cumulative earth topographic change over hundreds of millions of years, thus accounting, for example, for fossilized tropical life forms in present polar regions and, conversely, for glacially striated rock and erratics in low latitudes (Stein and Freymueller, 2002). Slow global shifts in landmasses were also accompanied by altitudinal change, in some cases by regional epeirogenic uplift, and in other cases by mountain folding and local uplift. These tectonically driven changes in the earth's landscape affected the major oceanic currents, the movements of large air masses, the distribution of lakes, streams, mountains and mineral resources, and hence dramatically impacted the worldwide distribution of paleo-ecosystems, such as those that must have existed on Pangaea, Gondwana, Laurasia and other ancient Paleozoic and Mesozoic landmasses (Cox, 1973; Gordon and Stein, 1992; Kearey and Vine, 1990; Unrug, 1997; Zonnenshein et al., 1990).

3.   MEDIUM-TERM TECTONIC IMPACTS

On timescales of 10^5-10^6 years, global tectonics have shaped most modern-day topography and hence ecosystem distribution. For example, at the start of the Cretaceous the Colorado Plateau in the western United States was at sea level. It now stands at ~2-km in elevation, a product of isostatic uplift averaging ~40 m/my for the first 40 my, but apparently increasing to about 200 m/my over the past 5 my (Sahagian et al., 2002; Pederson et al., 2002). It was also during the late Cenozoic when epeirogenic uplift of the Colorado Plateau accelerated incision of the Colorado River, giving rise to its present steep walls, to dramatic local changes in topography and microclimate, and inevitably to the evolution of innumerable small but complex interacting ecosystems (Cushing et al., 1995; Fule et al., 2002; Grim et al., 1997).

Similarly, continuing northward motion of the Indian Plate into the Eurasian Plate over the past 15 my has driven the Tibetan Plateau to elevations averaging 5 km, thus affecting weather worldwide, in particular, the monsoons of Asia and India (Bilsniuk et al., 2001; Bullen et al., 2002). The opening of the Rift Valley of Africa, driven by movement along and between the African and Arabian Plates also started at this time and likewise dramatically controlled the evolution and distribution of flora and fauna, particularly in and around the Rift Valley lakes, as recorded in paleo-anthropological sites (Scholtz and Contrearas, 1998; White, 2003; Zeyen et al., 1997). In addition, the European Alps, the Apennines of Italy and the high mountain ranges in southeast Asia are still undergoing global-tectonic-driven uplift, manifest mostly by earthquake-generating subduction, compression, and transformational movement between the African, Eurasian and Australian-Indian Plates respectively (Cloetingh et al., 2003).

So-called medium-term tectonic processes, particularly along plate spreading centers or subducting margins, produce volcanic flows and ash, which, in turn, host many complex natural habitats. Moreover, it is long- and mid-term tectonic movement that leads to exhumation of rocks once buried deep in the crust and thus to weathering and soil-substrate formation (edaphic control) on ophiolites, serpentinite, carbonates, siliceous dikes and other rocks that support unique ecosystems (Brozovic et al, 1997; Dobson et al., 2002; Jenny, 1989; Kruckeberg, 2002).

4. NEAR-TERM TECTONIC IMPACTS

Tectonic activity during the past few million years has similarly profoundly impacted ecosystems, particularly those of local extent and of relatively small scale. For example, continuing uplift of the Sierra Nevada and Cascade mountain ranges in North America, and the Andes in South America, has produced pronounced altitudinal zonation of soils and biota, as well as climatic rainshadows that control a host of regional-scale ecosystems (Alley et al., 2003; Bartlein, 1997; Hermanns et al., 2001).

A manifestation of near-term tectonism in high latitudes and in nearby coasts and subpolar high mountain chains is post-glacial rebound, the continuing uplift of land following periodic, world-wide deglaciation during the Pleistocene. In Fennoscandia, for example, the continuing post-glacial uplift is an estimated 830 m and this has been accompanied by episodic shoreline uplift and tilting, by faulting, and often by rock shattering, thus locally producing the "substrate" for endemic ecosystems (Morner, 1979).

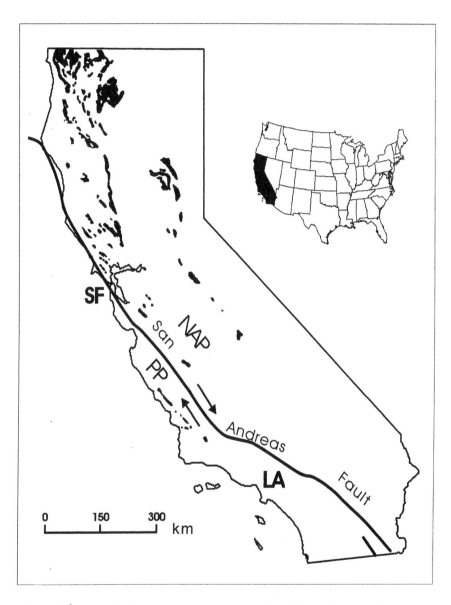

Figure 5-1. Map of California in the southwestern United States showing: (1) the San Andreas fault, the major boundary zone between the North American Plate (NAP) and the Pacific Plate (PP); and (2) the general distribution of subduction-induced terrane (dark color) hosting serpentine-controlled "ecological islands" in the North Coast, Klamath and Sierra Nevada mountains. SF=San Francisco; LA=Los Angeles.

In the Hudson's Bay area of Canada and in the adjacent northern United States, deglacial uplift is similar in extent and character, thus continuing drainage-pattern deformation and affecting evolution of estuarine ecosystems that began in early Pleistocene time (Denton and Hughes, 1981; Hughes et al., 1985).

On-going tectonism is also expressed by regional subsidence and ultimately by changes in regional drainage systems and local hydrology. This is exemplified by the 1811-1812 swarms of intraplate earthquakes in the central United States, which demonstrated that short-term tectonic activity can violently change the pattern of even large rivers such as the Mississippi (Schumm, 1986; Schumm et al., 2002).

The impact of near-term, and indeed recent, tectonism is particularly well illustrated by floral and faunal distribution and evolution in relatively closed natural systems. We provide examples of these, mainly from the State of California in the western United States (Fig. 5-1). With its present ~35 m population, an area of about 425^3 km^2 and a range of climate, topography and endemic species more numerous and diversified than any other region of comparable size in the United States, California supports more than 5000 species, subspecies and varieties of native flowering plants, conifers and ferns (Hickman, 1993). On-going study on taxonomy, nomenclature, distribution, ecology, relationships, and diversity of vascular plants will likely add many hundreds of native species to the flora of California (Jepson Flora Project, 2005). Additionally, the State contains 11 of the 12 world's major soil orders, and some 1200 different soil series, many of which owe their origin to the interplay of mid- and near-term tectonism, climatic and hydrologic change, and landscape evolution (Soil Survey Staff, 1999). This is particularly evident along and near the San Andreas fault system, the major component of the North American-Pacific Plate boundary in the central and southern part of the State, and along subduction faults in the northwest, where continuing tectonism juxtaposes a mélange of different rock units, produces sag ponds, vernal pools, linear estuaries and coastal embayments, hot springs, and serpentine terrane that support unique endemic floral and faunal ecosystems (Figure 5-1). These habitats often occur as isolated "ecological islands" and California is thus well known for its large number of endemic organisms. Indeed, more than 30-percent of the native plants in California are endemic to the State, and occur naturally nowhere else in the world (Hickman, 1993).

Figure 5-2. Typical tectonically induced sag pond along the San Andreas fault on the Carrizo Plain capable of hosting unique, wet-season, endemic floral and faunal ecosystems. Site about 150 km northwest of Los Angeles.

Here we present two examples of California ecosystems that occur on geologic terrane differing in scale and origin: sag ponds and vernal pools produced by the interaction of seismicity, regional uplift, weathering and bioturbation; and endemic serpentine ecosystems caused by subduction, uplift, and exposure and weathering of ophiolites, serpentinite and other deep-mantle lithologies

5.   SAG POND AND VERNAL POOL ECOSYSTEMS

Tectonism often produces or significantly contributes to the development of micro-relief that entraps seasonal "wetlands" and hence leads to formation of endemic ecosystems. All wetlands have several features in common and typically three main components: (1) water at the soil surface or within the root zone; (2) clayey (smectitic) soils usually typified by mottling and seasonal expansion; and (3) vegetation adapted to wet conditions and the related absence of flood-intolerant species (Mitch and Gosselink, 2000).

Figure 5-3. Remnant vernal pool and surrounding Mima-mound topography (foreground)
on the eastside of the Central Valley of California. Orange grove (background), other
agricultural practice and urbanization now destroying the endemic ecosystems (see text).
Western slope of the Sierra Nevada Mountains shown at the skyline.

Sag ponds are the classic examples of relatively rapid, tectonically
produced, micro-relief wetlands (Ferren et al., 1995). These features range
in size from a few m to tens of km in length, are usually somewhat elliptical
to linear in shape and, in most cases, are formed by relief-induced,
transtensional movement along major transform faults such as the San
Andreas in California (Figures 5-1 and 5-2).

The San Andreas zone is typified by mainly right-lateral slip of about
~30 mm/yr, about two-thirds of movement occurring along the deformed
boundary of the North American and Pacific Plates. The San Andreas, with
late Quaternary recurrent movement of up to ~7 m/event, has produced many
landslide dams, sag ponds and other local zones of water accumulation
(Earnst, 1981; Yeats et al., 1997). Within these areas of local hydrological
disruption unique biotic communities have often formed, as exemplified by
~10- to 1000-m long, tectonically produced sag ponds and vernal pools (a
type of seasonal wetland) on the San Andreas fault in southern and central
California.

Vernal pools are generally small, usually only a few m in diameter,
shallow and complex seasonal wetlands initially produced by regional uplift.
Developing on coastal and fluvial terraces, mesas, plateaus and alluvial fans

tens to hundreds of m above local base levels, the vernal pools are a manifestation of micro-relief often associated with relatively stable landscapes. These surfaces, tens of thousands of years old, may produce strongly developed soil profiles with poorly drained substrates, often clay (argillic) or silcrete (duripan) horizons (Jenny, 1989). Usually bordered by a ring of rodent-occupied "Mima-mounds", the vernal pools are typically hydrologically isolated from perennial inflow, and hence have internal drainage. Sediment input is generally limited to minor eolian influx, where coastal dunes or silt-laden floodplains are near, and to local colluviation derived from rodent-induced bioturbation on the adjacent mounds (Figure 5-3).

The California vernal pools harbor endemic ecosystems that are rare and now threatened (Ferren and Fiedler, 1993). California vernal pools are closely allied to more than 100 vascular plant species (Keeley and Zedler, 1998). Additionally, vernal pools support a unique fauna, including endemic species such as fairy shrimps (Eriksen and Belk, 1999). Other animal species dependent on vernal pools are two amphibians, the California tiger salamander (*Ambystoma californiense*) and the western spadefoot toad (*Scaphiopus hammondii*) (Keeler-Wolf et al., 1995).

The vernal pools and surrounding Mima mounds are inexorably intertwined, for there are no mounds without intervening depressions and without depressions there are no vernal or temporary pools (Figure 5-3). Though most California mound and vernal pool topography result from long-term uplift and resulting weathering processes, there is no consensus about a possible single cause for their origin. Hypotheses for California mound origin range from construction by fossorial rodents, usually pocket gophers (geomyidae), by former periglacial processes and, more recently, by seismic activity (Arkley and Brown, 1954; Berg, 1990a, 1990b; Cox, 1984, 1990a; 1990b; Riefner and Pryor, 1996; Shlemon et al., 1997). The seismic hypothesis suggests that some mounds may be initiated by earthquake-induced sand blows and similar forms of seismically induced liquefaction that cause local accumulation of surface sand, and thus an initial habitat for burrowing and mound-building pocket gophers (Berg, 1990a; Kuhn et al., 1995).

Along the San Andreas Fault in southern and central California, vernal pools are associated with micro-relief produced by either transtensional displacement (local grabens) or through temporary damming by seismically induced landslides (Figure 5-2). Unfortunately, the ecology of such fault-produced vernal pools is still poorly known (Keeler-Wolf et al., 1995).

Whether stemming from slow uplift or from more rapid movement, tectonism is the initial producer of terrane suitable for vernal pool development. From an ecological standpoint, vernal pools harbor a highly endemic flora and fauna closely allied to Mediterranean climates (Zedler, 1987). Vernal-pool wetlands and their endemic ecosystems are now increasingly endangered owing to rapid urbanization (Ferren and Fiedler, 1993). What tectonism originally produced over thousands of years may now, unfortunately, be lost in a few tens of years.

6. TECTONICS AND FORMATION OF SERPENTINE-CONTROLLED ECOSYSTEMS IN CALIFORNIA

Since about late Mesozoic time, subduction of the Pacific Plate under the North American Plate has accreted volcanic arc and offshore trench sediments into a broad, 7000-m thick, assemblage of ophiolites, graywackes, shale, limestone and silica-rich chert known as the "Franciscan Group" (Atwater, 1989). The Franciscan Group crops out primarily in the western part of northern and central California (Bailey et al., 1964; Figure 5-1). Its "greenstone" is more properly deemed "serpentitite," but is also often called "serpentine," a lithology so unique and extensive that it is designated as "The State Rock of California" (Kerrick, 2002).

Plate-boundary tectonism caused accretion of "exotic terrane", uplift and exhumation of the serpentinite. Subsequent weathering, however, produced soil parent material unusually rich in magnesium and iron, and locally in nickel, cobalt and chromium, but inherently deficient in calcium, sodium and potassium (Alexander, 2002). The resulting soils are therefore unusual and often support edaphically controlled "ecologic islands", typified by plant and animal species endemic to serpentine terrane (Schoenherr, 1992). Indeed, the California Floristic Province supports 215 species, subspecies and varieties of native flora restricted entirely or in large part to serpentine soils (Kruckeberg, 1984). Many plants are rare taxa because of their narrow geographic distribution (Wolf, 2001). Serpentine habitats also support endemic invertebrates, including species and subspecies of butterflies (Schoenherr, 1992), the best known of which is the threatened Bay Checkerspot Butterfly (*Euphydryas editha bayensis*), now restricted to local patches of California grassland developed on serpentine soils (Murphy and Weiss, 1988). Accordingly, the initial environmental factor determining the distribution and abundance of this butterfly is geological; namely, the uplifted and now-exposed serpentine terrane (Murphy et al., 1990).

Factors other than soil chemistry also impact serpentine ecosystems. Serpentines are inherently unstable; slope failures are common and the flora is thus also constrained by slope steepness and local microclimates. Additionally, some taxa occur solely on serpentine over a wide range of climates; yet others are found on adjacent rocks that bear different soil morphology and chemistry. Nevertheless, serpentine endemism occurs in many conifers, woody and herbaceous dicots, monocots, and ferns totaling 198 taxa, which represents approximately nine-percent of the total endemic taxa known to occur in the California Floristic Province (Kruckeberg, 1992). The most widespread and strict endemic control is probably expressed by distribution of the Leather Oak (*Quercus durata durata*), which is a near-constant indicator of serpentine soils throughout the Coast Ranges and foothills of the Sierra Nevada Mountains (Kruckeberg, 2002).

7.   SUMMARY AND CONCLUSIONS

Tectonic processes are but one element in ecosystem development. Long-term plate tectonic movement and resulting accretion of unlike terrane have produced land uplift, formation of consequent drainage, orographic barriers and other topographic influences on local climate and vegetation, and differential weathering of rocks and minerals. More recent tectonic movement is geomorphically expressed by seismically induced surface displacement and related secondary features such as fault-related scarps, landslides and sag ponds. Tectonism is thus the initial influence on natural ecosystem formation and extent, but it is the inexorable interaction of all climatic, hydrologic, weathering and anthropic processes that ultimately controls survivability of endemic ecosystems.

The vernal pools of California illustrate evolution of a very restricted but highly complex ecosystem, caused by initial regional uplift followed by relative landscape stability, local tectonic displacements, soil formation, hydrologic change and the activity of nearby fossorial rodents. In contrast, the more extensive serpentine terrane illustrates uplift and exhumation of deep mantle rocks that weather into unusual soils and hence strongly influence the distribution and evolution of other endemic species.

Tectonism, on a broad scale, may be the originator of both land and seascapes that "house" ecosystems, but tectonics in time, rate and space, change naturally, and so ecosystems, regardless of scale, likewise evolve. The process of change, however, whether good or bad, is now greatly accelerated by anthropic processes. Man is indeed a geomorphic agent. Only a fraction of former vernal-pool terrane still exists in California owing to urban development; and much serpentine area has since been logged, grazed or otherwise dramatically stripped of its natural vegetation. The impact of on-going tectonic processes, however, will still be felt: on a large-scale, uplift continues, albeit slowly, and faults continue to move, especially along plate boundaries; on a small-scale, subsidence, liquefaction and a host of other tectonic proces will leave their imprint on the Planet, thus likewise continuing to affect ecosystem development.

8.   ACKNOWLEDGMENTS

We thank Drs. J. Ridgway and B. Marker (UK) for review of a draft manuscript; and Professor I. Zekster (Russia) for encouraging us to develop the topic of "Tectonic Processes in the Interaction Between Geology and Ecosystems," and to submit this manuscript for review and publication.

Chapter 6

KARST AND ECOSYSTEMS

JULIUS TAMINSKAS[1], RICARDAS PASKAUSKAS[2], AUDRIUS ZVIKAS[2], JONAS SATKUNAS[3]
[1]Institute of Geology and Geography, Vilnius, [2]Institute of Botany, Vilnius, [3]Geological Survey of Lithuania, Vilnius

1. ABSTRACT

Karst is a type of landscape found on carbonate rocks (limestone, dolomite, marble) or evaporites (gypsum, anhydrite, rock salt) and is typified by a wide range of closed surface depressions, a well-developed underground drainage system, and a paucity of surface streams. Karst is, therefore, a particular kind of landscape where landforms due to solution processes prevail over other kinds of landforms. It is estimated that karst landscapes occupy up to 10% of the Earth's land surface, and that as much as a quarter of the world's population is supplied by karst water. The karst system is sensitive to many environmental factors. Karst aquifers contain the most important water resources of mountainous countries: these resources have often a good quality because karst areas have currently a low technogenic pressure.

Karst landscapes could be regarded as very conservative because the action of surface runoff is low. Moreover in karst areas we find underground systems of caves: these are probably the most conservative environments of the Earth's surface. In them we can find the records of the main environmental changes that occurred during the evolution of deep karst. In karst environments we often find important biotopes with endemic and relict species.

Upper Devonian gypsum and dolomite occurs beneath Quaternary sediments in North Lithuania. Karstic sinkholes appear frequently where the Quaternary cover is particularly thin and underlain by gypsum. The karstification is intensifying, perhaps related to climate change. The sinkholes and karstic lakes have a major significance for water exchange in the region under investigation. Sinkholes and karstic lakes play the role of 'windows' to underground aquifers and in the natural drainage network. Most sinkholes are dry or filled by water only in spring or after heavy rainfall. The sinkholes, the bottoms of which reach the groundwater level, form karstic lakes. The hydraulic connection with groundwater, which permits these to become permanently filled by water after some time, can be broken by sedimentation due to slope erosion leading to hollows that are dry or only temporarily filled by water.

In all karst regions of North Lithuania specific hydrological and hydrochemical conditions affecting dissolution of gypsum and migration of its products make the inorganic compositions of lakes diverse and often peculiar, to lakes of karst origin. In mid summer anoxic hypolimnions frequently develop such that hydrogen sulphide of biogenic origin is evolved from intensive sulfate reduction. While toxic to most aerobionts, it can support quite diverse microbiological processes of the sulfur cycle. On numerous occasions during the summer season, dense mats of purple sulfur bacteria occupy even the littoral zones of lakes.

It has become evident recently that intensive human activity in the region (e.g. pollution and/or use of fertilizers) may destroy the unique and fragile ecological balance. Gypsum karstic lakes of North Lithuania are considered as an important habitat the European scale. The Directives for Natural Habitat, Wild Flora and Fauna (92/43EEC) foresee their protection and inclusion into the network of EU protected sites. However, to implement the Directives a more detailed study of habitat characteristics is required.

2.   INTRODUCTION

Karst is a type of landscape found on carbonate rocks (limestone, dolomite, marble) or evaporites (gypsum, anhydrite, rock salt) and is typified by a wide range of closed surface depressions, a well-developed underground drainage system, and a paucity of surface streams. Karst is, therefore, a particular kind of landscape where landforms due to solution processes prevail over other kinds of landforms (Ford & Williams, 1989).

The highly varied interactions among chemical, physical and biological processes have a broad range of geological effects including dissolution, precipitation, and sedimentation and ground subsidence. Diagnostic features such as sinkholes (dolines), sinking streams, caves and large springs are the results of the dissolution action of circulating groundwater, which may exit to entrenched effluent streams. Most of this underground water moves by laminar flow within narrow fissures, which may become enlarged above, at, or below the water table to form subsurface caves, in which the flow may become turbulent. Caves contain a variety of dissolution features, sediments and speleothems (deposits with various forms and mineralogy, chiefly calcite), all of which may preserve a record of the geological and climatic history of the area. Karst deposits and landforms may persist for extraordinarily long times in relict caves and paleokarst (Jennings, 1985; Tools..., 1996).

Karst is most common in carbonate terrains in humid regions of all kinds (temperate, tropical, alpine, polar), but processes of deep-seated underground dissolution can also occur in arid regions.

The scale of karst features ranges from microscopic (e.g. zonation in chemical precipitates) to entire drainage basins (with caves that drain hundreds of square kilometers) and broad karst plateaus.

It is estimated that karst landscapes occupy up to 10% of the Earth's land surface, and that as much as a quarter of the world's population is supplied by karst water (Gvozdeckij, 1981). Karst systems are sensitive to many environmental factors. The presence and growth of caves may cause short-term problems, including bedrock collapse, disparities in well yields, poor groundwater quality because of lack of filtering action, instability of overlying soils, and difficulty in designing effective monitoring systems around waste facilities. Instability of karst surfaces leads annually to millions of dollars of damage to roads, buildings and other structures in North America alone. Radon levels in karst groundwater tend to be high in some regions, and underground solution conduits can distribute radon unevenly throughout a particular area.

Because the great variety of subsurface voids and deposits are protected from surface weathering and disturbance, karst preserves a record of environmental change more faithfully than most other geological settings. Temperature, rainfall, nature of soil and vegetation cover, glaciation, fluvial erosion and deposition, and patterns of groundwater flow can usually be read from cave patterns and deposits. This record can be resolved on an annual scale in the case of certain fast-growing speleothems.

In the case of Italy karst areas cover almost 20% of the whole territory; excluding the alluvial plains, the percentage is about 50%. Groundwater from karst aquifers plays a major role in the water supply of large regions in Switzerland and karst water is often of high enough quality that only simple water treatment is needed before its use (Tripe *et al*, 2000).

Karst areas represent an important environment for various reasons. Karst aquifers contain the most important water resources of mountainous countries: these resources have often a good quality because karst areas have currently a low technogenic pressures. Karst landscapes are sometimes referred to as very conservative because the action of surficial runoff is low.

Moreover in karst areas we find underground systems of caves: these probably are the most conservative environments of the earth surface. In them we can find records of the main environmental changes that occurred during the evolution of deep karst. In karst environments we often find important biotopes with endemic and relict species.

Karst responds with great sensitivity to environmental changes, and karst features (especially speleothems) contain many clues to past climatic and hydrological events and changes at a variety of time scales. It is uncertain whether future conditions can be interpreted from karst features, because many changes tend to be abrupt and discontinuous (Tools…, 1996).

3.   KARST OF NORTH LITHUANIA

In the north of Lithuania and in much of Latvia, Upper Devonian gypsum and dolomite occur under a thin cover of Quaternary deposits (Figure 6-1). Where the latter are 1–5 m thick, sinkholes abound and their number is ever increasing. Ground collapse usually affects the Quaternary cover and permits ready recharge of surface water into the Upper Devonian aquifers (Paukstys *et al.* 1997). Areas with such intensive water circulation in open gypsum systems are referred to as the intensive karst zone (Drake, 1980). However, even beyond the limits of this zone, where the cover sediments are thicker and not disrupted, intensive dissolution and removal of gypsum may also take place via underground cavities. Here calcium and sulfate ions predominate among the denudation products discharged with groundwater runoff, which later maintains river's runoff. If extreme changes of environmental conditions such as significant fall of groundwater level or pollution occur, sinkholes may develop in this outer region.

The boundaries of the karst region are determined by occurrence of carbonaceous and sulphatic rocks overlain by Quaternary cover. The zone of active carbonate (dolomite and limestone) karst covers occurs in Latvia but not in Lithuania (Figure 6-1).

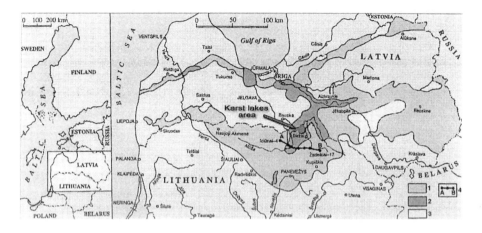

Figure 6-1. Lithuanian–Latvian karst region. 1 Pre-Quaternary carbonates; 2 zone of active sulphatic karst; 3 zone of active carbonate karst; 4 geological cross-section

The total area of active sulphatic karst where sinkholes appear frequently covers about 500 km² of North Lithuania. This is surrounded by a roughly equal area without recent sinkholes, though even here many springs indicate intensive underground denudation of gypsum. Both these areas are included in the zone of active sulphatic karst Figure 6-1. Sinkholes form when rock bridges collapse into underground cavities formed by long term gypsum dissolution followed by widening fissures, with larger cavities formed where fissures cross. The underground cavities are determined by the structural features of the rock bridges, humidity and static load or by the fall of the water level in the cavities. In the northern part of Lithuania the active recent karst development takes place in the Tatula and Suosa formations of the Upper Devonian (Table 6-1), represented by gypsum, dolomite–gypsum, gypsum–dolomite with clay, dolomite and dolomitic marl interlayer and lenses (Figure 6-2).

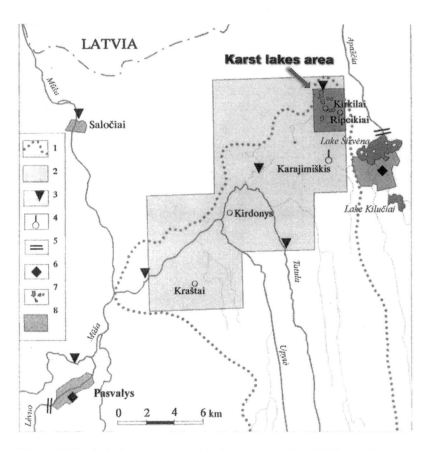

Figure 6-2. Geological cross-section of the karst region of North Lithuania (Iciunai–Tetervinai–Pervalkai–Stasiskiai–Gaiziunai–Sviliai–Zadeikiai): 1 groundwater level; 2 Quaternary deposits; aquifers: 3 Istras–Tatula; 4 Kupiskis–Suosas; 5 Sventoji–Upninkai; Aquitards: 6 Lower part of Tatula Formation: 7 Jara; 8 clay; 9 glacial loam; 10 gypsum; 11 dolomite; 12 dolomtic marl; 13 stratification of gypsum, dolomite and dolomitic marl; 14 limestone; 15 dolomitic flour; 16 sand with siltstone and clay interbeds and lenses; 17 borehole and tested extent of aquifer and piezometric level

Table 6-1. Middle and Upper Devonian stratigraphic scheme

SYSTEM	SERIES	STAGE	SUBSTAGE	REGIONAL STAGE	Group	Formation	Member	Geological index	Thickness, m	Lithology
DEVONIAN	UPPER	FRASNIAN	UPPER	PAMUSIS		Pamusis		D_3pm	3–2.5	clay, dolomite, mare
			MIDDLE	DAUGUVA		Istras		D_3is	2–4.5	dolomite
						Tatula	Nemunelis	D_3t^{nm}	6-20	gypsum, dolomite, dolomite marl
							Kirdonys	D_3t^{kd}	4-8	dolomite marl, dolomite
				DUBNIK			Pasvalys	D_3t^{ps}	6-25	gypsum, dolomite, dolomite marl
			LOWER	PLAVINAS		Kupiskis		D_3kp	5-10	dolomite
						Suosa		D_3s	12-19	Dolomite with gypsum
						Jara		D_3j	4-8	dolomite marl
				SVEN TOJI	Sventoji			D_3sv	90–100	sandstone, siltstone, clay
	MIDDLE	GIVETIAN		BURTNIEKI	Upminkai	Butkunai		D_2bt	100–130	sandstone, siltstone, clay
				ARUKULA		Kukliai		D_2kk		

GLOBAL STRATIGRAPHIC SCALE		REGIONAL STRATIGRAPHIC UNITS				
EIFELIAN	NARVA	Kernave		D_2k	6–20	clay, siltstone, sandstone,
		Ledai		D_2ld	60–80	dolomite, marl dolomite, clay

The karstic rocks are mostly covered by Quaternary deposits – Upper Pleistocene tills (boulder clay with lenses of sand and gravel), glaciolacustrine clays and Holocene sediments – alluvial and lacustrine sand, gyttja (Marcinkevicius & Buceviciute, 1986). The rocks of the Tatula Formation are dissected in various directions by fissures of different size. Karst cavities 10 and more meters in diameter are found through the entire vertical section of the Tatula Formation. Numerous boreholes detected them. Some cavities are partly filled with loam or sandy loam with debris of dolomite, dolomite marl and sometimes gypsum. Moving deeper, the number of karst cavities decreases. The deepest karst cavity filled with water has been found at a depth of 96.4 m. The vertical size of this cavity was 2 m. Active karst process also takes place in 0.8–3.8 m thick gypsum layers of the Upper Devonian Suosa Formation (Table 6-1) (Marcinkevicius, 1998).

Karst pits, karst wells and kettles, sinkholes, etc. comprise the surface karst landscape. Underground forms, which are unevenly distributed, include fissures, channels and cavities of various shapes. More than 8500 surficial sinkholes were counted in the sulphatic karst region of Lithuania (Marcinkevicius & Buceviciute, 1986). In some places more than 200 sinkholes of different size can be counted over an area of 1 km², comprising in total more than 0.3 km², as in Ripeikiai and Karaimiskis villages. Some of sinkholes are permanently filled by water. Therefore they are called karstic lakes. Often several karstic lakes are connected between each other and make rather large (up to 4 ha) karstic lakes. Nearby to these lakes territories with abundance of old generation lakes and sinkholes occupied by bogs and overgrown by trees are located (Figure 6-3). There the sinkholes and karstic lakes play a role of drainage system of groundwater. Precipitation and surficial runoff get into the sinkholes and further seeps through cracks in the gypsiferous rocks, enlarging them slowly, even forming caves up to several hundred cubic metres in volume. Occasionally the roof of a cave collapses, creating a sinkhole.

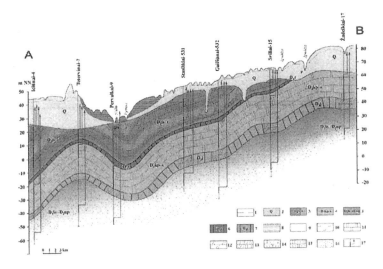

Figure 6 -3. Karst lakes area: 1 new generation karst lakes area; 2 old generation karst lakes and sinkholes area

3.1 Hydrology and hydrogeology

The karst region is located in the eastern part of the Baltic artesian basin. The active water exchange zone is up to 270 m thick and includes aquifers in Quaternary, Istras–Tatula, Kupiskis–Suosa and Sventoji–Upininkai Formations (Table 6-1). This series of aquifers is underlain by the 60–100 m thick regional aquitard of the Narva Formation. The main source for recharge of the Istras–Tatula and Kupiskis–Suosa aquifers is shallow groundwater and surface water. Variations in level and chemical composition of groundwater are determined by the rate of infiltration of precipitation. The thin cover of Quaternary deposits and the active karst features allow substantial pollution of groundwater. The Kupiskis–Suosa aquifer is the main source of water for single householders. The Sventoji–Upninkai aquifer is the source for centralised water supply in Birzai, Pasvalys and Panevezys towns.

The sinkholes and karstic lakes mainly are concentrated in the watersheds of the karstic region. The water collected there discharges mainly by springs directly to water courses or on the banks of streams. However, only one disappearing stream is known in this region. This stream (catchment – 11 km²) flows into a sinkhole, and continues flowing underground until reappearing at the surface as a spring at the village of Likenai. This spring is the source of a surface stream – Smardone.

The river runoff coefficient (ratio of runoff depth to precipitation depth) is 0.23–0.25 in the karst region. The average annual depth of runoff in the region varies from 126 to 221 mm. The greatest discharge occurs during spring floods. The minimal runoff occurs in the drought periods from July–October (warm season) and December–February (cold season).

The groundwater discharge into the karst rivers of the region makes up from 25% to 40% of annual runoff whilst in non-karst rivers it is only 8–16%. The runoff of karst rivers in drought periods always exceeds that of non-karst rivers because, during the spring flood, the karst system regulates the runoff and accumulates c.a. 15% of spring floodwater. This water significantly increases the minimal river runoff of summer drought period (Taminskas, 1994).

A great part of the Tatula river basin (hydrological monitoring network) is in the active sulphate karst zone (Figure 6-4). The normal runoff value is 193 mm (or 6.1 l s ^{1} from km²). It is 5% higher than the normal runoff value of neighbouring nonkarst landscape rivers. The normal groundwater runoff value of Tatula basin is 2 l s^{-1} from 1 km² (Taminskas, 1994).

Figure 6-4. Monitoring network of karst region: 1 Tatula river basin; 2 study area in active sulphatic karst zone: gauging and sampling site; 3 wells; 5 dams; 6 waterworks; 7 karst lakes; 8 urban area

3.1.1   Ecology of karst lakes

Karstic lakes are sinkholes permanently filled by water. Most of the sinkholes of the region under consideration are dry or are temporarily filled by meltwater in spring or by heavy rainfall. These lakes form where the base of the sinkhole is below the groundwater level. Karstic lakes maintain a hydraulic connection with the groundwater and are permanently filled by water unless sedimentation due to slope processes removes this connection leading to a dry depression or one only periodically filled by water. The sedimentation rate in the karstic lakes can reach up to 1.8 kg per m² per year. Two thirds of the sediment consists of mineral products of slope erosion and the remainder of organic detritus. The major part of the organic materials are leaves and remains of trees growing on the lake shores.

The mean depth of the karstic lakes varies from 3.5 m to 7 m; the area from 170 m² up to 4 ha.; and the water volume from 100 up to 42000 m³. The largest karstic lakes are formed from several connected sinkholes (Figure 6-5). The water level does not drop below the base of Quaternary deposits beneath which (2–5 m below surface) occurs an artesian (D_3t) aquifer consisting of fissured rocks.

Most of the karstic lakes have small basins with a very low surface inclination and rather impermeable soils. In arable lands, the karstic lakes often are isolated from their basin by soil rimwalls. Due to this the surficial runoff is minimal. The main flow of water into the lakes is from precipitation falling onto the surface of the ground (up to 920 mm per year, average – 640 mm per year) and groundwater inflow. Evaporation from the karstic lakes depends on the temperature regime, morphometric properties and density of surrounding trees. During warm periods of the year, evaporation is from 0.6 to 3.3 mm per day (on average 460 mm per year). The intensity of evaporation from deep lakes with steep slopes overgrown by trees is less by 2–3 times, compared with shallow, low and treeless shores. The mean residence time of water in these lakes is 2–5 years.

Karstic lakes are located in river watersheds of the unified recharge basin of the D_3t aquifer. The highest water levels in karstic lakes are reached during the spring thaw, with much lower rises during the summer and autumn rainfalls. However, sometimes, water levels of the karstic lakes rise in winter thaws. Fluctuations of water level can be up to 2.5 m. In long periods without precipitation, the water levels drop fairly evenly by from 4 to 8 mm per day. These water level fluctuations are identical to, and reflect, the groundwater fluctuations of the D_3t aquifer.

Because of high isolation and small area, water convection in the lakes is slow. The main water mass convection takes place in spring and late autumn due to the change of thermic gradient. Therefore, clear thermic stratification persists for long periods.

The hydrochemical regime of the karstic lakes is mainly determined by water exchange with groundwater and biochemical processes in each lake. Karstic lakes are characterised by particularly high values of sulfate and calcium ions and total dissolved solids (TDS). Values of other chemical elements differ less and are similar to the other water bodies of the region (Table 6-2). High values of oxygen demand (BOD_7 6.4–14.08 mg/l; COD – 25.2–2320 mg/l) and comparatively high concentrations of phosphorus and nitrogen determined in karstic lakes indicate pollution by organic and inorganic compounds that, in turn, have a direct influence on types and intensities of processes of primary production and destruction. One of the most influential and persistent sources of biogenic pollution, in addition to agricultural runoff, are the leaves of arboreal plants, growing densely on the shores of the karstic lakes.

Karstic lakes have characteristic abiotic parameters – water transparency, thermal and salinity regime, dissolved O_2, H_2S, CH_4. Gradients of these parameters define stratification of the water mass, which can be divided into two different, often overlapping, zones where favorable conditions occur for existence of specific microorganisms. Due to particular hydrologic and hydrochemical properties, that define the solubility of gypsum and migration of its products, the karstic lakes are characterized by very varied sulfur cycle processes and related microorganisms.

Table 6-2. The hydrochemistry of a karst lake (Kirkilu Lake, year 2002)

	Min.	Max.	Average		Min.	Max.	Average
pH	6.16	7.75	7.36	N_t, mg l^{-1}	0.66	3.37	1.33
SO_4^{2-}, mg l^{-1}	405	972	849	P_{min}, mg l^{-1}	0.00	0.02	0.005
Cl^-, mg l^{-1}	16	24	21	P_{org}, mg l^{-1}	0.01	0.13	0.05
HCO_3^- mg l^{-1}	207	356	246	P_t, mg l^{-1}	0.01	0.15	0.05
NO_3^- mg N l^{-1}	0.065	1.45	0.438	Ca^{2+}, mg l^{-1}	203	497	398
NO_2^- mg N l^{-1}	0.00	0.122	0.03	Mg^{2+}, mg l^{-1}	18	49	33
NH_4^+, mg N l^{-1}	0.00	1.027	0.342	Na^++K^+, mg l^{-1}	0.00	14	3.48
N_{min}, mg l^{-1}	0.08	2.60	0.81	TDS, mg l^{-1}	923	1843	1550
N_{org}, mg l^{-1}	0.05	0.88	0.52	BOD_7, mg O l^{-1}	6.4	14.08	9.93

It has been determined that processes of terminal anaerobic decomposition of organic matter (sulfate reduction) takes place in both the bottom sediments of different lakes (1.02–11.40 mg S^{2-}/dm^3 per day) and in anaerobic near-the-bottom layers (0.012–0.049 mg S^{2-}/dm^3 per day). The concentration of hydrogen sulphide at the bottom usually reaches values of only up to 5 mg/l and under the anaerobic conditions is a hydrogen donor for anoxic photosynthesis reactions of colored sulfur bacteria (Paskauskas *et al.*, 1998; Zvikas *et al*, 2002). On numerous occasions during the summer season dense mats of purple sulfur bacteria occupy even the littoral zones of the lakes.

Karstic lakes can be distinguished by a wide variety of aquatic microorganisms but even water bodies that are in close proximity vary in composition of species of phytoplankton. A high diversity of green algae, diatoms, and golden-brown algae species as well as cryptomonads is characteristic of karst lakes. The highest diversity of phytoplankton species (88) was determined in the oldest water bodies such as Lake Kirkilu. Many fewer species (up to 15) were determined in recently formed small sinkholes. All in all 223 algae species, forms and varieties have been identified (Kalytyte *et al*, 2002). The abundance of phytoplankton in certain layers of lakes reaches up to 8.0×10^6 cells/l, while planktonic heterotrophic protists comprise 30800 ind./l. The characteristic species structure of planktonic heterotrophic protists in these karst water bodies is subjected to a distinct change in abiotic parameters, in the course of a year leading to an abundance of certain species (*Spirostomum teres, Caenomorpha medusula*) that tolerate low oxygen concentrations. Chlorophyll-bearing mixotrophic ciliates *Prorodon viridis, Coleps hirtus* are also abundant in karst lakes. Furthermore, formation of an anaerobic community of certain cyanobacteria and heterotrophic protists species in sulphide bearing strata is common in most of the investigated lakes.

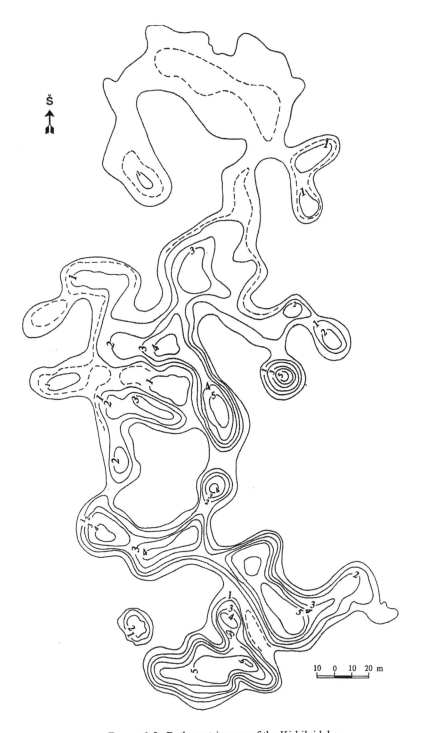

Figure 6-5. Bathymetric map of the Kirkilai lake

Thus, specific and exceptional abiotic conditions such as photic microaerobic metalimnions and anoxic hypolimnions with hydrogen sulphide, high salinity, etc., markedly influence the spatial and temporal structure of plankton communities in karst lakes as well as the formation of anaerobic photosynthetic and heterotrophic/mixotrophic populations of planktonic microorganisms.

4.   CONCLUSIONS

Karst landscapes, particularly in intensive karst zones, are very vulnerable environments that must be monitored to avoid hazards and to reconcile ecological and economic interests. Sinkholes can damage buildings, communication system and agricultural fields, as well as creating pathways for pollution of groundwater.

Most of the recently formed sinkholes are located in the watersheds. The majority are dry or filled by water only in spring time or during the heavy rainfalls. If the bottom of the sinkhole is below the groundwater level, karstic lakes form.

Sinkholes having a good hydraulic connection with groundwater aquifers and filled permanently by water (karstic lakes) can lose their hydraulic connections due to sedimentation from slope erosion and become dry or only periodically filled by water.

Sinkholes and karstic lakes perform the role of the drainage network of the territory. Precipitation and surficial runoff enters sinkholes and karstic lakes and further seeps through cracks in the gypsiferous rocks, enlarging them slowly, and even forming caves up to several hundred cubic metres in volume.

The increasing water exchange forms favourable conditions for the occurrence of the new sinkholes.

The largest water flow into the karstic lakes comes from precipitation on the surface of the lake and recharge by groundwater. Karstic lakes differ very much in their morphometric parameters, such as the time of the forming of the kettle, degree of overgrowing of shores, and connections with groundwater. This leads to differences in balance, and physical, chemical and biological indicators. Investigations of these indicators and processes in lakes demonstrates that even lakes in close proximity can be very different, closed water systems in which natural process are rather isolated from the wider environment. This peculiarity of karstic lakes, as well as capability to change their properties, is a basis for investigating these karstic lakes as "independent natural aquariums".

The karstic lakes are notable for their increased values of sulphates, calcium ions and total dissolved solids due to recharge by groundwater. High levels of biogenic elements indicate pollution by organic and inorganic matter that, in turn, modifies the character of intensive production-destruction processes in lakes. Leaves of arboreal plants, densely growing on the shores of the karstic lakes, are one of the potential sources of biogenic pollution. The leaves fall in autumn and decay in the water and bottom sediments.

Due to these, and other, unique hydrobiological properties, the gypsum lakes of North Lithuania have been recognized as habitats of European significance. They are proposed for inclusion in the EU network of protected sites of natural conservation under the Directives for Natural Habitat, Wild Flora and Fauna (92/43EEC).To implement the provisions of the Directives effectively, a more detailed study of habitat characteristics will be necessary (Rasomavicius, 2001).

PART II:
ENVIRONMENTAL IMPACTS OF THE EXTRACTIVE INDUSTRIES

Chapter 7

GEOENVIRONMENTAL PROBLEMS OF MINERAL RESOURCES DEVELOPMENT

A. I.KRIVTSOV
Central Research Institute of Geological Prospecting for Base and Precious Metals, Moscow, Russia

An idea of balanced development of the world, brought forward by the United Nations Organization and supported by 182 countries (Rio-92), implies satisfaction of today's demands for mineral resources without any damage for future generations. Possible ways to achieve this goal were analyzed during recent years in a number of international projects taking account of political, economic, social and environmental factors. It is obvious that balanced development requires achieving a balance of interests in the system "state (S) – mining companies (MC) – community (C) – habitat (H)". The first two elements in this system possess the strongest influences, whereas possibilities for cooperation community are considerably limited by the potential of democratic institutions for protest movements. The habitat is, in general, defenseless and passive, but only until critical situations creat a threat to mankind's existence. Preventive measures for such situations comprise at the present only a small part of nature-protective actions; mankind struggles chiefly with adverse effects already accumulated and current influences upon the environment.

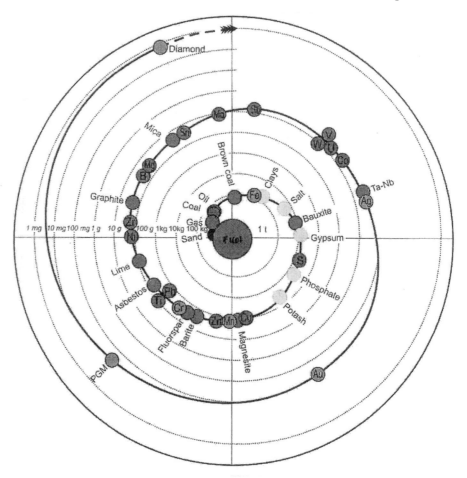

Figure 7-1. Masses of basic useful minerals per 1 of a thermo-electric plant in the world consumption of 1997 (Krivtsov, 1999).

Intensively developing processes of globalization in the mineral resources sphere have depended the division of countries into suppliers of raw minerals (ores) and consumers of them (in many cases these are consumers of already enriched ores). Suplier countries increase the load on their own geological environment and, hence, create ecological hazards for themselves. This is accompanied by improvement of the ecological situation in consumer countries, including the benefit that technological wastes are remain outside their territories. In any case, one should take into account the increasing division of countries according to "ecological cleanliness".

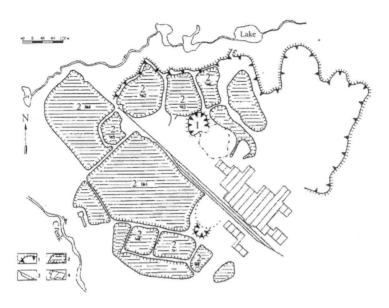

Figure 7-2. Location of dumps, dams and settling ponds of the mine "Yun-Yaga" in the Pechora Coal Basin. 1 – rock dumps; 2 – ash- and slag-settlers; 3 – access roada; 4 – dressing mill, boiler room and other manufacturing spaces (Subsurface Resources of Russia).

If scientific and technical progress is to be achieved, in the observable perspective mankind cannot refuse to consume mineral products belonging to non-renewable (in scales of mankind's history) natural resources. Repeated utilization of some metals and minerals (recycling) is not much help for controlling their initial production, whereas heat- and energy-generating products (oil, gas, coal) are consumed at the first stages of their use, with incomplete utilization of their combustion wastes.

The welfare of mankind depends on production and consumption of fuel and energy resources. A basic scheme – production of metals and materials for generation, delivery and consumption of energy - dominates in a modern technological infrastructure of the world. In turn, energy is consumed not only for production of metals and materials, but also for production of itself.

Possibilities for increasing production of mineral raw products in the far future are limited by depletion of natural resources in the subsurface of the Earth on continents, decrease of a finite and only slowly replenishable supply of useful minerals to technologically and economically sustainable levels (the so-called mineralogical barrier), as well as almost impossible technological barrier of increasing depth of production. The pressure for pursuing the latter is obvious from intensification of oil and gas extraction in closed and open seas, and in shelf zones.

In the present-day world, there exists a mineral resources balance within a global technological infrastructure, which is proved by the metal capacities (material capacities) of energy consumption, i.e. mass of raw mineral product per a ton of conventional fuel consumed (Figure 7-1).

There is no doubt that in the foreseeable future mankind will consume more and more mineral resources. To maintain the level of habitat comfort that has been achieved, it would be necessary to increase consumption proportionately with the growth of population, which cannot provide that level of the majority of people in the world.

According to the data of many researchers, today's world population amounting to about six billion people could grow during the next half century by 1.5% per year and may be doubled by the year 2050. At the present time, available resources for one inhabitant of the Earth is (Subsurface Resources of Russia): 2.32 ha of land, 0.26 ha of plentable soil, 0.8 ha of forest land, at water supply of 7.2 ths. m^3; the specific provision of already discovered fuel- and energy reserves is: 26.2 t of oil, 25.4 ths.t of gas and 300 t of coal for each inhabitant of the Planet.

American researchers consider that each inhabitant of the USA for his entire life will require 1620 t of basic useful minerals (including about 370 t of oil, over 160 ths. m^3 of gas – compare with specific reserves) – (Figure7-2).

Long-term average-world tendencies of extraction of basic mineral resources, as predicted by 2025 (Krivtsov, 1999), generally correspond to high rates of consumption, but do not reach the level in the USA (Table 7-1).

The degree of impacts of development and use of mineral resources on the environment directly depends on amounts of useful minerals extracted from the subsurface and subjected to different types of treatment for their final usage. It is inescapable that all these factors produce and accumulate anthropogenic wastes, and the greater the waste, the lower is the useful content from the initial raw resources (ores). Though such materials as sand and gravel can be utilized without wastes, it is quite another matter with production of relatively poor gold ores when extraction of one g of this metal produces not less than one ton of "waste" rocks.

Twenty-seven basic minerals extracted from the subsurface in 2003 are conventionally subdivided into the following groups (by initial data of the USA Geological Survey): I. 2-1 billion t (cement - 1.86; copper ore – 1.4; iron ore – 1.12; ores for abrasives – 1.1; gold ore – 1.0); II. 210-100 mln. t (zinc ore – 210; rock salts – 210; bauxites – 144; nickel ore – 140; phosphates – 138; platinoid ore – 120; limestone – 117; lead ore - 100); III. 55-27 mln. t (clay – 55; diamond ore – 50; silver ore - 40, tin ores, peat, potash salts – 27 each); VI. 16-10 mln. t (manganese ores – 16; chromium

Table 7-1. Prediction of world extraction of basic useful minerals (except Russian Federation and China) Absolute values and normalized for the levels of 1950 (1)

Useful minerals, measuring units	Factual, 2000	Prediction, 2025	Accumulated extraction
Oil with condensate, mln.t	2893,02	4730,00	95287,75
	5,84	9,56	
Natural gas, billion m³	2327,69	3710,00	75471,13
	12,93	20,61	
Coal, mln. T	3116,45	4200,00	91455,63
	2,69	3,62	
Total fuel- and energy resources recalculated	9777,94	15010,00	309849,20
	5,10	7,83	
Iron ores, mln. t	669,80	1236,00	23822,50
	3,40	6,27	
Phosphates, mln. t	99,51	195,00	3681,39
	5,21	10,21	
Copper, ths. t	12320,00	15524,00	348050,00
	5,60	7,06	
Chromium ores, mln. t	12,24	17,00	365,50
	7,20	10,00	
Nickel, ths. t	833,00	1304,00	26712,50
	7,00	10,96	
Molybdenum, ths. t	94,50	186,00	3506,25
	6,75	13,29	
Antimony, ths. t	8,50	26,40	436,25
	0,25	0,78	
Tin, ths. t	132,84	127,00	3248,00
	0,81	0,77	
Lead, ths. t	2325,00	2930,00	65687,50
	1,55	1,95	
Zinc, ths. t	6475,00	8690,00	189562,50
	3,50	4,70	
Mercury, ths. t	0,80	3,10	48,75
	0,20	0,78	
Gold, ths. t	2,08	3,10	64,75
	2,60	3,88	
Silver, ths. t	15,08	20,40	443,50
	2,90	3,92	
Manganese ores, mln. t	15,40	12,50	348,75
	4,81	3,91	
Diamonds, mln. carat	7,80	179,00	2335,00
	6,00	11,70	

Platinoids, t	265,00	402,00	8337,50
	17,55	26,62	
Bauxites, mln. t	120,75	189,00	3871,88
	17,50	27,39	

ores – 14; molybdenum ores – 13; holystone – 13; talk - 10); V. < 10 (barites – 7; fluorites – 5; silica - 4). In total, the measured quantity of hard minerals extracted in the world in 2003 is equivalent to a rock mass of not less than 10 billion t.

In comparison, the quantity extracted annually of fuel and energy raw resources is considerably greater than the first group of hard minerals. Over three billion tons of oil, almost four billion tons of coal, over three trillion m^3 of natural gas were extracted in the world in 2003.

In evaluating mineral ores consumed, one should take into account "handcrafted " extraction of so-called popular minerals that is typical in many countries around the world but cannot be exactly measured.

Totally, by minimal estimates, not less than 20 billion t of basic mineral resources are extracted annually at the present time. In excess of this, great amounts of so-called stripped rocks are extracted from the subsurface, blocking access to accumulations of useful minerals. Centralized accounting of such rock masses is not made. According to data from (Mezhelovsky ans Smyslov, 2003), each ton of extracted coal is accompanied by extraction of 1.5 m^3 of stripped rocks.

Integral estimates of mineral amounts that may be extracted by 2025 are given in Table 7-1

Based on the data from this table and through averaging the contents of useful minerals in ores accumulated by 2025, the wastes from utilized basic mineral resources are estimated at 150 billion t, including iron – 36; molybdenum – 35; gold – 31.6; phosphates – 20.8; copper – 34.5; zinc, lead, nickel, platinoids - from 4.5 to 2.5 billion t.

Respectively, growth of extraction and utilization of mineral raw products reduces sizes of initial geological space for production, primary and final processing, as well as disposal of wastes from mining, enriching and metallurgic manufactures.

In the most general case, usage of mineral raw products corresponds to a cycle of conjugate processes beginning with geological works on prospecting and exploration of mineral deposits, with achievement of success only in a limited number of objects. And only those discovered deposits that seem attractive for investment undergo further development. The latter includes extraction of mineral reserves from the earth's subsurface and is accompanied in major cases by their primary processing, i.e. enrichment with separation of useful components from so-called dead rocks.

After crushing and appropriate technological operations, these rocks are disposed into dumps. In most cases, concentrates of minerals are subjected to reprocessing at other plants, including metallurgic ones, where, in turn, new wastes are produced. Metals are turned into hardware, a part of which after their useful life serve as secondary sources of metals (scrap). All these processes affect the environment by degrading the state of its different elements.

It is gratifying to consider that geologic-exploratory works exert a minimum influence on the environment, and their consequences are easy to eliminate. In this respect, an exception is prospecting and exploration of oil and gas deposits where deep boreholes are drilled with the use of technological systems that cover large areas. Consequences of laying access roads and communications, construction of sites for drilling rigs and so-called storehouses for wastes and other geomechanical disturbances of the earth's surface after completion of borehole sinking are not liquidated fully. In Russia during drilling one m^3 of water is used for each meter of borehole sinking; drilling waste waters, drilling mud and sludges are accumulated. In West Siberia, the latter amount, respectively, to 0.24, 0.20 and 0.18 per each meter of borehole sinking. Annual volume of such wastes contaminating, first of all, topsoils, is estimated at 25 mln. m^3 with a content of chemical reagents equal to 1.7 mln. t – recalculated for hard substances. Scales of such influences depend on drilling amounts and concentration of drilling works. Volumes of drilling accumulated in Russia exceed 160 mln. m^3 (Mezhelovsky and Smyslov, 2003). The same factors influence the environment during construction of exploitation boreholes, depending on specific loads upon a unit of area.

Oil and gas extraction leads to considerable geomechanical disturbances of the earth's surface occupied by production and transportation infrastructures, contamination of soils and water by oil products, and multiple increase of discharged contaminated waste waters into surface water streams. By data from publication "(Mezhelovsky and Smyslov, 2003), during 1991-1999 the volume of contaminated waste waters in the Russian oil industry amounted to about 200 mln. m^3.

A specific problem of the domestic oil extraction is complete utilization of co-extracted gas, which in Russia of today does not exceed 80%(Mezhelovsky and Smyslov, 2003). Over 5.5 billion m^3 of gas are burnt annually in so-called "gas torches", with outbursts of about 400 ths. t of hazardous substances into atmosphere. Burning torches generate convective heat fluxes of air and create local meteorological media within a radius of to five km from a borehole mouth.

Annual leakage of methane to atmosphere at oil- and gas extraction is estimated at 560 mln. t (Mezhelovsky and Smyslov, 2003,); scales of accidental gas outbursts during borehole gushing and breaking of pipelines are not accessible for reliable integral estimation.

Table 7-2. Content of microelements in oil and stratum waters (Subsurface Resources of Russia)

Elements	Content of microelements			Elements	Content of microelements		
	in oil, g/t	in oil ash, %	in stratum water, g/m^3		in oil, g/t	in oil ash, %	in stratum water, g/m^3
Vanadium	0,03- 1170	to 20,0	0,003	Copper	0,1-20,0	0,03-10,0	to 29,0
Nickel	2,0-350,0	to 6,0	0,06	Tin	0,001-0,6	10^{-3}	-
Aluminium	1,0-75,0	to 2,5	-	Gold	to 0,001	-	0,01
Zinc	0,1-35,8	to 1,1	0,1-28,0	Silver	10^{-3}-9,8	10^{-4}	to 2,0
Cobalt	0,03-42,7	10^{-4}	0,004	Boron	to 10,0	to 0,3	6,0-2054
Iron	3,2-165,0	to 3,7	0,127	Gallium	to 0,001	10^{-3}	-
Selenium	0,03-4,0	-	—	Indium	to 0,5	10^{-3}	-
Manganese	0,1-30,0	10^{-3}	0,004	Germanium	0,015-0,69	10^{-5}	0,002
Arsenic	0,05-8,8	0,005-0,03	to 0,03	Titanium	to 3,4	10^{-3}	0,01-1,3
Scandium	to 0,01	10^{-4}-10^{-5}	-	Cesium	-	-	0,3-15,0
Chromium	0,1-2,4	to 0,08	to 0,7	Cadmium	0,02-12,7	-	-
Antimony	0,03-0,11	-	-	Uranium	to 0,001	10^{-4}-0,05	0,4-0,7
Mercury	0,02-30,0	-	0,02-0, 18	Radium	to 10^{-8}	-	10^{-9}
Barium	0,01-0,14	to 0,35	to 60,0	Thorium	to 10^{-4}	-	-
Bromine	1,0-10,0	to 1,9	51-4107	Selenium	0,1	10^{-6}	-
Iodine	1,0-10,0	to 3,2	2,0-120,0	Rubidium	to 0,28	-	to 8,7
Strontium	0,2-19,0	0,03-1,51	$(8*10^{-3})$-8	Rhenium	0,05-0,2	_	to 0,01
Lithium	-	to 0,14	2,0-32,0	Antimony	0,05-6,0	10,0-5,0	-
Molybdenum	30,0	0,02	0,001	Beryllium	to 0,1	-	0,0002
Lead	0,01-10,0	to 0,68	to 84,0	Cadmium	to 0,66	-	-

Exploitation of oil- and gas deposits is accompanied by strong geochemical processes: huge masses of a whole number of elements, including radioactive ones, are transported with oil, gas and oil waters to the surface. It is reported that during past 20 years, the oil-producing enterprises of the USA have accumulated about eight mln. t of radioactive wastes.

According to scales of elements transport and increase of their concentrations, oil-producing enterprises are comparable with natural ore-generating systems where exploited oil-bearing seams and layers serve as recharge zones, and technological facilities and, first of all, pipelines serve as accumulating zones (Table 7-2).

Considerable transformations are also caused in the subsurface space. Changes in geochemical characteristics in combination with drastic alterations of hydrodynamic regimes bring principal changes in hydrogeochemical and hydrogeological states of the subsurface space, which in combination with inter-stratum water seepage leads to irreversible changes in water-saturation degree, disturbing, hence, underground water balance and creating complicated problems of current and prospective water supply in large oil-producing centers.

The most threatening signals of excess in redistribution of hydrocarbon extraction from the subsurface admissible by nature are earthquakes anthropogenically induced in oil- and gas-producing regions due to a disturbance of geodynamic equilibrium in exploited deposits.

Development of hard minerals differs from oil development not only by techniques of extraction, but also by much larger amounts of technological wastes per unit of a useful component, as well as by much larger territorial spreading of mining enterprises (excluding some of the large system-forming mining and metallurgical plants).

In Russia 330 basic enterprises are engaged in coal development - 212 mines (150 mln. t per year) and 118 opencast collieries (210 mln. t per year). It is reported in (Mezhelovsky and Smyslov, 2003) that each one mln. t of extracted coal is accompanied by discharge of 3.22 mln. m^3 of water, stripping of about 1.5 mln. m^3 of rocks, outburst (into air) of almost three ths. t of hazardous substances, and by disturbance of over 10 ha of lands. Wastes from development and concentration of coal are added by products of their burning (ash and slags) containing multiple increased contents of many elements including also some that are toxic and radioactive.

Extraction of hard minerals from the subsurface both at open-cut and underground quarrying is inevitably accompanied by rock deformations and disturbances in geodynamic equilibrium, which often leads to subsidence of the earth's surface, appearance of anthropogenically induced landslides, and in some cases to local earthquakes affecting also urbanized centers. Mined holes cause within their influence areas a change in hydrogeological regime due to pumping of groundwater discharged to surface water streams.

Use of explosive substances to crush extracted ores and rocks is accompanied by pollution of air and surface areas adjacent to mining enterprises. Influence of mass explosions is especially notable when they are carried out in quarries to cut out and crush rocks and ores. Basic pollutants of atmospheric air in areas of mining and metallurgical enterprises are: sulphur dioxide and dust (annually 200 mln.t for the Russian Federation), nitrogen oxides (60 mln.t), hydrocarbon oxides (8 billion t).

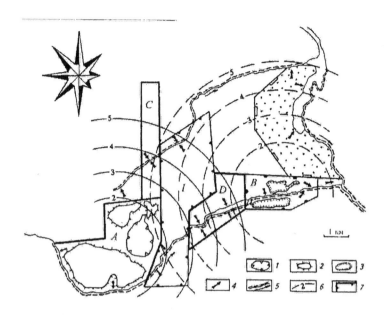

Figure 7-3. Map of runoff and dust loads upon the area of Sibai City (Bashkorstan) (Authors: A.I.Krivtsov, A.N.Gerakov) 1 – mining quarry; 2 – dumps of waste rocks and balanced ores; 3 – settling ponds of dressing mill; 4 – directions of industrial, ore and domestic waste runoffs; 5 – areas of the hydro-network contaminated by wastes; 6 – distribution of dust load from its basic sources (figures show a distance in km); 7 – industrial-urban zones (territories): A – mining, Б – dressing, В – railway transport, Г – local industry, Ж – dwelling, Р – recreational zone.

Generally, a mean-productivity mining enterprise requires a land area of 2-3 ths. ha. Under action of wind and water migration, the influence zone of a single plant increases by 10-15 times (Figure7-3). Representative is the example of ore-dressing plants in Krivoy Rog Region, which cover over 20 ths. ha and dispose annually 90 mln.t of tailings from iron ore dressing.

All types of anthropogenic wastes have an increased toxicity that depends both on initial properties of useful minerals and a complex of chemical reagents used for ore dressing. Of significance is the impact of exogenic factors and, first of all, atmospheric precipitation and oxygen, upon changes in mineral forms and mobility of ore-generating elements that determines their possible migration in the habitat sphere.

The publication (Mezhelovsky and Smyslov, 2003, "Subsurface Resources of Russia") gives an overview mining and metallurgical enterprises of Russia that have both long and relatively short functioning histories.

For over two centuries the Russian Federation has been and still remains today actually the only country with high extraction of gold from placer deposits, most of which are located in especially unstable landscape zones. During the entire history of the Russian gold production, over 4000 placer deposits, not taking into account numerous small objects, were involved in development.

Mining works are carried out in valleys of water streams; already before the start of processing gold-bearing sands, huge masses of stripped rocks ("peats") begin migrating, covering river flood plains and terraces, which leads to full removal of vegetation and formation of anthropogenic landscapes, in response to requirements of mining works. During washing of gold-bearing sediments, dead rocks are accumulated in dumps that grow annually with an decrease in the already limited contents gold content. Thus,-, the amount of migrating and washed rocks in 1995 reached 1 mln.m^3 against 500-600 mln.t in 1970.

Loosing of sediments and washing of gold-bearing sands are accompanied by contamination of water and silting of river channels for many kilometers from placer deposits, which influences the state of water biocoenoses.

At dredge-aided development of placers, artificial water bodies are formed that radically change hydrological regimes with dewatering of below-lying flood-plain valleys.

Changes in structure of drainage systems exert additional influences upon thermal regime of permafrost with a change in rock state.

Areas with long gold development periods have accumulated huge volumes of stripped and dumped rocks: in Lensko-Bodaibinsky area – 10.5 billion m^3, Verkhne-Kolymsky area – 9.0 billion m^3, Aldan area – 4.0 billion m^3, in areas of Middle Urals, South Urals, Yenisey River, Upper Undigirka River - from 2.4 to 2.0 billion m^3. In 10 other areas these volumes amount to about one billion m^3. Some areas of gold placer development possess an ability of self-purification and self-recovery with a velocity of 7-20 to 15-30 and 15-50 years. However, these processes are interrupted in cases when wastes of earlier development are again involved into developing process.

In case of development of hard minerals, main part of unusable and deteriorated lands appear to lie near mined holes, rock dumps, tailing- and slag-storages. Areas of deteriorated lands are considerably different in size: extraction of 1 mln. t of iron ores in Russia requires from 14 to 500 ha of land; limestone - from 60 to 1200 ha; phosphorites - from 20 to 80 ha.

The book (Mezhelovsky and Smyslov, 2003) describes environmental influences of the largest Russian territorial-industrial plant complexes on development, enrichment and metallurgical processing of raw ferrous and non-ferrous metals.

The Kursk Magnetic Anomaly (KMA) integrated complex on development of iron ores is located in a dense-populated Chernozem zone of Russia. Three quarries and one mine of the complex extract annually about 65 mln.t of ore and dump over 25 mln.m^3 of stripped rocks. Each of these mining enterprises has functioning ore-dressing mills accumulating tailings of ore dressing. Treatment of concentrates is carried out at the Starooskolsky electrometallurgical integrated plant; a plant of iron hot-briquetting is under construction at the Lebedinsky plant.

Since beginning of the KMA complex operation, Pre-Cambrian rocks are stripped over an area of 200 km^2; over 100 ths. ha of chernozem are made unusable for agricultural production; wastes of ore-dressing amount to about 60% of the ore mass and together with metallurgical - about 80%; share of used rock mass is estimated at 10% of the extracted amount; volume of dumps and tailing-storages exceeds 1.5 billion m^3. An anthropogenic relief is intensively formed - the valleys and ravines are filled with wastes (tailings), dumps of stripped rocks have a height of 60 to 100 m; Lebedinsky quarry with an area of 10 km^2 reached a depth of 350 m, Stoilinsky quarry (7 km^2) - 200 m.

About 100 mln. m^3 of water are annually pumped from the quarries; the cone of depression at the Lebedinsky quarry covers 380 km^2 with a water level decline by 10-15 m which has led to drainage of dug-wells and water wells, as well as streams and small rivers. Water losses from tailing-storages estimated at thousands of m^3/hour, lead to contamination of sources of water supply.

Cutting out of rocks in quarries with the aid of mass explosions with a recharge capacity of over 400 t of explosive substances is accompanied by outburst of dust and gas in amounts of 15-20 mln. m^3 to a height of up to 300 m with subsequent distribution and outfall at a distance of up to 10 km.

A specific problem of the KMA complex is the growth of radon emitted from rocks during crushing, and from stripped fissured structures.

Pollution of the atmosphere and dust loads on soils cause accumulation of many elements, including toxic ones. Within a radius of 15-17 km from the Lebedinsky complex it is not recommended to use for food a number of agricultural crops.

Additional plants to process mining wastes have been created at the KMA complex. In turn, such plants create, in one or another way, additional loads upon all the components of the geological environment.

The above-cited publication (Mezhelovsky and Smyslov, 2003) contains the characteristics of the Norilsk mining complex, extracting and processing ores of non-ferrous metals. The complex emits great amounts of sulphur anhydride to atmospheric air (to two mln.t per year); copper accumulations in soils near smelting enterprises exceed the primary concentrations in natural ores. Norilsk City is blocked from all sides by metallurgical plants and is subjected to action of multi-component smoke.

In general, the main factors affecting the geological environment of hard mineral extraction are determined by processes that form hollows formation in the subsurface and accumulation of wastes on the surface. The main factors disturbing equilibriums are created by technological processes of breaking and treatment of ores with the use of explosive substances and different reagents.

Mining works and metallurgical plants possess a high-energy capacity; basic amounts of energy resources are produced by burning processes that consume atmospheric oxygen. Not obtained yet are appropriate estimates of losses of this element for transformation of disintegrated rocks and ores, which gradually come in chemical (and geochemical) equilibrium with conditions of the earth's surface.

The problem of mankind's oxygen supply was discussed in the publications of A.T. Pikhlak (Pikhlak, 2005) who showed in particular that burning of all the fuel and energy reserves ready for treatment would require 1637 billion t of oxygen. At a constant level of consumption, this resource may be depleted already by the year 2038. Losses of atmospheric oxygen for oxidation of huge rock masses extracted to the surface are not included in the above estimate, but it is obvious that they are considerable. It is impossible to ignore the opinion of A.T.Pikhlak who is sure that only forest covers can save the Planet not only against oxygen lack, but also against the greenhouse effect.

The expected doubled growth of population on the Earth by 2050 will lead to the fact that the earth's area per one inhabitant will reduce almost to one (!) ha. The same growth rates of production and consumption at the expense of waste accumulation on the earth's surface will further speed up this reduction. In spite of intensifying nature-protective actions, it is far from assured that mankind will manage to "lay back" all the wastes accumulated from mineral production, because this will require gigantic expenses not comparable, perhaps, with cumulative expenses for production and treatment of mineral ores.

It follows from the mineral resources balance of our modern technological structure (metal- and material capacity of consumed energy) that in 2000 consumption of one t, obtained from oil, gas and coal, was accompanied by accumulation of not less than 6.5 t of wastes produced from extraction of 25 basic useful minerals (not taking into account metallurgical wastes). An illustrative example is such a fact that each golden ring of a weight of nine g with a diamond of three carats is "shadowed" by production of 4-5 t of wastes, all of which lies outside the habitat of the owner of this adornment.

A growing threat of habitat area exhaustion with deterioration of state and quality of air balance, surface and ground waters can be weakened through wider use of non-traditional sources of energy (in the first turn, geothermal) and creation of basically new systems (alternatives) of energy supply.

The worst alternatives are demonstrated in fantastic films diverting people from the disfigured surface of the Earth into the subsurface space emptied by unreasonable extraction of its natural riches.

Chapter 8

GROUNDWATER AS A COMPONENT OF THE ENVIRONMENT

IGOR S. ZEKTSER
Water Problems Institute of Russian Academy of Sciences

1. ENVIRONMENTAL IMPACT OF GROUNDWATER EXTRACTION

It has become obvious during recent decades that any type of anthropogenic activity (civil engineering, agriculture, deforestation, extraction of surface and ground waters, etc.) inevitably has an impact upon ecosystems. The main task in solving this problem is to predict possible changes in the geological environment and to provide scientifically sound recommendations for the prevention of negative anthropogenic influences on marine and terrestrial ecosystems.

As is widely acknowledged in many countries, one of the most important natural resources is groundwater. Fresh groundwater plays a significant and growing role in drinking water supply and irrigation in many countries. This is because when compared with surface waters, groundwater has a number of advantages as a water source: it is of better quality with less risk of pollution and infestation; and it has a lower susceptibility to seasonal and long-term fluctuations. In the majority of cases, the use of groundwater does not require expensive water treatment. In many European countries (i.e. Austria, Belgium, Germany, Hungary, Denmark, Rumania, Switzerland) groundwater use exceeds 70 % of the total water consumption. In the USA the share of groundwater in municipal and drinking water supply is about 50 %, in Russia it is about 45 %. According to data from the European Economic Commission, groundwater is the basic source of urban municipal and drinking water supply in the major European countries. The water supply of large European cities such as Budapest, Vienna, Hamburg,

Copenhagen, Munich, Rome, Vilnius, Minsk and others is completely, or almost completely, based on groundwater. In countries with an arid or semi-arid climate, groundwater is widely used for irrigation. About one third of all irrigation is at the expense of groundwater.

When dealing with problems of groundwater use, it is always necessary to recognise that groundwater is not only a useful mineral, but also an important component of the environment and, hence, of surface ecosystems. Therefore, any changes in the groundwater regime and resources are inevitably manifested to some degree within the other components of an ecosystem. Groundwater extraction can force mineralised waters to move from deep layers upward to well fields, and in coastal zones, – salt water intrusion into near shores aquifers. All these circumstances should be taken into consideration when studying the stability of ecosystems and their resistance to anthropogenic impacts.

Zektser et al. (2005) review a few examples of the most notable impacts of intensive groundwater extraction upon ecosystems in the northwestern USA. In particular, the authors focus their attention on the negative influences of lowering shallow groundwater levels, resulting in stream flow reduction and the decline of lake water levels, stunting   of vegetation, land subsidence, and sea water intrusion into coastal aquifers.

Some examples of methods and studies of the prediction and elimination of impacts from intensive groundwater extraction upon ecosystems, both within and outside water bodies, are given below.

2. INFLUENCE OF GROUNDWATER EXTRACTION ON RIVER RUNOFF

One of the most serious environmental consequences of groundwater extraction, along with water resource depletion, water level lowering and the formation of cones of depression, is a change in the interconnection between surface runoff and the subsurface. This is especially important in the assessment of water reserves and environmental consequences caused by the filtration recharge of aquifers from rivers. Such recharge areas are located along rivers and their exploitable reserves are formed completely by water filtration from the river. It is obvious that the operation of such extraction wells has a direct influence on river runoff. It is thus natural that the study of the interaction between surface and ground waters under conditions of groundwater exploitation is the subject of numerous scientific investigations, international conferences and symposia.

Changes in surface runoff due to groundwater extraction were discussed for the first time by Theis (1935). Theis analysed the exploitation of single water well with a constant yield near a perfect river, assuming that the water-bearing layer was homogenous and spread infinitely directly from the river. Most of the problems concerning the interaction between surface and ground waters were first given a theoretical grounding by Khantush (1965), who showed that computations of water-intake productivity should take into account the influence of the boundaries of water-bearing layers, seepage from neighbouring aquifers and imperfections in the hydraulic link between aquifers and rivers. Later, other authors showed that without reliable assessment of changes in the surface runoff caused by groundwater exploitation, it is impossible to make a realistic estimate of the productivity of well fields

Possible changes in river runoff due to groundwater exploitation depend on a range of natural and anthropogenic factors, among which the most important are:

- type of a hydraulic link between the aquifer to be assessed and the river in different seasons, controlling regime and dynamics of subsurface runoff to the river;
- seasonal variability of river runoff over the annual and long-term cycles;
- type and amount of an aquifer recharge and discharge, including possible changes in evaporation from the shallow groundwater surface due to an exploitation-induced lowering of its level;
- water- extraction yield and distance of pumping wells from river channels;
- duration and regime of water- extraction exploitation;
- hydraulic capacity of the exploited aquifer.

It is necessary to take into consideration all the factors listed above in the quantitative assessment of prospective productivity of water extraction wells, as well as in determining possible changes in river runoff due to groundwater exploitation.

In most cases, exploitation of groundwater that is hydraulically connected to rivers causes changes in the river runoff. However, the types, trends and scales of such changes can vary widely at different stages of water well operation. Owing to the inherent stability of water-bearing systems, changes in river runoff do not become visible immediately: they tend to be smoothed out and there is a time lag in their appearance. Moreover, in some cases river runoff can actually be increased through the discharge of waste groundwater that has been extracted from deeper aquifers not hydraulically connected to the river under study. The methodologies and models currently available enable prediction of possible changes in river runoff due to groundwater exploitation with a sufficient degree of accuracy. This, in turn, provides the possibility of establishing a system for regulating extraction wells productivity that would prevent a catastrophic or unacceptable (for different spheres of man's activities such as fishing, shipping, recreation purposes etc) decrease in river discharge. In some cases when a considerable decrease of mean river runoff due to groundwater exploitation is predicted, special compensating measures can be put in place, such as the construction of regulating dams, etc.

Assessment of changes in river runoff makes wide use of hydrodynamic calculations, including analytical, numerical and analogue methods that are described in detail in the specialist literature. The present chapter, therefore, presents only some examples and results for particular assessments of the influence of intensive groundwater extraction upon river runoff.

A detailed analysis of the influence of groundwater exploitation upon river runoff was carried out by (Zeyegofer et al, 1991) and co-workers in connection with the preparation of "A general scheme for a joint water supply system for Moscow City and the Moscow Region using subsurface water sources". Calculations of changes in particular components of the water balance for the defined river basins showed that extraction of groundwater would cause an increase of its inflow to water wells from aquifers located outside the Moscow Region. At the same time, it was possible that subsurface recharge to rivers would decrease during exploitation, perhaps down to 30 % of its natural value. However, the real damage to the average river runoff at a 50%-water provision would be considerably lower.

Greater damage would occur in those rivers that drain the lower layers of the fresh water zone in the Moscow Region composed of Carboniferous carbonate rocks and in which the exploited aquifers are located.

According to the data of Zeyegofer et al. (1998), the subsurface river recharge from these aquifers during their intensive exploitation (60-80 years) could be reduced almost by 50% compared with earlier natural conditions. However, this statement concerns large rivers such as the Klyazma, Moskva, Oka which have a rather high natural and transient runoff, that in absolute terms is many times higher than the predicted reduction in subsurface inflow. Therefore, if the decrease in subsurface recharge is compared with the average annual river water discharge, the damage to the river runoff appears to be rather low. It should be emphasised that this optimistic and important (in practical respects) conclusion applies only to large rivers and average annual river discharges. However, in periods when the mean water level is low, the decrease in river runoff due to intensive exploitation of the aquifers could be very considerable, and should be taken into account when designing different constructions on rivers (e.g. installations, recreational structures, etc.).

The above, whilst valid, does not take into account the return runoff, i.e. discharge of groundwater back to rivers after its use. If the considerable amount of used fresh groundwater that is usually discharged back to the river is taken into consideration, then the damage to river runoff will probably be negligible. As the used water is often not discharged in the same place that extraction occurred, a redistribution and change of the mean runoff regime, in comparison with the natural state, may be expected in some areas of the river. This phenomenon is further complicated by the discharge of used surface waters to rivers, which makes it difficult to analyse changes in the river runoff due to groundwater extraction. In some areas of rivers, especially in the headwaters of small rivers and streams, a discharge increase is observed because of the return of waste (used) groundwater into them (Zeyegofer et al, 1998). A considerable reduction of river discharges in areas where large groundwater extractions are located has been recorded by English specialists.

Cherepansky (1999) carried out a great deal of work on the use of modelling methods to predict possible changes in the runoff of small rivers in Byelorussia. These studies made it possible to distinguish the river areas that are most at risk from runoff reduction and to establish measures for the prevention of the impact of groundwater extraction on the river runoff.

Kovalchuk and Kropka, (1993) report interesting data on the influence of long-term groundwater exploitation upon the water regime of the Drama river basin in Poland. Due to the formation of a deep and wide cone of depression within the exploited aquifer in this area, changes have occurred both in the hydrodynamic boundaries of the basin and in the general nature of the river runoff.

Considerable reduction in river discharge is observed in the Tokyo area, caused by a decrease in the yield of springs or their disappearance under the influence of intensive shallow groundwater extraction.

Intensive groundwater extraction from the upper aquifers in the southeast of the Coastal Valley (USA) has led to a decrease in the subsurface recharge of rivers and lakes in this area (Testa, 1991).

3.   INFLUENCE OF GROUNDWATER EXTRACTION ON VEGETATION

As has been mentioned above, in those regions where there is a strong interconnection between exploited aquifers and overlying unconfined shallow groundwater, water extraction causes a lowering of the shallow groundwater level. This lowering can affect the state of the landscape. The most sensitive landscape element that reacts to changes in the level of the shallow groundwater table is vegetation.

How the lowering of the shallow groundwater level affects vegetation depends on which regime of water recharge of plants is predominant – automorphic or hydromorphic.

In the automorphic regime of recharge, the plant roots do not reach the subsurface water table or the height of the capillary rim and, hence, the plants get water only through atmospheric moisture infiltration into the root zone. The depth of plant roots added to the height of the capillary rim is often called the critical depth. If the depth of shallow groundwater occurrence is deeper than the critical depth, the plant recharge regime is automorphic, i.e. mainly at the expense of infiltrating precipitation. But if the depth of shallow groundwater occurrence is less than the critical depth, a considerable contribution to plant recharge comes from shallow groundwater, and such recharge regime is called hydromorphic.

The results of numerous pilot and experimental investigations (Zhorov, 1995; Zeyergofer, et al, 1998) show that the root depth of the majority of plants does not exceed 5 m. Thus, in the near-Oka River reserve area of the Moscow Region, the maximum root depth for pine is generally less than 3 m (only in one of 17 measurements did it reach 3.9 m), for oak – 5.1 m, for linden – 2.5 m, for birch – 3.4 m, for aspen – 4.4 m, with the major mass of roots located in the upper 0.5 m thick layer, i.e. in the zone of intensive precipitation infiltration. In the Reserve Area the major mass of roots for the 10-15 –year old and 60-65 –year old pines were at depths of 0.4 m and 0.8 m respectively.

Grass and cultured plants have the same general order of root depths. Kovalevsky (1994) indicates that the relationship between ecosystem productivity and shallow groundwater table depth is parabolic. The extremum of such parabolas correspond, on average, to optimal depths of shallow groundwater table during the growing period.

Various experimental investigations have shown that, on average, the optimal depth of shallow groundwater occurrence, during the growing season, is 1.2 to 1.5 m for the cotton-plant, from 0.7 to 1.5 m for most vegetables, and from 2 to 3 m for fruit gardens. The highest productivity of the coniferous forests that are widely developed in the humid zone of the East-European plain is found in sandy loams with a shallow groundwater table at depths of 1.5 – 2.0 m.

The height capillary rise of moisture above the shallow groundwater table depends on the lithological composition of the unsaturated zone. For sands of various granulometric compositions, it varies from 0.1 to 0.5 m, for light loam and peat bogs from 2.0 to 2.5 m, for heavy loam from 3.0 to 4.0 m.

The data given above allow the following principal and practical conclusion to be drawn: if the natural shallow groundwater table, connected hydraulically with an exploited aquifer, lies at a depth lower than the critical depth, groundwater table decline caused by water extraction will have no impact upon vegetation. In practice, this means that if the shallow groundwater level in sands is deeper than 5 m and in loams deeper than 7 m, then no intensive exploitation of groundwater can affect vegetative communities. However, this conclusion is true only for typical vegetation in the humid zone and the zone of moderate wetting. For plants of arid climates such as, for example, eucalyptus, the critical depths will vary and the influence of water extraction on vegetation will be different.

In the hydromorphic moisture regime, the best conditions for growing plants, both natural phytocenoses and cultured plants, are observed when the depth of shallow groundwater level is in the range of 0.5 to 2.0 m. This is the optimal depth interval where the basic mass of plant roots are usually located., A decline or rise of groundwater levels relative to the optimal depth may thus have an adverse influence on vegetation.

Some examples of the negative influence of groundwater extraction on vegetation can now be considered.

In the mountainous regions of the Balaton Lake basin, Hungarian scientists have reported the disappearance of a number of forms of vegetation along the banks of streams, caused by the lowering of the level of the karst aquifer due to intensive pumping of mine waters. In the Netherlands, the lowering of shallow groundwater levels in the lowland areas during recent decades has led to losses in both flora and fauna in nature reserves (Zhorov, 1995). Proposals have been put forward to reduce groundwater extraction from the sediments of the coastal zone of the Netherlands by using external water supplies. In the Gvadalkvivir River estuary of southern Spain, intensive groundwater extraction has resulted in a decrease in the area of wetlands from 200,000 ha to 27,000 ha, thus greatly reducing rest areas and feeding grounds for birds migrating from Europe to Africa and back.

Kovalevsky (1994) reports that an increase in the thickness of the unsaturated zone and shallow groundwater depth, due to groundwater exploitation in the Severny Donets River valley (Russia), has led to the disappearance of drying out of flood-plain lakes, reduction in the forest canopy due to desiccation soils, and changes in plant species. Similar exploitation has caused drying-out of the cedar groves in the Reserve Area of the Urals and oak woodland in the Lebedinsky quarry area of the Kursk

Magnetic Anomaly, and gardens in the area of Krasnodar City in the North Caucasus.

Groundwater extraction has lead to the formation of extensive cones of depression in river valleys of many arid zones causing the death of hydrophylous plants (hydrophytes) and a notable depression of phreatophyte numbers. Considerable damage was brought about by intensive groundwater extraction (about 800 l/sec) in the valley of the ephemeral Karakengir River (Central Kazakhstan). There, according to Khordikainen, vegetation dried out and died, and its transpiration discharge was sharply reduced. A change in the moisture regime due to the lowering of the shallow groundwater level led to drying-out of meadow grass in areas of the river valley close to extraction wells and river runoff decreased sharply.

There are numerous cases where groundwater extraction has led to the drying-out and even disappearance of lowland swamps, causing depression of the swamp fauna and flora, the death of hydrophylous swamp plants, or changes in their species.

In the former FRG in the mid 1970s, a period of high groundwater extraction coincided with a number of dry years, which led to a significant drop in the groundwater level over large regions and, as a result, to a change in the landscape and vegetation. In the same period a negative influence upon vegetation was exerted by a decrease in the shallow groundwater level due to water pumping in West Berlin.

However, it should be noted that in a number of cases where groundwater exploitation leads to draining of wetlands, this exerts a positive effect on the crop productivity of grass in near-river water meadows and its species composition. Thus, in some areas of the Prisukhonskaya Lowland that are artificially drained by groundwater extraction and irrigation, the stiff, low-productivity sedge-type vegetation has changed into rich high-quality meadow grass with a higher crop productivity (Kovalevsky, 1994).

Analysis of the influence of groundwater level changes upon forest crop-productivity shows that in a very humid zone, an additional draining of territory due to intensive groundwater pumping can increase the growth of forest whereas in arid areas it may decrease it to the point of destruction of the forest.

Numerous examples of the influence of intensive groundwater pumping upon landscapes and vegetation in different regions of Germany are described in the detailed analytical overview of A.A. Zhorov "Groundwater and Environment" (1995). The most representative of these examples are given below. It should be noted that Germany is a country where some of the most complex and intensive in-situ investigations, observations and assessments (experimental work, modelling, predictive computations, etc.)

of the influence of intensive groundwater extraction on different environmental components have been carried out.

In the territory of the Furbergsky Field in Low Saxony, the lowering of the shallow groundwater level due to exploitation caused a disturbance in the water regime of the vegetation over 470 ha of forests and 1400 ha of meadows, which amounts to 5.5 % of the total area of the Furbergsky Field. The dependence of the amount of root-occupied layer wetting on the lowering of the shallow groundwater table due to water extraction is illustrated by graphs showing soil moisture changes for 105 botanic test sites. To the east of Furberg City, a small-sedge swamp with 15 plant species changing into a clover meadow due to a decline in the groundwater level by 4m after 25-years of exploitation. During the years 1964-1982 the Rhine River valley (Gessen fluxes) had the lowest groundwater level for a long time period. The increased groundwater extraction during this period coincided with a number of dry years. The consequences of this were the dying out of forest vegetation and subsidence of the land surface, in turn causing damage to buildings. The total damage, including destruction of houses, streets and railways, damage to fruit trees, expenses for irrigation and surface levelling in agriculture, and damage to forests, was estimated at 16 million DM.

Detailed data on the connection between surface ecosystems, in particular vegetation, and depression of the groundwater table by water extraction are reported by Zhorov (1995) for the other regions of Germany, where long-term observations of a variety of environmental parameters (i.e. soil moisture, shallow groundwater levels in natural and anthropogenically disturbed conditions, composition and state of vegetative communities, river runoff regime, etc.) are being carried out. The author notes that the methodology for assessing landscape sensitivity to lowering of the groundwater level and the risk of landscape changes due to water extraction was developed during the design of a new water extraction system in the former FRG. This took place over a period of 10 years, when the Working Group on "Ecology and Environment" implemented several projects using a unified technique. The methodology is based on analysis of the water regime in the topsoil-vegetation layer and determination of the role of the soil-moisture factor in it. Special attention was paid to studying the maximum height of capillary rise for different soil types, determination of field moisture capacity and depth of the root-habitable layer, plus rock lithology and density, organic matter and soil moisture content. Schemes were worked out to assess the sensitivity of agricultural and forest areas to a lowering of groundwater levels. These are being successfully used in assessing water-extraction influence upon the environment and, what is more important, serve as a basis for different nature-protection measures.

It can be concluded that the basic measures which would prevent or eliminate the detrimental impact of groundwater extraction upon the landscapes, vegetation and ecosystems include: regulation of large groundwater water extraction schemes (in many cases, this is involves decrease in water pumping), artificial replenishment of groundwater reserves and regulation of surface runoff by means of special water-management actions.

4.   INFLUENCE OF GROUNDWATER EXTRACTION ON LAND SUBSIDENCE

Groundwater movement is accompanied by mass-transfer and redistribution of substances inside the earth's crust. Transfer of dissolved substances with groundwater flow is one of the most important processes of chemical element migration, determining scales of subsurface. Quantitative assessment of subsurface chemical erosion is achieved by computing the amounts of dissolved substances removed with groundwater with time, or the time taken for the earth's surface to settle by a certain value (for example, by a metre) because of removal of dissolved substances by groundwater. In natural conditions this process cannot be clearly seen. Thus, for example, in the Baltic artesian basin, groundwater removes 30 t of dissolved substances per km^2 annually, which causes the earth's surface to settle by 0.008 mm/year. In general, under natural conditions the earth's surface is subsiding slowly and imperceptibly due to subsurface erosion (less than 1 meter per 100,000 years). In spite of the importance of this process for the geological evolution of the Earth, it exerts no significant adverse consequences on man's every-day life.

Subsidence of the earth's surface can also be caused by the extraction of groundwater, oil or gas. It is well known that in areas of intensive groundwater pumping declines of piezometric levels (cones of depression) covering areas of tens or hundreds of square kilometres are often observed

Lowering of groundwater piezometric levels and changes in stratum pressures cause a change in rock stresses and in velocities, and sometimes directions, of groundwater flow. All this intensifies suffosive and karst processes.

Under some conditions, groundwater level declines lead to subsidence of the earth's surface; under the others - to the formation of collapses. Earth surface subsidence is most commonly met in regions where groundwater is enclosed in high-permeable sandy-gravelly rocks with a low compressibility and interbedded with clayey rocks that are of low permeability, but have a high compressibility. During pumping, the groundwater head pressure is lowered, which increases the effective pressure upon the lithological structure and leads to consolidation of compressible sediments and, hence, to subsidence of the earth's surface.

Depending on the type of sediment, consolidation can be predominantly elastic, recovering with a water level rise, or predominantly plastic, leading to irreversible reconstruction of the granular structure of the sediment. Hard rock aquifers are virtually incompressible. Pebble, gravel and sand aquifers are slightly compressible, but their consolidation occurs rapidly and has an elastic character, i.e. with groundwater level rise these rocks largely recover their former state. Essentially, surface subsidence is connected with consolidation of low-permeability clayey sediments.

A decrease of head pressure in aquifers creates a hydraulic gradient from overlying clays and other low-permeability rocks inside an aquifer to high-permeable rocks. In a system of aquifers, a decrease in head pressure creates a hydraulic gradient from high in clays and low permeability strata to low in highly permeable rocks.

Pressure in a low-permeability stratum increases firstly in the pore water, and only with the removal of water is it gradually passed to the granular framework. Because of a slight hydraulic conductivity in such rocks, vertical water transport and the subsequent decrease in pore pressure proceeds slowly.

Karst-suffosive processes often develop in carbonate rocks containing fresh groundwater of a good quality. The mechanism of these processes can be represented in a simplified form as follows. Carbonate rocks commonly well jointed and under the action of physical and chemical weathering, they are usually penetrated to great depths by numerous hollows and caverns of various sizes and configurations, mainly filled by loose sediments. With long and intensive pumping of confined waters from such rocks, filtration becomes much faster (by tens or hundreds of times). This leads firstly to redistribution of the loose filling and then to its complete removal. Finally, when the roof of the hollow space so formed is unable to carry the burden of overlying, water-saturated, sandy-clayey sediments, slow surface subsidence occurs.

Earth surface subsidence and collapse often leads to hazardous consequences such as flooding and water-logging of territories, deformation of highways, railway lines, water pipelines and other communicative facilities, changes in river channels and slopes, and deformation of industrial and civil buildings.

Surface subsidence occurs widely in the USA. Polland (1981) reports that the depth of subsidence varies from 0.3 m in the area of Savanna City (Georgia) to 9 m in the west of the San-Hoakin valley (California). The total area affected by subsidence in California reaches 18,000 km^2. One of the largest environmental changes caused by groundwater pumping is seen in the San-Joaquin valley. Almost 1.5 million ha of irrigated lands cover the valley, half of which is involved in subsidence. In this region, the decrease in groundwater level due to intensive extraction reached a few tens of meters, causing land subsidence of up to 9 m in some areas. Subsidence cracks of up to 3.0-3.5 km long are recorded in California. In some places, irregular subsidence causes disturbances to canals, water pipelines and water wells, the restoration of which requires considerable financial expense. In San Francisco, lowering of the earth's surface by 2.4 m necessitated the construction of special dams in order to prevent water from the Bay invading the shore. In the coastal plain of Los Angeles, the earth's surface has been settling at a velocity of 0.7 m/year for a few decades. This is caused by the integrated action of a number of anthropogenic factors, such as oil-, gas- and water pumping, injection of fresh water in areas of seawater intrusion, as well as by the influence of modern neotectonics.

Subsidence within Mexico City during the last 70 years has reached 10.7 m. This has caused damage to buildings, bridges, water pipelines and sewerage networks. The Palace of Fine arts has subsided to more than 3 m below the level of surrounding streets. In order to stop the process of subsidence, groundwater pumping was reduced and instead, surface waters were involved in the supply of water to the city. Special measures for the regulation of groundwater extraction and use of water resources are currently under development.

Lind (1984), in his book "Water and a City", presents examples of the close interconnection between all the environmental components, including general water resources, groundwater, the earth's surface, etc. In particular, he describes in detail the geological-hydrogeological conditions of Venice where a slow and catastrophic submergence of the city into the sea is taking place, caused by extreme groundwater extraction from the water-bearing layers in combination with the influence of the Adriatic Sea tides. The counter-measures undertaken were directed at a considerable reduction of groundwater pumping in the Venice City area and use of surface water from the Sile River. Whereas in the period 1952-1968 subsidence was on average 5-6 mm/year, the decrease in water pumping (the number of acting water wells were reduced by 60%) resulted, by the year 1975, in uplift of the earth's surface in Venice by 2 cm (Lind, 1984).Similar problems are typical of many large coastal cities that intensively exploit groundwater.

There are also other examples that demonstrate how intensive groundwater extraction influences subsidence. Thus, according to data from the Technological Institute if Indonesia, beginning in 1994 intensive groundwater exploitation caused land subsidence of up to 0.5 m in Djakarta City. This resulted in seawater being drawn to the surface, causing in turn, flooding and, in some places, damage to building basements.

At present, there are known to be over 150 regions in Mexico, Japan and USA, where surface subsidence due to intensive groundwater pumping and mineral deposit development has reached 10 m.

Intensification of groundwater extraction in particular seasons, for example in summer for irrigation, can cause subsidence of loose soils.

Surface settlement on an alluvial plain (Japan) was detected in 1973. In the 1980s and early 1990s, the rate of subsidence, caused by a decrease in groundwater head pressure due to pumping, reached 2 cm/year.

The total area of Japan that has subsided to below sea level because of groundwater extraction amounts to 1200 km^2. In Tokyo, in the period 1900-1975, land settlement reached 4.7 m. The Water Law, accepted in Japan, is aimed at reducing groundwater extraction and increasing the use of surface waters. At the present time, in a number of regions in Japan, the rate of subsidence has decreased owing to more rational groundwater pumping and an increase in the use of water from surface storages (Yamamoto Soki, 1986).

In Bangkok City (Thailand) considerable settlement of the earth's surface due to groundwater exploitation has been observed since 1960. Now, about 15,000 water wells pump water from a confined sandy-gravel aquifer overlain by clays.

In China, intensive groundwater pumping led to the formation of cracks of to 0.5 m wide on the earth's surface (Foster, 1995). Land surface subsidence caused by groundwater extraction is also documented in Malaysia (Haryomo, 1995).

In some regions subsidence involves huge areas. Thus, on Taiwan, land subsidence caused by uncontrolled groundwater extraction covers a territory of 250 km^2, with the maximum subsidence reaching 2.5 m. Along the coastal plains of Taiwan, sea water intrusion can be seen. In some areas measures to reduce groundwater extraction are in place (Chian Min-Wu, 1992).

In a number of regions, earth subsidence, caused by intensive groundwater extraction, has lead to the destruction of existing civil and industrial installations. Such cases are known, for example, in Calcutta City (India) where there is damage to buildings caused by subsidence at a rate of 1. 4 mm/year (Mishra and Singh, 1993).

The danger of seawater intrusion, alongside subsidence, gives rise to the necessity to reduce groundwater extraction in many coastal cities. Such a hazard is observed in the coastal areas of California, on the Cape Verde Islands, on the coast of the Gulf of Mexico in the USA, in a number of cities in Italy, and many other regions. Sea water intrusion led to negative changes in the groundwater quality on the coast of Israel (Mellout et al, 1994). Many coastal areas of England have a problem of salt-water intrusion into the aquifers. Thus, in the eastern coastal zone of England, seawater intrusion into a Cretaceous aquifer intensively exploited for drinking and industrial purposes has been recorded. Specialists have recommended measures for stabilization of water extraction with the aim of preventing further seawater intrusion (Spink et al, 1990).

Intensive seepage of salt water into exploited aquifers is known in the coastal zone of the state of Queensland (Australia) where the rate of water extraction is almost twice the rate of present-day infiltration recharge (Hiller, 1993).

One important and, perhaps the only, way to deal with this type of surface subsidence is to reduce groundwater extraction from exploited aquifers and introduce constant monitoring of shallow groundwater levels and the state of the earth's surface.

Only the development of strict rules for the control of groundwater pumping will allow the prevention or reduction of the negative consequences of water extraction-in areas where the geological-hydrogeological conditions favour surface subsidence and/or salt water intrusion.

The most vivid example demonstrating the battle against subsidence and salt-water intrusion is the work carried out in Texas (USA). There, in Houston City, after 40 years of groundwater exploitation, subsidence reached 4 m, which caused flooding of a large region by sea waters. In 1976 the authorities of the state of Texas took measures to artificially replenish the groundwater. After this, subsidence decreased considerably. It should be noted that Texas is one of the few regions in the world where the government is carrying out regular long-term investigations (including stationary observations, modelling and predictions) in order to assess the influence of intensive groundwater extraction upon surface subsidence.

The formation of extensive cones of depression in areas of exploitation of the Potomak-Magoti aquifer (New Jersey, USA) led to seawater intrusion into the coastal zone, which made it necessary to halve groundwater extraction by large users (Kroop and Nocido, 1988).

The problem of surface subsidence due to groundwater pumping (similar subsidence occurs due to the pumping of oil and gas) attracts the attention of scientists in many countries, who direct their efforts to studying the features and regularities of subsidence. The main tasks are to predict the development of this process under different geological-hydrogeological conditions, develop a rational regime of groundwater exploitation, especially in areas subject to subsidence and karst-suffosive phenomena, and to prepare recommendations for the prevention or reduction of negative consequences from intensive groundwater extraction in such areas.

This brief analysis of the possible influence of intensive groundwater extraction upon the basic components of surface and water ecosystems (river runoff, vegetation, eco-landscapes) enables the following important conclusion to be made.

Resolution of the widespread conflict between the desire, and often necessity, to draw as much groundwater as possible from the subsurface and the likely negative consequences of this action on different ecosystem components should be based on integrated environmental monitoring data, including long-term assessment of large-scale groundwater extraction programmes. Such studies will allow rational scales and regimes for water extraction schemes to be substantiated, taking into consideration the requirements of environmental protection.

Chapter 9

GROUNDWATER USE AND PUBLIC HEALTH

LEONID I. ELPINER, ANDREY YE. SHAPOVALOV
Water Problems Institute of Russian Academy of Sciences

Processes of continental water quality degradation stemming from the powerful anthropogenic pressure continue to figure prominently amongst current environmental problems. Particular attention has been paid to groundwater because it has been considered that it may show better quality indices compared to surface water sources. At the same time, rapidly accumulating water chemistry data indicate that such views may need fundamental revision, especially in areas where, due to a variety of circumstances, the natural condition of groundwater formation and factors influencing its quality are disturbed.

The processes of terrestrial water quality degradation linked to strong anthropogenic impact are among the most important of modern ecological problems. It should be noted also that the growing store of medicoecological data indicates an obvious and direct cause-and-effect relationship between the infectious and non-infectious sickness rates in the population and the degradation of water quality in groundwater sources used for drinking purposes.

Even in countries featuring advanced development of municipal water supply systems and effective environmental protection agencies, the role of water factor in the spread of infectious diseases still persists, in particular as a result of the use of contaminated groundwater from the topmost aquifer. For instance, most (76%) of the 34 outbreaks of water-related infections recorded in 1991 and 1992 in 17 states of the USA were associated with the use of water from wells for drinking. A total of 17464 persons were affected. Outbreaks were caused by dysenteric bacteria, hepatitis A viruses, Lamblia or cryptosporidia (Moore et al., 1993). The detection of groundwater infected with cryptosporidia is more and more frequently mentioned in publications. This is of great importance, because this agent causes an infectious disease that undermines the immune system of organisms. Twelve outbreaks of this disease have been detected in the United States since 1985. The agent of the disease was detected in groundwater sources (Rose, 1997).

The use of poorly protected subsurface water sources can be a cause of outbreaks of viral intestinal infections. An example is the outbreak of acute gastroenteritis (up to 3000 affected persons) that occurred in Finland in 1994 because of use of water from wells infected by adenoviruses A and C, rotaviruses, and SRV viruses (Kukkula et al., 1997). Water-related outbreaks of intestinal infections associated with the consumption of contaminated groundwater are recorded in a number of Russian regions. Substantial viral contamination of drinking water and its supply sources is established: in 1993, 1994, 1995 the average percentage of tap water contamination with enteric viruses was 1.6%; 1.4% and 1.28%, respectively; with hepatitis A virus antigen - 7.6%; 7.8% and 5.8%, respectively; with rotavirus antigen - 3.6%; 3.7% and 7.68%, respectively (On the state ..., 1996).

Investigators pay more attention nowadays to data on medical ecological problems associated with chemical groundwater pollution, primarily with the incidence of cancers.

Studies conducted in the USA by the National Cancer Institute (Cantor, 1997) showed an increased risk of the development of cancer pathology in the population groups consuming groundwater containing elevated concentrations of nitrites, asbestos-containing products, radionucleides, arsenic, and secondary products of water chlorination. Studies conducted in the Argentine confirm a correlation between an increased mortality due to bladder cancer and the presence of inorganic arsenic in drinking water (Hopenhayn-Rich et al., 1996). The importance of this problem for groundwater is noted also in other papers (Haupert et al., 1996).

Japanese researchers (Tohyama, 1996) established a positive correlation between uterine cancer and the fluoride content of drinking water in 20 areas over that country.

Recent publications pay increasing attention to data on penetration of carcinogens into groundwater due to fuel leakage. For instance, in the course of their studies of the risk of cancer due to groundwater contamination with MTBE, M.C. Dourison and S.P. Felter (1997) established that this compound remains toxic when it enters humans from drinking water. The authors pointed out that the effect of MTBE on cancer incidence requires further investigations. At the same time, according to B.R. Stern et al. (1997), nearly 5% of the US population consumes water with high concentrations of this substance (700 to 14 000 ppm).

The risk of an increase in the incidence of cancer is also associated with carcinogenic organic compounds of anthropogenic origin in subsurface sources.

For example, there are suggestions of correlations between the increased incidence of breast cancer in Hawaii and the 40-year consumption of groundwater containing chemicals such as chlordan, heptachlor, and 1,2-dibrom-3-chloropropane. The National Cancer Institute (Allen et al., 1997) drew particular attention to the pesticide contamination of groundwater. Studies at the Technical University of Denmark (Bro-Rasmussen, 1996) showed that a number of pesticides (DDT, Lindane, dieldrine, and others) can persist in groundwater for a long time. Croatian researchers (Goimerac et al., 1996) detected contamination by the widely used carcinogenic herbicide -- atrazine. Dogdeim S.M. at al. (1996) detected, in an Egyptian region, pesticide contamination in groundwater, foodstuffs, and soil, on the one hand, and in women's milk, on the other hand. These are warning data in the light of earlier work that revealed the mechanisms of the pesticide reaching the breast milk via the chain: soil → drinking water → human being.

Other studies have confirmed a correlation between cancer diseases and the presence in drinking water of trihalomethanes that form when water that is poorly cleaned of organic compounds is treated by chlorine (Guidelines…, 1993).

Studies of other toxic substances in drinking water showed a positive correlation of Al with Alzheimer is disease (McLachlan et al., 1996), As - with bladder cancer (Hopenhayn-Rich et al., 1996), and NO_3 - with gastric cancer (Yang et al., 1998). A negative influence of Al on children's central nervous systems was also detected.

A closer study of groundwater pollution processes in the Moscow Region indicated that pollution is invariably preconditioned by the fact that most water intake structures (together providing 80% of the water produced in the region) are positioned near industrial sources of groundwater pollution. In some of the aquifers that are currently being used extensively, the standards for manganese, fluorine, chlorides, arsenic, selenium and lead are observed to be exceeded several times. One cause is a violation of rules for livestock farm manure storage and application (Elpiner et al., 1998).

Several investigations have been aimed at the study of the influence of groundwater polluted by toxic organic and non-organic substances from municipal landfills on human morbidity and found adverse health effects among local population. (Bruner M.A. et al. (1998); Najem B.R. et al. (1994)).

Of growing importance in assessing health risks associated with groundwater are studies of the effect of a complex of contaminants from the Colorado State University.

Considering the possible adverse effects of chemical pollution in drinking groundwater, it is worth mentioning data on the role of natural environmental components. In recent studies Taiwanese researchers found a correlation between colon cancer and low levels of drinking water hardness in municipal water supply systems (Yang, Hung, 1998). Japanese researchers N. Sakamoto et al. (1997) established a significant positive correlation between death rates from cancer of the stomach with the Mg^{2+}/Ca^{2+} ratio in the water of subsurface sources and in tap water. In Poland, B. Zemla (1980) revealed a positive correlation between the low death rates from stomach cancer and high water hardness.

High concentrations of several elements of natural origin also exert an adverse influence on the population's health. For instance, in Russia it was found that the consumption of groundwater with "extremely high" natural levels of boron and barium led to the increased digestive morbidity among children (Rakhmanin Yu.A. et al 1996).The traditional notion of water hardness as an index affecting its organic properties and the suitability for domestic needs has recently been revised as a result of a series of studies aimed at establishing a correlation between drinking water hardness and the incidence of cardiovascular diseases (Rubenoviz 1996; Zielhuis 1981).

Further development of these studies with the use of ecological--epidemiological methods confirmed these observations and yielded some more accurate and important data. Evidence of the correlation between the coronary disease and the consumption of soft water (in particular, with Mg deficiency) was also obtained in Finland (Punsar, Karvonen, 1979), Italy (Masironi et al., 1980), Spain(Gimeno Ortiz et al., 1990), Germany (Sonneborn, Mandelkow, 1981), Russia (Plitman, 1988), the United Kingdom (Lacey, Shaper, 1984) and Taiwan (Yang et al., 1996).The published data show the significance of the drinking water composition as a whole rather than the specific role of each separate element. It should be noted, however, that the mechanism by which the above considered elements affect the cardiovascular system are still to be studied. The observed effect might be a secondary manifestation associated with the impact of substances contained in the water on the nervous, regulatory, hormonal, and other human systems. In any case, we have to agree with the principal conclusion concerning the existence of relationships between the conditions of the cardiovascular system the any composition of drinking water consumed.

It has also been reported that the frequency of urolithiasis increase when using groundwater with the hardness of more than 10 mg eq/l, which also contained 300-500 mg/l of calcium (Elpiner, 1995).

The developing environmental situation calls for new approaches to the assessment of groundwater quality and intensification of measures aimed at groundwater protection and safe use.

It can be literally that the new problem has been formulated in terms of modern hydrogeology. Response to this problem needs the development of fundamental, practical scientific investigations to determine effective measures of safe groundwater use under intensive technogenesic pressure.

An example of the application of new approaches to water-relative problems is the research based on a medical geographical approach to of large-scale ecological epidemiological investigations (Smolensk oblast, Russia). It consists in the analysis of demographic, medical demographic, hydrological, hydrogeological, geographical, and ecological data of official statistics in order to draw possible links between water quality and the population's health (Shapovalov, 2002).

It has become an obvious necessity to take into account modern medical ecological interpretations of hydrogeochemical information when assessing the conditions of drinking water use and choosing new groundwater sources. To guarantee effective administrative measures for health protection in the population, it is necessary to obtain enough reliable data to properly characterize the medical ecological situation and hydrogeological conditions for groundwater use. It is impossible to make an informed choice between possible measures without being able to predict the development of the situation. It is, therefore, most important that medical studies should be based on full hydrogeochemical information. This problem can be solved only if specialists are properly acquainted with the prerequisites of the necessary work.

These emerging problems call for a unified interdisciplinary approach incorporating techniques of hydrodynamics, hydrogeochemistry, geoecology, and preventive medicine. The development of the medical ecological line of investigation on hydrogeological problems is related to the use of up-to-date hygienic, ecological, social demographic and medical geographical approaches.

The circumstances call for urgent development of the theoretical basis and techniques of medicoecological zoning in terms of the relationships between public health and the quality of groundwater that is either used or planned for use. One more, and no less important, line of investigation is the development of methods of forecasting variations in those medicoecological conditions that are related to the hydrogeological conditions of water use.

Chapter 10

METHODS OF GROUNDWATER POLLUTION RISK ESTIMATION FOR ECOSYSTEM SUSTAINABILITY

ANNA P. BELOUSOVA
Water Problems Institute of Russian Academy of Sciences

An important environmental problem is that of estimate risks and hazards of groundwater pollution in conditions of increasing negative impact on them from natural and anthropogenic factors. Risks and hazard are estimated in relation to groundwater as an ecosystem life support system and, more generally, to their economic significance for sustainable development of the investigated regions.

The sequence of action in estimation of danger, risks and hazards to groundwater from pollution is (Belousova 2003, Dzektser 1992, Ragozin 2001):

I. Identification and definition of ecological HAZARD (Figure 10-1)

- Sources of pollution of the saturated and unsaturated zones;
- Establishment of the locations of probable and prove sources of pollution;
- Mapping of groundwater with identification of areas that are vulnerable to hazardous processes;
- Identification of the possible processes causing pollution of groundwater from revealed dangerous sources of pollution;
- Preliminary forecast estimates of pollution of groundwater from potentially dangerous sources of pollution and use of indicators and indexes for estimation of complex hydrogeochemical conditions in the underground part of the hydrosphere;
- Estimation and mapping of vulnerability of groundwater to the identified dangerous processes;

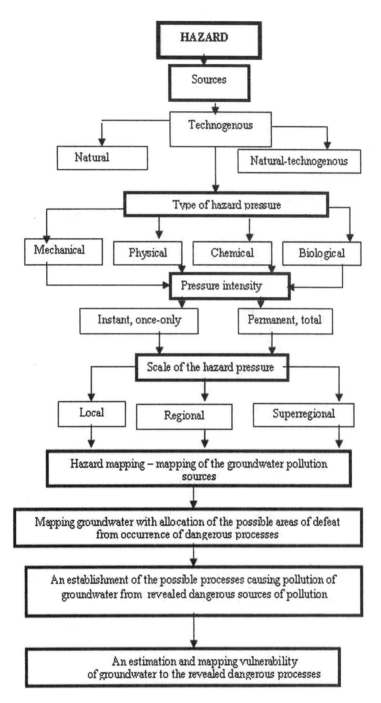

Figure 10-1. Hazard structure

II. Estimation of RISK and DAMAGE to groundwater from pollution (Figures 10-2, 10-3, 10-4):
- Definition of areas within which risks are estimated, according to definition of vulnerability to pollution and damage to groundwater;
- Estimation of impacts;
- Identification of the range of risks: material, economic, social, ecological;

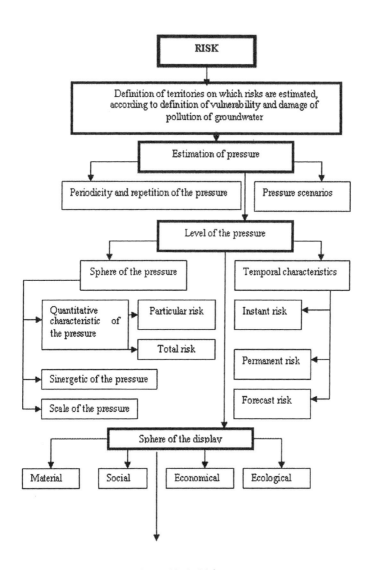

Figure10 -2. Risk structure

III. The characteristic of RISKS and DAMAGE to groundwater from pollution:
- Assessment of probable risk;
- The determined risk assessment.

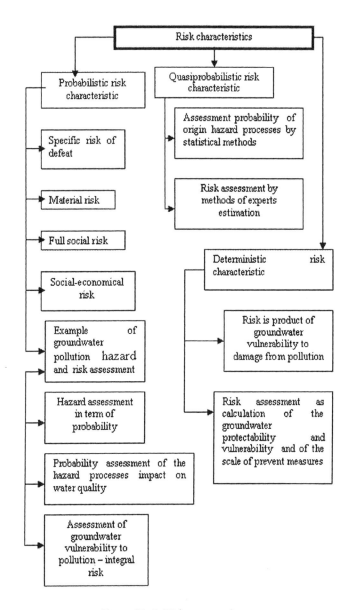

Figure10 -3. Risk structure2

IV. Classification of RISKS and DAMAGES:
- Classification of risks: comprehensible, significant, extreme and catastrophic;
- Classification of damage: direct and indirect hazards;

- Construction of maps of risks of pollution of groundwater and ecosystems;

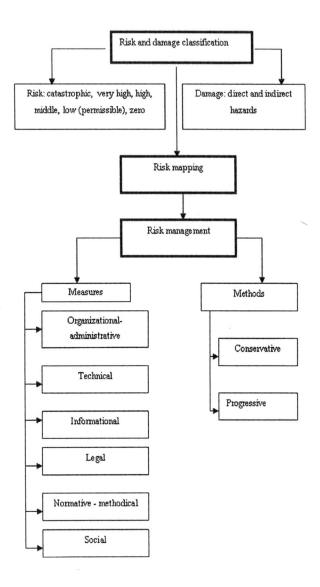

Figure 10 -4. Risk structure3

V. Management of RISKS:
- Actions to reduce risks and the potential for damage to groundwater (organizational, technical, information, legal, methodical, social);
- Methods for the management of risks: conservative (old methods), progressive (new technology).

Risk level	Level of the groundwater vulnerability to pollution					
Level of groundwater protectability and (travel time of pollutants)	Catastrophic (MPC > 100)	Very strong (MPC = 50-100)	Strong (MPC = 10- 50)	Moderate (MPC = 1- 10)	Light (< MPC> backgro und concentrat ion)	Not vulnerable (background concentration)
Extremely weak (0-5 years)	Catastrophic risk	Extreme risk	High risk	Moderate risk	Low (permissible) risk	
Weak (5-10 years)	Extreme risk					
Moderate (10-25 years)	High risk					
Conditionally (25-50 years)	Low (permissible) risk					
Protected (> 50 years)	Low (permissible) risk					Zero risk

Figure 10 -5. Managment of risks

The quantitative estimation of risk of groundwater pollution differs from methods of estimation of various natural and technogenic risks, because groundwater are not always in direct contact with possible sources of pollution.

Consider various approaches of the determined method of risk estimation. The forecast of risks for a case when the source of pollution is on the ground surfaceand through the protective zone separating groundwater from superficial pollution (soil and unsaturated zone) can be assess, at a first approximation, by:

$$R = \sum_{i=1}^{n} Vl_i \times D_i \qquad (1)$$

where R is risk of pollution of groundwater; i is the polluting substance; n is the quantity pollutants; Vl is vulnerability of groundwater to pollution in respect of a certain pollutant, defined by the ratio of technogenic loadings of that pollutant to "protectability" from it   (i.e. the ability of a protective zone to interfere with penetration of a pollutant into groundwater);

$$Val = k_{MPCi} \ t_c / (t_m + t_{MPCi}); \qquad (2)$$

where k is the frequency that the maximum permissible concentration of the pollutant is exceeded; ; t_m is time of i- pollutant migration; t_{MPC} is time of achievement i-pollutant MPC in groundwater; t_c is characteristic time which can be defined for carrying out of rehabilitation actions or time (critical) for which pollutant will penetrate on critical depth in a protective zone and after which expiration of action will not bring expected result.

Damage (D_i) to groundwater from a pollutant is determined by calculation of the specific cost (D_c)of remediation of 1 m^3 polluted infiltration water or 1 m^3 of the polluted soil and rocks (in this case in formula 2 we replace values of porosity with volumetric weight of the rock mass) from pollution on 1MPC.

$$D_i = S \cdot h_c \cdot D_c \qquad (3)$$

where S is the area of defeat (pollution), h_c - critical depth on which pollution which should be liquidated by rehabilitation actions can penetrate.

Estimation of direct risks of pollution to groundwater can be carried out for processes of pollution and types of pollution of the groundwater observed up to 1995, when industry in Russia was still functioning fully and strong pollution of groundwater was observed. Risks of this type can be termed "postponed" pollution.. That is, it is possible to use known tendencies of development of processes of pollution and to calculate the "postponed" risks. Quantitative expression of such risks can be carried out using formula (1, 2, 3) by replacing the thickness of the protective zone, thickness of the aquifer (or thickness of the polluted layer). This method of estimating risks and potential damage can be used successfully in relation to rehabilitation of groundwater using chemical and biological methods. Where hydrogeological methods of remediating polluted groundwater are used, the estimation of damage can be related to the expense of pumping the polluted water from the aquifer, and subsequent transportation and recycling.

Risk of groundwater pollution can be assessment with method of the expert's estimation, using for its results of the groundwater vulnerability estimates (Figure 10-5).

The proposed method supports complex estimation of the ecological condition of groundwater in various area and time scales with the purpose of securing sustainable development of the underground hydrosphere as component of an environment.

As a whole, the considered technique for estimation of risks should provide the basis for a subsystem of forecasts and estimates in the general system for ecological monitoring of an environment.

PART III:
ANTHROPOGENIC DEVELOPMENT, GEOLOGY AND ECOSYSTEMS

Chapter 11

URBANISATION AND THE GEOENVIRONMENT

BRIAN MARKER
Minerals and Waste Planning Division, Office of the Deputy Prime Minister, UK

1.   INTRODUCTION

Most of the chapters in this book are relevant to urban areas to at least some extent and the geological characteristics of urban areas are usually broadly similar to those of nearby non-urban settings. However, urban areas are extreme cases where human activities have the greatest impacts on the local environment, and environmental processes can affect the greatest numbers of people. There is thus good reason to give them separate attention.

Decisions made by developers and urban managers have major implications for health, safety, the economy and the environment. In addition, deposits resting on the bed-rock beneath many urban areas are often extensively modified, or created, by human activity ("made ground") (Simpson, 1996) and may be of archaeological value (Moseley, 1998; Oxley, 1998). Since ground processes and the nature of the underlying materials influence the occurrences of geological resources and hazards, it is important that these should be properly understood so that appropriate actions are taken to safeguard natural resources, reduce risks to people (Fell, 1994) and the environment, and ensure that good opportunities for conservation (Barker, 1996) and, or, development are not missed when making urban planning and management decisions (Marker, 1996). However it often requires professional geoscience training to anticipate problems and opportunities. If urban managers are trained in other disciplines, important issues may not be identified and information may not be sought from earth science specialists at the right time, at the required level of detail. It is important that specialists provide adequate clarification for the non-specialist, Otherwise the result may be damage or delays to development, impaired health, injuries and, sometimes, loss of life. Therefore good communication is needed between urban authorities, developers and geoscientists.

This chapter briefly examines impacts of the environment on cities (and their inhabitants), and impacts of cities (and their inhabitants) on the environment. It also considers how the issues can be addressed through better communication.

2.   GROWTH AND RENEWAL IN URBAN AREAS

The geographical origins of towns and cities are diverse (Mumford, 1961). Most develop in response to natural and local conditions as, for example:

- Convenient stopping places on routes such as passes between hills, fordable points on rivers, or at water sources in arid areas;
- ports on sea coasts, rivers or lakes;
- locations in farming areas where routes meet allowing the development of markets;
- defensive sites, often on rocky prominences with areas below that could be used for building dwellings; or
- administrative centres, normally at prosperous settlements developed for other reasons but sometimes separately, for instance to refocus activity by removing administration from a coastal town to a site in the interior to promote development.

All but the last of these are associated with specific geomorphological settings that reflect underlying geology. Therefore most urban areas are subject to problems and opportunities associated with these. Urban areas therefore reflect their origins, history, and contemporary environment.

The early development of individual urban areas often accommodates to, and makes use of the landscape. As towns grow, however, development often extends progressively onto less suitable land (Figure 11-1). Land is used for development with limited regard for environmental matters because the process is driven primarily by social and economic considerations (Simonds, 1978). Hills are quarried and hollows are infilled. Nearby embayments of the sea or shallows of lakes may be "reclaimed" by infilling. Wetlands are drained and streams are diverted or culverted. The aim of accommodating more development is achieved by engineered solutions, re-

Figure 11-1.

designing the landscape rather than working in harmony with it (Legett, 1973; Robertson & Speiker, 1978).

Further growth leads to amalgamation of earlier settlements through suburban sprawl into conurbations and then into "megacities". Alternatives to lateral spreading are widespread use of high-rise buildings or, less commonly, extensive development of underground space. These types of development inevitably require a thorough understanding of the geological environment to ensure that they are properly engineered. However the concentration of more population and facilities in a limited area inevitably increases the risks arising from natural and manmade hazards (Figure 11-2). Conversely, dispersal leads to the need for more travel between centres and increased environmental impacts of traffic. Engineering of transport infrastructure, therefore, becomes a key environmental issue in most extensive urban areas.

Figure 11-2.

But, in time, urban areas become outworn. Old industries shut down and sections of the work force become unemployed. Neighbourhoods become depressed. Industrial sites are left derelict and, often, contaminated (Figure 11-3). Amongst the dereliction, however, there are often individual buildings and sites that are worth preserving as part of the cultural heritage. Sooner or later attempts are made to rejuvenate such areas through land reclamation and investing in new industries. Again the "drivers" for this process are essentially social and economic but the environmental aspects include investigation and treatment of contamination and identification of the most appropriate uses of land (Roberts & Sykes, 2000).

The construction and maintenance of any urban area depends on large supplies of natural resources. The presence of a city has major impacts on its surroundings in terms of the demand for minerals (Figure 11-4) and water (Figure 11-5). Urban areas also generate large quantities of waste. Although attempts are being made in many countries to reduce the amounts of wastes that are produced and to recycle or recover value from these, much still goes, and will continue to go, into landfills situated outside the urban region (Figure 11-6). These demands have an inevitable effect on wildlife although, in general, the impacts from large-scale agriculture often outweigh other source of impacts. The urban setting is often seen as one that is hostile to wildlife even though some suburban areas display a greater biodiversity than nearby areas of intensive agriculture (Figure 11-7).

Figure 11-3.

Although geosciences have a strong part to play in planning and achieving sustainable redevelopment and conservation, this is not widely understood amongst the general public. The following section examines aspects of management of the urban environment. It is followed by some examples of urban issues in the form of scenarios so that a wide range of possibilities can be included within the limitations of space.

3. MANAGEMENT OF THE URBAN ENVIRONMENT

In the past, many urban areas grew in an uncontrolled manner, thus urban populations experienced increased risks, a poorer quality of life, and increased health problems. This trend continues where migration from rural to urban areas is very rapid (UNESCO, 1997). However, in general, there has been increasing management of many urban environments through a variety of regulatory regimes (Marker et al, 2003), for instance:

- land use planning – which is concerned essentially with locating development in the most appropriate places, while protecting areas worthy of conservation, so that development is undertaken at the best balance of social, economic and environmental costs;

Figure 11 -4.

- building regulations – which aim to ensure that built structures are properly constructed and are fit for use;
- environmental regulations – regulating the operation and occupancy of sites, ensuring that wastes and emissions are properly controlled and dealt with and that air and water are protected, so as to minimise social and environmental damage, including risks to health, conserving natural habitats and maintaining biodiversity;
- health and safety regulations – to ensure that employees, and other people entering places of employment, are safeguarded from unnecessary risks;
- environmental health regulations – to ensure that people are not exposed to potential hazards to health; and
- management of emergencies – setting out and disseminating procedures for protection, evacuation and relief, making provision for emergency services in relatively low hazard locations, and minimising the likelihood of major accidents.

Provisions vary significantly in detail from country to country so it is necessary in a short chapter to generalise. However regimes should be broadly complimentary so that, overall, they provide comprehensive opportunities and safeguards. All need to draw, to varying extents, on sound social, economic and environmental information. However there has long been a tendency to emphasise socio-economic factors and natural habitats, while overlooking other environmental issues, including geoscience matters.

Figure 11-5.

Land use planning, potentially, has a key role in setting the framework for sustainable development and conservation proposals and decisions (Marker, 1998). The main steps in most planning systems are:

- the preparation of some form of regional spatial strategy setting out broad policies for development, conservation and environmental protection;
- the preparation of a development plan that allocates or "zones" particular areas for development or conservation; and
- the consideration and determination of planning applications and, if development is approved, the formulation of planning conditions to control the way in which it takes place.

All steps require sound social, economic and environmental data – ranging from generalised information for regional planning, to detailed data on specific planning application sites. However some important data may not be readily available and, even if they are available, they are often not in a form that a non-specialist can use easily.

Figure 11-6.

Thus a geological map of the area will inform the geologist of much that needs to be taken into account but may be incomprehensible to the planner unless he or she has had a geological training (Marker & McCall, 1990). But geological maps do not usually contain information on geomorphological issues which may be equally, or more, important for land management (Brook & Marker, 1998). Normally the planner will rely on widely available public sources of data for regional or local planning and on the developer to provide data on a proposed development site. The developer's consultant will, similarly, rely on readily available sources for general contextual information when a site investigation is designed or for use in an environmental impact assessment (EIA). However, widely accessible public sources of data often have limitations on scope and content (DTLR, 2002). Therefore important issues may be overlooked.

Figure 11-7.

However environmental testing of policies is becoming increasingly formalised through, for example, environmental statements prepared under EIA of major development proposals (DETR, 2000), or of plans and programmes ("strategic environmental assessment"), or sustainability appraisal, which extends to social and economic issues. In the case of the environmental impact assessment, a wide range of data needs to be brought together but there will be some elements that are poorly covered by readily available information and the consultant may not be in a position to obtain detailed data given time and cost constraints (Tucker et al, 1998). Since EIA commonly reflects the quality of available data rather than a balanced view of the key data, some key factors may be overlooked or underestimated because little of the guidance on undertaking such assessments relates to geological or geomorphological characteristics. Indeed European Directives on these matters explicitly mention soils but not geological or geomorphological characteristics (Cendrero et al, 2001). Where guidance is directed towards themes such as housing, retail, and transport there is commonly nothing included to alert the reader to geoscience issues because these are often thought to be specialist matters, of limited relevance, by decision makers. Therefore much published guidance and advice to planners makes little or no reference to geoscience issues except when concerned with minerals, landfilling by waste or, sometimes, geological conservation.

Although the planning authority requires a site investigation, or where appropriate an Environmental Impact Assessment (EIA), to be submitted with the planning application, there may be no expert within the department to judge whether the report is thorough and whether the implications of the findings have been fully explored (Thompson et al, 1996). Therefore, commonly, the report will be accepted at face value if it appears to have been prepared by reputable consultants. Alternatively another department within the authority may be consulted, for instance that of the building control officer or the municipal surveyor/engineer. While these people are likely to have more geoscience background that in many planning officers they may not have the necessary depth of expertise in some geological or geotechnical matters. While advice might be sought elsewhere this is often not done because of cost.

Building regulations aim to ensure that built structures are properly constructed and are fit for use. Therefore the official responsible for building control is responsible only for the structure itself, including matters such as seismic or wind loading, and the adequacy of the foundations. Consideration normally applies to the development site only. They do not, in general, take account of matters outside the site such as the possibility of a landslide on a slope above the site affecting the building (Lee et al, 1991). However, in some cases regulations might identify certain types of construction that should be used in particular settings such as a flood hazard zone. The provisions should, therefore, compliment land use planning measures within which the broader context of the proposed development can be considered.

Environmental regulators are charged with preventing, and organising responses to, pollution incidents, making sure that those which do occur are dealt with effectively, and ensuring that the legacy of past contamination and pollution are cleared up safely and to appropriate standards (Bolton and Evans, 1997). This function may extend to developing strategies for dealing with affected areas and for setting affordable priorities. Such agencies may prepare land management plans that are complimentary to development plans. These might include, for example:

- shoreline management plans (Defra, 2003) that examine priorities for coastal protection and conservation, including matters such as sea defences, beach replenishment and, in some cases, managed retreat in which man-made defences are abandoned allowing, protection by salt marshes and dune systems to regenerate; and
- catchment management plans (Dunn & Ferrier, 2003) in which characteristics of a river basin are considered in terms of land and water management activities and the groundwater management is taken into account alongside that of surface waters.

Certain environmental protection agencies are charged with caring for natural habitats and securing improvements to biodiversity. Much of the focus of this work is in undeveloped areas and safeguarding from the impacts of development. However it is increasingly recognised that urban areas also have a part to play and that introducing "semi-natural" habitats into such areas may help to improve the quality of life in cities. Opportunities exist in city parks and relict patches of habitats left in, for instance, areas between transport corridors, or can be planned into urban developments and redevelopment (Barker, 1996). It is important to consider the potential for habitat creation when reclaiming damaged land.

The health and safety inspector focuses on risks arising in the workplace associated with the use of machinery, access to parts of the works, and potentially hazardous processes, but may not identify potential hazards from the ground such as accumulations of radon gas in a poorly ventilated basement (Appleton & Ball, 1995). Similarly the environmental health officer's time may be taken up largely by risks to health at poorly managed restaurants and shops and not by similar gaseous accumulations in dwellings. However there may be responsibilities for ensuring that hazards are not created by, for instance, disturbing contaminated land during construction or remediation (Pratt, 1993).

Emergency planning officers are charged with setting out contingency plans for emergency services, administrators and the public. These include plans for evacuation and emergency relief. Such plans need to set out and disseminate procedures widely so that all participants including the general public are aware of the actions that should be taken. However it is the land use planning system that should make provision for placing emergency services in low hazard locations and environmental protection and health and safety measures are those that strive to reduce the risk of major hazards and accidents (Alexander, 2002)

The systems are usually administered by separate authorities, which need to act together to secure sustainable urban management. All need to draw on common sources of information but many have limited access to geoscience expertise or awareness that this might be needed. Therefore actions and responses are frequently disconnected and are, sometimes, conflicting.

4. ILLUSTRATIVE EXAMPLES OF SOME URBAN ISSUES

4.1 Scenario 1.

An industrial centre grew on the basis of local coal and ironstone mining that supported heavy industry. Groundwater was abstracted extensively to supply industrial processes, the growing population and to de-water the mines. As time passed the mineral resources beneath the city were exhausted but mining continued nearby. The growing city expanded across formerly mined ground and, after some years, development was affected by intermittent localised subsidence of the mines. While there were good records of the later stages of mining, the earliest mines were not mapped in any detail. Eventually the coalfield was exhausted and the remaining mines were closed. Mine closure led to cessation of pumping. Consequently groundwater began to rise towards its original levels. This led to flooding of basements, resurgence of marshy ground such that some clay formations underwent swelling, damaging foundations, and the restoration of ancient discharges from springs and into surface water courses. The discharged water was of poor quality due to high concentrations of iron hydroxides derived from chemical reactions within the mined areas. Mineral extraction, groundwater abstraction, and lateral expansion of the city therefore combined to leave a legacy of subsidence, foundation problems, contamination and increased flood risk. However the effects were not confined to the city. The contamination affected downstream areas leading to the disappearance of most benthic organisms within nearby streams and rivers and discharges and drainage of mine waters led to increased downstream flood risk.

In this example, the administrative focus is likely to be on reducing the very visible effects of flooding, and improving stream water quality. However it is unlikely that the complexity of the situation would be fully appreciated. The legacy of ground problems needs to be addressed through mapping of the nature and extent of mined ground, investigation of groundwater quality and behaviour, examination of historical patterns of springs and drainage, and environmental monitoring, as a basis for planning of new land uses and environmental protection. In particular, the civic authorities need to develop strategies to reduce pollution, and to determine areas in which ground investigations prior to development should take account of mined ground. The implications affect a number of management systems including land use planning, environmental protection and environmental health regimes.

4.2   Scenario 2

Small quarries were opened for the extraction of limestone in the 14th century. The stone was used in the construction of a castle, and major local buildings. The quarries were abandoned long ago. The buildings are now a recognised part of the cultural heritage and are subject to heritage conservation designations. Over the years, the town expanded greatly and the old quarries, now surrounded by buildings, re-vegetated naturally. There are few other nature conservation sites in or near the town so these are an important educational resource as well a focus for recreation and community involvement in site management. These regenerated sites are recognised as a valuable wildlife habitat and designated as a Site of Special Scientific Interest. More recently some have become designated as international sites of conservation under the European Union Habitats Directive, a more restrictive type of designation. The historic buildings were damaged by industrial air pollution and the passage of time and now require restoration. This needs to be done using stone that matches the original materials in appearance and behaviour on weathering. Because the stone resource area has been completely built over, apart from the original quarries, a planning application was submitted for limited additional extraction of stone from the site specifically for conservation purposes. The planning authority refused the application on the grounds that the nature conservation interest would be damaged. An alternative was to secure stone that is a less good match from a greenfield site 50 kms away, a proposal that was opposed by local residents at that site because they felt that it might set a precedent for more quarrying at a larger scale in a previously unaffected area and because of additional traffic on narrow lanes, even though the proposed output of the quarry was small. This proposal is refused. In the event, a poorly matching limestone is imported from another country. The use of this is strongly criticised by architectural and historical interest groups.

In such circumstances public authorities commonly focus on the nature and cultural conservation issues but, to address these issues properly, information is required on characteristics of past and potential building stone, the geological extent of resources, nature conservation potential of sites, and evaluation of environmental impacts of quarrying. The issues have implications both for land use planning and environmental protection regimes.

4.3   Scenario 3

A major manufacturing city was linked to a sea-port by an aging railway that had been constructed through a range of hills, along a river valley and across coastal lowlands. Maintenance had not kept pace with wear due to traffic and the capacity of the system had been exceeded due to the development of a major airport beside the line. There were speed restrictions on the line, limited capacity for additional traffic, and relatively narrow tunnels. These limitations led to rapidly increasing use of road transport with consequent air pollution problems. It was determined that a major new rail link should be constructed. Major stretches of the route, especially on the coastal lowlands, were to be on a newly constructed embankment. The modernised railway brought the expected benefits in terms of rapid transit from the airport and bulk haulage to and from the port. However, a few years later, there was an exceptionally wet winter followed by spring thunderstorms. Water levels in the river rose rapidly and both the river channel and drainage pipes beneath the railway embankment proved inadequate to carry the peak load. The water pooled behind the embankment, which acted like a dyke, leading to major flooding of the farmland behind it as well as parts of the manufacturing city before it was breached by the floodwaters thus breaking rail communications for some time and causing considerable industrial disruption and financial losses.

The traffic capacity issue is the one most likely to be concentrated upon initially by administrators until flooding proves to be costly. It is essential, however, that the responsible authorities should evaluate the potential for flooding through modelling of the effects of proposed changes, and establishing ground conditions for construction at the earliest opportunity. If the issue had been recognized, then the embankments could have been interspersed with viaducts and drainage channels thus greatly reducing flood risk. Again land use planning and environmental protection, especially catchment management, are involved and protection from flooding requires emergency planning.

4.4   Scenario 4

A city is located beside a bay with a narrow coastal plain backed by mountains. It was a port but expanded along the coast as housing, industry and tourism developed. Eventually travel times from the extremities of the city to the centre were becoming inconvenient. Therefore it was determined that future development of the city should be more centralised. This would require reclamation of part of the bay. The most economical approach was to build a series of dams and infill the areas behind them with sand dredged from the sea. The dams were progressively constructed and filled until a sufficient area had been reclaimed. At each stage, the dredged sand was pumped into the compartment through a pipe directly from the dredging vessel. The area was then built upon, including high commercial buildings, and high-rise flats, with the foundation suitably piled down to bed-rock. Several years later the city experienced a major earthquake. Parts of the infilling of fine sand, still waterlogged, liquefied when subjected to the tremors. Buildings foundered into the sediment or tilted because of liquefaction, and had to be demolished. The event caused major damage, financial losses and some loss of life. A seismic risk study was commissioned so that losses from any future event could be reduced. However considerable public concern arose because the consultants concluded that much of the reclaimed area was at high risk if a significant tsunami were to occur. The public authorities introduced stronger controls on building but were fully aware that there was no real defence from a major tsunami. However such was the level of occupancy, the value of investment in the area, and lack of alternatives, that it was not practical to evacuate the population gradually to a more suitable location.

In this example, public concern is likely to lead to pressure on politicians to provide effective evacuation plans because of the costs of treating the made ground and of developing more suitable buildings. Evaluation of seismic (including tsunami) hazard and risk would help with understanding the problem but it would be very difficult to resolve it. It might have been better not to reclaim the part of the bay in the first place, but now there is no alternative to relying on emergency procedures.

4.5   Scenario 5

A city had been a major manufacturing centre. More recently, traditional industries had declined and a thriving service sector was now the main employer. As manufacturing industry declined, so did the need for groundwater abstraction. Less abstraction led to the groundwater table rising towards its former level. It was decided that the city should make better use of its space without encroaching further on the surrounding scenic countryside. In addition to building higher structures, it was decided that more use should be made of underground space for travel, storage, shopping and recreational facilities. Work commenced using abandoned former limestone mines, extended by excavation and tunnelling. This led to flooding of basements and increasingly costly precautionary and maintenance works to prevent water ingress into the new underground developments. The final cost of the developments was significantly greater than the original estimated budget leading to assertions that the municipal authorities had mismanaged the initiative.

Exploitation of the opportunities offered by existing and potential underground space are dependent on a good understanding of the extent and stability characteristics of existing underground space, the bulk rock properties of ground that might be excavated, and the monitoring and modelling of groundwater movement (rising, falling and lateral flow) before a sound strategy can be identified and remedial works designed. This example relates to land use planning, environmental protection and building regulations, but also brings in health and safety at work issues during stabilisation.

4.6   Scenario 6

A major industrial area developed in the 19th century. The complex was powered by coal and by manufacturing of gas and coke. Ash was tipped on areas between the factories and gave rise to localised contamination. Liquid effluents from the gas works led to the accumulation of contaminants such as phenols, cyanides and heavy hydrocarbons. Eventually most of the factories closed due to cheaper overseas competition, and the gas works closed and became derelict. The land remained vacant for many years before an urban regeneration initiative was proposed. There was, by that time, limited recollection of the precise processes that had been carried out in the works or the nature of the materials that had been tipped, or the nature and distribution of the foundations beneath the buildings, some of which were no longer standing. Some of the level and tipped areas were re-colonised by plants while the site remained vacant. During the site survey it was found that rare plant species were present because of the unusual chemical conditions in the substrate. It was proposed that these plant should be protected as a significant contribution to local biodiversity. This provided a significant constraint on reclamation of the site by digging out of the contaminated materials and either disposing of them elsewhere or by washing and replacing the engineering soils. However seepage of contaminants into the groundwater and surface water had to be prevented leading to costs of development appreciably greater than had been estimated. The remaining industrial plants were still burning coal but, because of exhaustion of local sources, this was now imported and contained relatively high concentrations of sulphur. This gave rise to sulphur dioxide that was deposited as acidic rain in nearby hills covered with pine forests. These were being harmed by the emissions, and acidification of lakes and rivers has given rise to release of aluminium into the waters with adverse environmental effects on fishing interests. In parts of the city the topography was such that, on days with little wind, fogs or smogs built up with adverse effects on the population in respect of increased risks of respiratory diseases and heart ailments. Industrial sources had also given rise to relatively high concentrations of heavy metals in certain soils in the urban area and also in the open countryside downwind of the city. These were concentrated in some types of vegetation and were a potential hazard especially in respect of crops grown in urban allotments and used locally.

Addressing these issues requires consideration of past (potentially contaminating) land uses as a context for site investigation and remediation, possible strategic geochemical analysis of urban soils, ecological survey, evaluation of groundwater behaviour and quality, air emissions inventories, and use of meteorological records as well as direct action such as emissions controls. As well as being wide ranging in terms of environmental information requirements, this example has implications for land use planning, environmental protection, building regulations, health and safety at work, and environmental health. It requires a fully integrated approach by the regulatory authorities.

4.7   Scenario 7

Porous sandstone beneath part of a city is the major source of local water supply. This aquifer is overlain in places by sand and gravel and elsewhere by low permeability clay. It is determined that the city should be extended by the creation of a major industrial area including chemical works. Some decades later it was discovered that toxic chemicals occurred in low concentrations in water taken from boreholes into the sandstone both within and outside the urban area. Monitoring proved that the concentrations were increasing as time passed. A major investigation was undertaken. It was established that leakage from the chemical works had given rise to a pollution plume because the works was built on permeable gravel deposits immediately above the sandstone. An old unlined landfill has also given rise to pollution. Cleaning up the effects was difficult, if not impossible, making it likely that the water resource could no longer be exploitable for domestic purposes. The alternative was a pipeline to bring water from 50 kms away and the construction of major reservoirs.

The problem could have been reduced if the authorities had been aware of it earlier. Mapping of bedrock and superficial deposits, and assessment of groundwater behaviour and quality could have been used to define aquifer protection zones. Potentially polluting activities could have been kept away from the most vulnerable areas. The nature and extent of previous pollution could have been determined, and monitored, and remedial works could have been instituted earlier. Relevant regimes include land use planning, environmental protection, especially catchment management, environmental health and health and safety at work issues.

4.8   Scenario 8

An area of uplands was rich in timber of high commercial value. Because of the need to earn foreign currency the Government licensed major logging concessions. The logging company wished to maximise its profits and undertook rapid tree felling over a few years. This increased the rate of surface run-off of water and led to soil erosion that carried sediment into local stream channels. The sediment load reduced the diversity of benthic stream organisms. Sediment was carried downstream to be deposited lower down the river channel, decreasing channel capacity, while the increased run-off led to a greater peak load of water. Farmers were keen to increase crop yields on the flood plain, not least because of increased demand for cheap food in a nearby city. They undertook extensive land drainage in the former wash lands and water meadows beside the river thus reducing capacity to store floodwater. These factors combined to cause increased incidence and severity of downstream flooding in urbanised areas that had not been affected previously. In addition, the increases of pore water pressure in engineering soils in the highlands led to new and reactivated landsliding with the result that vital communications routes were blocked and the logging industry was, itself, disrupted each time an event of this sort took place. High costs for losses and emergency and remedial action resulted.

The key problem is that a variety of actions undertaken distant from the city, and within another administrative area, led to direct impacts on the urban area. The issues can be addressed only by looking at management of the river catchment as a whole. This requires geomorphological mapping; assessment of landslide hazard, erosion potential, and storm magnitude and frequency in the uplands, the extent of tree cover that could be removed in more sustainable forestry operations, rates of run-off and sediment removal, channel capacity and deposition in the downstream areas, and surveys of land drainage capacity, leading to an assessment of flood potential. However the results will only be of value if industry, farmers and the authorities collaborate at all levels. Thus the issues involved include land use planning, environmental protection, including agricultural land and catchment management practices, and emergency planning in relation to flooding.

4.9   Scenario 9

A low-lying coastal area, lined by dunes, marshes and areas of mangroves was located to the west of a major port that had been designated for expansion. This involved the building of a breakwater to the east of the undeveloped coast and the dredging of a deeper access channel. Part of the dune complex was found to be rich in monazite sand and permission was given for extraction and processing. Because of tourist potential, a resort was developed on part of the marshes. Following construction of the breakwater, erosion of marshes and dunes has increased significantly. On investigation, it was established that deposition that formerly maintained the marsh and dune system was now being intercepted by the breakwater and dredged channel, also significantly increasing the cost of maintenance dredging. Eventually erosion reached the area in which mineral extraction had been undertaken and the dune line was breached to erode valuable farmland, areas of marshes and mangroves of conservation value, and part of the tourist resort. A public debate on whether the farmland should be protected by sea-defences at the expense of conservation areas began. However it was recognised that coastal defences would be costly and it was uncertain whether it would be possible to afford these in the medium to longer term because of rising sea levels and an increasing frequency of major storms. Removal of the breakwater was not considered an option because of the potential impacts on trade.

This is another instance of be "wise after the event". There should have been careful analysis of coastal currents and sediment budgets to establish the implications of intervention in the natural system at the outset. A sound understanding of the existing geomorphology and habitats should have been secured and, in particular, their sensitivity to change should have been assessed. Temporal trends should have been considered, particularly the implications of rising sea levels. The nature and value of existing and proposed land uses should have been subjected to cost-benefit analysis in relation to coastal defence options. Evaluation of the mineral resources needed to be set within the context of the potential environmental implications of extraction, especially increased flood and erosion potential if the dune system was removed. Collaboration was needed on environmental protection, especially shoreline management, land use planning and emergency planning actions.

4.10   Scenario 10

Major population growth took place in a city before mains drainage was constructed. Severe pollution of surface water and wells by sewage occurred. Following epidemics of typhoid and cholera, there was major investment in mains water supply and sewerage. The system was designed to the best standards recognised at the time. Water draining from roofs and streets, slightly contaminated water from domestic sources, industrial waste water, and foul water containing sewage were channelled into a single comprehensive system of sewers that carried the effluents to treatment plants. These discharged into a major river down-stream of the city, affecting water quality and habitats as the system deteriorated through age and inadequate investment. Maintenance costs grew rapidly. Leakage from pipes led to deterioration of groundwater quality, leading to supply problems. In some suburbs soakaways were used instead of mains drains. In some areas these led to mobilisation of sediments infilling solution pipes in limestones leading to sudden subsidence events and severe damage to buildings and roads.

Such problems can be tackled by encouraging new development and regeneration initiatives to include measures for more sustainable urban drainage. This includes the storage and use of rainwater, and separation of foul water from slightly contaminated water to reduce the burden on treatment works. In addition, slightly contaminated water can be treated by passing it through reed beds leading to the development of pleasant water features and improved biodiversity. It is necessary to carefully survey the drainage system to assess priorities for maintenance and potential for phased replacement, especially the maximum drainage load that can be accommodated at any point. In addition, the extent of limestones likely to contain karstic solution features, data on groundwater quality and behaviour, and work on the potential for sustainable urban drainage systems is needed. This example brings together the need for co-ordination by authorities responsible for land use planning, environmental protection including catchment management, health and safety at work, and environmental health.

4.11   Scenario 11

An important town grew several hundred years ago on the slopes of the mountain. It was initially a market centre for farming based on rich local soils and plentiful rainfall on the slopes. From time to time, there were volcanic eruptions but most of these were limited to small-scale ash falls. These events caused few problems for the original settlement. Subsequently, however, the town grew into a major city that now occupied much of the lower slopes and nearby plains. A small eruption led to the burial and collapse of some buildings because of localised down-wind ash fall in one area. While the damage was limited, the event gave rise to public concern because, although there were records of earlier eruptions, most people had tended to assume that either the volcano was extinct or that nothing would happen in their life times. The municipal authorities responded to this concern by commissioning studies by volcanologists who established that, although historic eruptions had been small, there was evidence of a number of episodes of major pyroclastic flows and lahars over the past millennium. Some of these extended well into the plains below the mountain. Mapping established that these were most likely to occur on the eastern slopes of the volcano and to be predominantly channelled down several valleys. It was found that many of the city's key emergency services and a key evacuation route for the population were located in the area that was at greatest risk.

Information is required on the volcanic history of the area, including the style of eruptions and the areas nearby that are most vulnerable in order to design an effective emergency response plan and to define a relatively safe evacuation route. Monitoring of the volcano for heat flow changes, gas emissions, seismicity and ground movements is required to help determine when the plan might need to be implemented. This has implications for both land use and emergency planning.

4.12 Discussion

Many more examples could be cited but these are sufficient to illustrate the diversity of issues that may affect urban areas and the nature of the information that is needed to address them. The information comes from many sub-disciplines, some of which are rooted in geoscience but others extend, for instance, to ecology, economics, sociology, architecture, and archaeology. Multidisciplinary studies are needed. The issues also relate to a wide range of administrative and regulatory systems and, in some cases, involve several of these. Therefore it is necessary for those who provide essential information to communicate with a variety of parties ranging from administrators to the general public, other geoscientists and researchers from other disciplines. In some cases communication may be straightforward but often there are barriers of unfamiliarity and technical complexity that can stand in the way of prompt and effective action.

Importantly, actions in urban areas can have implications for considerable distances away and events and management actions outside urban areas can have significant effects within them. Therefore management of urban areas cannot be considered in isolation from the surrounding environment – they are part of the ecosystem. This can present difficulties where separate administrations and regulatory agencies are responsible for different parts of the area and, especially, where distant and trans-boundary environmental effects may be relevant.

For satisfactory results to be achieved it is necessary to understand the implications of the geology and earth surface processes of the area and to explain them clearly in terms that can be understood widely. However that will not achieve the desired objective unless there is adequate, easily accessible, and readily understood information, advice, guidance and training on how to deal with these issues (Marker, 1998).

5.   INFORMATION, GUIDANCE AND COMMUNICATION

Regulatory regimes are often administered by different organisations or different administrative departments therefore most systems have requirements for consultation between specialists. However consultations do not necessarily include all of the issues that they should. Sound urban management requires ready access to the right information, in the right form, at an appropriate level of detail, and at the right stage in the administrative process. The requirements range from generalised information to alert developers, their consultants, and urban managers to the issues that need to be taken into account in development planning, control of development and in designing site investigation or EIA through simplified material for use by those who do not have training in the geosciences, to detailed information required by specialist geological consultants (Thompson et al, 1998; Waters et al, 1996).

The information necessarily comes from many disciplines but the main focus of this chapter is on the geosciences and the following are therefore particularly relevant (Brook & Marker, 1987; Smith & Ellison, 1999):

*   resources - minerals, water, soils, landscape, good building land – types, extent, quality and significance; safeguarding and potential
*   hazards and constraints - subsidence, slope instability, flooding, storms, natural and manmade contamination, erosion, volcanic activity, seismicity – types, extent and relative significance.

Information is needed on all environmental issues that are relevant to a particular area at both the regional and more detailed levels. This needs to be secured from strategic survey and synthesis of existing data. The results of environmental monitoring, individual site investigations and environmental impact assessments need to be taken into account to expand understanding of the issues, and because the environment is dynamic.

The availability of such information is generally sparse. There have been initiatives undertaken by many geological surveys and some university departments to provide engineering or environmental geology maps of specific urban areas. However, it is clear that while these have great value, urban areas cannot be considered properly in isolation from their broader surroundings. While some countries have promoted systematic applied geological mapping of wide areas, these are few in number and have not dealt comprehensively with all relevant themes (McCall and Marker, 1989). By and large national, regional and local governments have been reluctant to support systematic data collection though the cost is a very small fraction of the amounts of money that are spent each year on site evaluation, development, precautionary and remedial works to address "unexpected" ground problems, and intermittent disasters, notably those arising from flooding (Site Investigation Steering Group, 1993). This is undoubtedly partly because of a lack of appreciation of the significance of such data. In addition, many of these initiatives were undertaken some time ago, before access to modern GIS and databasing facilities. Maintenance and updating has, therefore, been difficult. There have also been individual studies to develop approaches to dealing with specific problems in individual locations but while the methods can be generalised to some extent, they are not always fully transferable. However it is important that adequate basic geological, hydrogeological, geomorphological and geochemical research is undertaken at the strategic and more local level in order that applied outputs and modelling procedures can be securely based (Walton & Lee, 2001).

In addition it is often difficult to secure funds for maintenance of the data system. Few public authorities are equipped to do this even though the costs are very small compared with development costs and the bill for putting problems right after the event. However there are signs of increasing interest in some areas, such as the European Union, as a result of the need for sustainability appraisal of plans and programmes and recognition of the need for geographic information systems.

Traditionally geoscience information has been presented on paper maps. More recently it has moved onto GIS (Nathanail & Symonds, 2001). However it has essentially been in two-dimensional representations that can be interpreted in three dimensions by specialist users but not readily by non-specialists. In addition the technical content and language used was not readily accessible. Environmental and engineering geology mapping was introduced to overcome this barrier with some success. However there is now an increasing need for three-dimensional representation, especially in connection with planning for the development of the subsurface (Paul et al, 2002). A number of geological surveys are now working on digital geoscience spatial models that can form the basis of such systems (Walton and Lee, op cit). These might, in future, be linked to thematic GIS layers and models that display possible changes as time passes in order to move towards prediction of possible outcomes of actions.

It is important, however, to provide information that can alert non-specialists to crucial issues and guide them towards the questions that need to be asked and where the answers can be obtained. There is a role in this for decision support systems (Alker et al, 2002).

6.   CONCLUSIONS

In summary, the urban area has often been regarded as if it intrudes within and is therefore separate from the ecosystem. However urban areas have major effects on their surroundings and processes in nearby areas have impacts on towns and cities. Therefore the urban area should be regarded as an integral part of the overall functioning of the ecosystem if environmental issues are to be properly analysed and resolved. Traditionally urban development has been undertaken with limited regard to ground conditions and earth surface processes, often leading to expensive problems and risks to property and people. It makes more sense to understand the nature of the ground and environmental processes at the outset since this can usually achieve development at less overall costs − to work with nature rather than against it − as well as to provide more effective responses to the legacy of past damage and unexpected events in the future.

There is a considerable diversity of environmental issues that have a geoscience element and many problems involve multidisciplinary consideration within a variety of management systems, often administered by different authorities. These authorities often need to draw on the same body of knowledge with varying degrees of emphasis but they can only do so if they are aware of the relevant issues and significance and availability of the necessary information. In practice, there is commonly a communication gap between various groups of administrators and specialists, as well as with the general public and politicians. This arises partly from limitations on education in the geosciences and a lack of dialogue and mutual learning. The gap can be bridged by the various interest groups working together to develop soundly based plans and solutions to problems. A key aspect is for stakeholders to know who to ask, when to ask, and what to ask when problems arise.

A key to sound urban management is the compilation and maintenance of databases not only for the urban area but also for its regional surroundings. These are particularly valuable if linked to GIS, environmental information systems and software for modelling potential outcomes of possible decisions. This, however, requires adequate funding at the outset and for maintenance. It is important that adequate strategic geological, hydrogeological, geomorphological and geochemical research is undertaken as a basis for applied outputs. Experience shows that authorities are often unwilling to invest in this way even though the outlay can lead to considerable savings in the future. There is a need to improve the understanding of administrations that this is an important element in securing more sustainable future development.

In addition there have been few attempts to date to develop software interfaces for delivery of environmental information to planners and others in forms suitable for decision support. This is not a simple task because it requires careful analysis of the procedures that are undertaken and the questions that are posed before it is possible to identify what information is needed and at what level of detail at each stage in the process.

Chapter 12

ASSESSMENT OF EFFECTS OF DISCHARGED WATERS UPON ECOSYSTEMS

Zhanna V. Kuz'mina[1], S.Y. Treshkin[2]
[1]Water Problems Institute, Russian Academy of Sciences, [2]Institute of Bioecologi, Uzbek Academy of Sciences, Uzbekistan

1. INTRODUCTION

The impacts of discharged waters upon ecosystems is very complicated, involving issues such as land degradation through secondary salinization, rising groundwater level and deterioration in the quality of drinkable water supply.

In 2000 it was decided that construction of the South-Karakalpakian main collector (SKMC), a project of the former USSR, in the Republic of Uzbekistan should be resumed. This will provide a waterway in the old course of the Akcha-Darya for collecting drainage waters in the Zhana-Darya - the old stream bed of Syr Darya (Figure 12-1). Construction is supported financially by the World Bank. The SKMC would lead to reduction of drainage in the Amu Darya river, within three agricultural regions of Karakalpatia, decreasing the mineralization of the river water about 0.2-0.3 g/l. Moreover, exploitation of water resources at the Kyzyl-Kum and Biruni pumping stations should be terminated (Figure 12-1). At present, the Kok-Darya old channel serves as a collector for waters from Biruni pumping station. It surrounds the whole area of Badai-Tugai nature reserve and enriches ground waters in half of this area. When the Biruni pumping station closes, the Kok-Darya channel will become completely dry (Figure 12-2).

The present chapter considers this problem in terms of a region of Central Asia and the South-Karakalpakian Main Collector (SKMC) in the Pre-Aral region in particular (Figure 12-1).

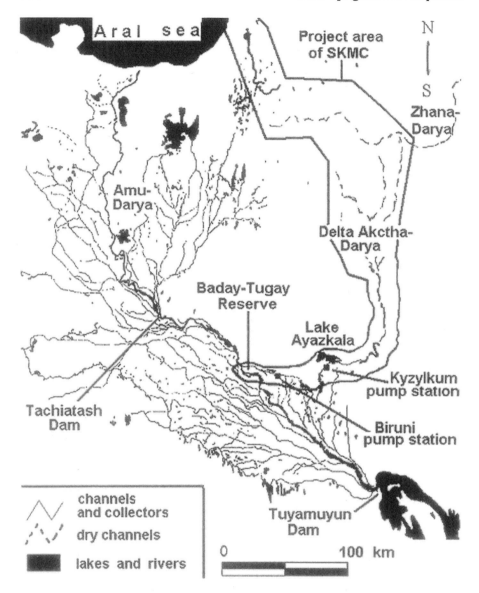

Figure 12-1. Location of South-Karakalpakian Main Collector (SKMC) project.

Research has been carried out in connection with the construction of the SKMC with the aim at assessing its effects upon the environment.

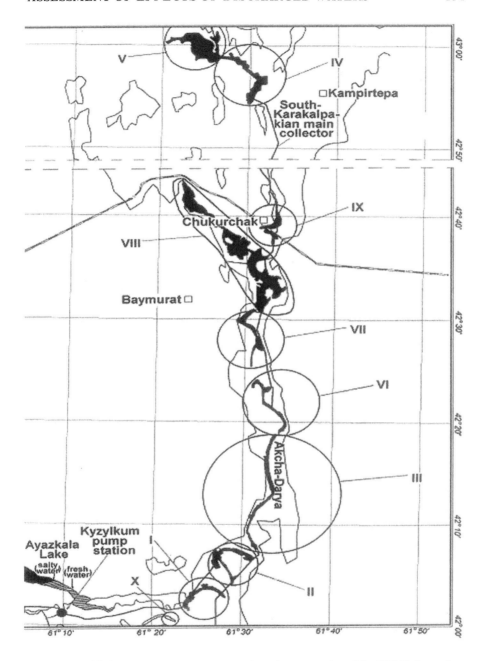

Figure 12-2. Location of planned wetlands under contraction of the SKMC. Legend: square with inscription – settlements, roman figures – planned wetlands on the route of the SKMC.

2.   THE INVESTIGATION METHODS

The vast area affected by the SKMC was divided into four regions according to physiographic conditions: (a) coastal plains of the Aral Sea including the dried sea floor, (b) the old Zhana-Darya and Akcha-Darya channels and delta, (c) the zone of influence exerted by discharged waters in the area near the Ayazkala lake and (d) the "Badai-Tugai" nature reserve with adjacent tugai[1] and agricultural lands.

Monitoring of salinization of surface and ground waters, groundwater depth, the state of plant and soil cover, and bird populations was undertaken to assess the impacts of the Collector upon the environment and ecosystems. Through examination of sociological and economic indices for a long period of time it was also possible to estimate the damage that would be caused to the native population. The data obtained in the course of long-term scientific observations in the Amu-Darya delta provided a basis for assessment of ecological and social consequences in all four of the defined regions.

3.   CHARACTERIZATION OF ECOSYSTEMS IN THE PROJECT AREA

In the zone of influence of waters discharged into the Akcha-Darya old channel (wetlands, Figure 12-2) and in the area near Ayazkala lake (lake banks and adjacent plain) hydrophytic[2], halophytic[3] and tugai vegetation is developed.

The unstable vegetation is localized, being mainly represented by ecotone communities: (a) hydrophytic vegetation – *Phragmiteta australiae, Typheta angustifoliae* and *Bolboschoeneta maritimae*, (b) halophytic vegetation – *Halostachydeta belangerianae, Salsoleta dendroidis, Lycium ruthenicum, Climacoptereta aralensis, Suaedeta salsolae,* etc. (c) tugai ecotone halophyte complexes – shrubs such as *Tamariceta hispidae* and grasses (*Karelinieta caspicae, Aeluropideta littoraliae*), tugai.

In the region of the Akcha-Darya old channel and delta and the Zhana-Darya old channel typical desert vegetation is dominant in   automorphic

---

[1] Tugai is a floodplain forest of river valleys of Central Asia. Tugai's are a unique type of ecosystem. It must be considered as a natural reserve of peculiar flora and fauna. In tugai vegetation prevail such species as: *Populus diversifolia, P. ariana, P. pruinosa, Elaeagnus turcomanica, E. orientalis, Salix songarica* and bushes from *Tamarix* genera.

[2] Hydrophytic vegetation (hydrophytes)- the vegetation growing on wet soils.

[3] Halophytic vegetation (halophytes) – the vegetation growing on saline soils.

territories: (a) psammophilous shrubs on desert soils of hilly and hilly-ridged sands (*Haloxyleta persicae, Ammodendreta conollis, Calligoneta caput-medusae, C.aphyllae, Salsoleta arbusculae*) together with grass species: *Carex physodes* and *Stipagrostis karelinii* (b) shrubs and grasses on takyr[4] soils of the old delta plain (in depressions): *Haloxyleta aphyllae, Anabasieta salsae, Climacoptereta lanatae.*

As a whole, the floral list includes 125 species of higher plants in the regions under study.

The Badai-Tugai nature reserve, with adjacent lands, occupies a rather small area but it is most important for preserving biodiversity in Central Asia. The nature reserve was formed to preserve the disappearing floodplain intrazonal tugai ecosystems. Because of regulated river flows everywhere in Central Asia and conflicts in Tajikistan and Afghanistan, the Badai-Tugai nature reserve (6 000 hectares) may be considered as one of the last remaining areas of tugai vegetation in desert plains. It is inhabited by bukhara deer, wild-boar, khiva pheasant, bustard, fox, jackal, badger, tolai hare, etc. There are more than 70 kinds of birds and 10 kinds of amphibian and reptiles (Kouzmina & Treshkin, 2001).

Within this nature reserve, four vegetation types, developed under hydromorphic and semihydromorphic conditions, can be distinguished: (a) arborescent tugai (Populeta arianae, P. pruinosae, Elaeagneta turcomanicae), (b) shrubby tugai with its distribution linked to salinization of soils and ground waters (Tamariceta ramosissima, Tamariceta hispidae), (c) grassy tugai (Alhagieta pseudalhagii, Aeluropideta littorales, Glycyrrhizeta glabrae, Phragmiteta australiae and Trachomiteta scabrii), (d) widespread solonchakous[5] vegetation: Halostachydeta belangerianae, Salsoleta dendroidis, Climacopterteretalanatae, Suaedeta salsolae, etc. (Kouzmina & Treshkin, 2001; Novikova et al., 1998, 2001; Treshkin, 2001).

The ecological state of the ecosystems may be divided into three types: relatively well preserved (20%), at different stages of desertification (50%) and completely desertified ecosystems (30%).

4. ECOLOGICAL SITUATION IN THE AREA NEAR AYAZKALA LAKE

Ayazkala lake has an anthropogenic origin (Figure12-2). In the 1980's it was artificially created among typical sand deserts as a reception basin for drainage water.

---

[4] Takyr – clay soil of deserts, often is saline.
[5] Solonchakous – soil having very higher salinization, more than 1% (in percentage of salt).

The banks of the western saline and stagnant part of the lake (Figure 12-2) are covered by sparse solonchakous vegetation of little use for cattle (*Suaeda salsa, Halostachys helangeriana, Tamarix hispida, Climacopters lanata*). This vegetation began to appear after construction of the lake. The typical productive pre-existing psammophytic vegetation was degraded to a considerable extent. Higher groundwater levels caused the disappearance of such typical desert grass-shrubby communities as *Haloxylon persicum+Ammodendron, conolly+Calligonum, spp.+Salsola arbuscula-Carex physodes,* and they were not replaced by productive tugai complexes. Due to higher salinity in this part of the lake the banks are covered by a thick salt crust, remain bare, and solonchakous vegetation is rare. Therefore this artificially formed lake serves as a focus for salinization. In addition to ground salinization of adjacent areas, aeolian salt transfer takes place, thus reducing the productivity of nearby agricultural lands.

The banks of the eastern influent and slightly saline part of the lake (Figure 12-2) are covered by *Phragmites australis* and *Phypha angustifolia* in strips (1-3 m wide), which are not of economic or environmental value.

The floristic composition of adjacent lands, including typical sandy, solonchakous vegetation with admixture of tugai is very poor and contains only 35 species of higher plants.

In connection with the construction of the South-Karakalpakian collector, the Kyzyl-Kum drainage pumping station will be closed and Ayazkala lake will be entirely drained. Therefore the degradation of typical sand desert ecosystems through high groundwater and salinization will be stopped and the regeneration of degraded sand desert vegetation will help the formation of rangelands suitable for some cattle. Moreover, there will be no salt accumulation in the lake and closing of the drainage pumping station will save money that can be diverted to other purposes.

Changes in the ornithological complex and the increase in salt transfer from the dried bottom of the lake may be considered as possible negative consequences of this transformation, because additional investment will be required for rehabilitation of the dried lake bed.

However, it is feasible to prevent salt transfer by means of biological rehabilitation of lands. As seen from data obtained in Ben-Gurion University (Israel) and in Turkmenistan, some salt-tolerant plant species can be intensively grown and developed on the dried lake beds if planted by special methods. For instance *Haloxylon* planted on saline soils under irrigation with the sea water (Nebitdag desert station in South-Western Turkmenistan) revealed good root formation. Thus, the problem of salt transfer in the dried bottom of Ayazkala lake may be successfully solved.

In view of the low water level in the Amu-Darya river (1999-2001) the local fishery declined by a factor of 20, and self-restoration is impossible in the near future. Within the period 1998-2002, catches decreased from 10 tons to 500 kg per month. Following the disappearance of fish in the lake 85% of the inhabitants have left Ayazkala settlement and only 10-15 people are now engaged in fishing. Therefore the consequences of the disappearance of this lake for fisheries will be minimal.

5.   ECOLOGICAL SITUATION IN THE REGION OF THE AKCHA-DARYA OLD CHANNEL AND DELTA

Along the Akcha-Darya channel (Figuress. 12-1, 12-2) are *Haloxylon* plantations of higher bonitet[6]. The plants are rather tall (1.5-4.5 m) and are 6-20 years old. *Haloxylon aphyllum* occupies takyr-like soils and solonchaks, while *Haloxylon persium* occurs only on sands. The collector water floods to form wetlands covered by hydrophytic vegetation (Figure 12-2). In the periphery of the wetlands there is a narrow unstable strip (1.5-7 m) of halophytic-solonchakous vegetation and/or tugai ecotone halophyte complexes.

The Akcha-Darya old delta and Zhana-Darya old channel display peculiar features, which are as follows: solonchakous vegetation consisting of *Halocnemum strobilaceum* and *Halostachys belangeriana* so specific to solonchaks in the southern dried part of the Aral Sea (region (a),Figure. 12-1); vast bare areas of takyr-like soils; and near absence of tugai complexes due to a high degree of soil salinization.

Stipulated by the plan, the average water discharges in the collector are 23 m3/sec and the forced water discharges – 40 m3/sec, being estimated to 23 m3/sec near Chukurkak settlement. In the central part of SKC (42°16' - 42°45',Figure. 12-2) the relief is highly dissected. When wetlands III, VI, VII, VIII, IX are formed by discharged water, the depressions between ridges in the adjacent sand desert will become flooded to form small lakes. As a consequence, the productive psammophyle vegetation suitable for grazing of sheep and camels will become degraded. Depressions with stagnant water will promote the reproduction of blood-sucking insects, which are dangerous for sheep and camels; as observed in this region in 1998.

---

[6] Bonitet of forests – qualitative assestment of economic productivity of a forest. Bonitet depends on natural conditions and antropogenic impact on forest. There are five classes of bonitet. The basic parameter of bonitet is average height and diameter of trunks of trees in the certain (comparable) age.

All of the wetlands are covered by hydromorphic vegetation of the same type. *Phragmiteta australiae* and *Typheta angustifoliae* can be of use for cattle, however the cattle-breeding is practically absent here. The region is scanty populated. The inhabitants of Baimurat, Chukurkak and Kempartobe settlements are engaged in farming sheep and goats as well as camels (more than 22000 head), but the above vegetation is unsuitable for these animals. Ayazkala settlement is also small, consisting of only 40 families. Therefore increased live-stock farming should not be expected after the formation of wetlands III, VI, VII, VIII (Figure 12-2). It is also impossible to develop the fishery because of the high content of salts in collector waters. It is evident from 1998 test flooding of the Akcha-Darya channel that water was unsuitable for drinking in five drainage holes, because its salinity was increased to 2.1 g/l. Also, five pastures located along the collector became degraded. Due to rising groundwater the typical productive psammophilous vegetation including *Ammodendron conolly, Salsola arbuscula, Calligonum caput-medusae, Calligonum aphyllum, Haloxylon persicum* dried, and bare solonchaks were formed.

To be able to prevent the sand accumulation in collectors, local wild plant species should be grown along them and adjacent territories in strips 300-500 m wide. Such plants reveal good root formation, and are more tolerant to salinization and the greater fluctuations of groundwater level compared to cultivated plant species. In every region the trees and shrubs for planting should be selected with regard to soil salinization and the groundwater depth preferences (Kuz'mina & Treshkin, 2004; Pankova et al., 1994). Planting to a distance of 500 m from both sides of collector channels will prevent sanding and seepage loss of water.

6.   ECOLOGICAL SITUATION IN THE BADAI-TUGAI NATURE RESERVE

Typical relict tugai ecosystems that remain here are practically absent in the lower reaches and delta of the Amu-Darya river (Kouzmina & Treshkin, 2001).

Created in 1973, the nature reserve contains nearly 167 species of higher plants, including 35 families and 120 genera. However, this list includes all the species of typical desert and solonchakous vegetation that are widespread only in the protected buffer desert zone (1362 hectares). In floodplain tugai areas, embracing 5929 hectares, there were only 90 plant species.

In 1985 the tugai flora list declined to 61 plant species. (Kouzmina & Treshkin, 2001). Hydrophytic and mesophytic vegetation disappeared as a result of anthropogenic xerohalophytism. After constructing Tuyamuyun (1980) and Takhiatash (1974) hydro-electric stations (Figure 12-1) the water level in the Amu-Darya river fell by 4.0-4.5 m over 25 years (Kouzmina & Treshkin, 2001; 2003). River water taken for irrigation and functioning of the Tuyamuyun storage lake was a destructive factor for the nature reserve, because the water intake was more than 30% (500-800 m3/sec) of monthly water discharge.

Changes to environmental conditions in the Badai-Tugai nature reserve include cessation of floods required for maintenance of water levels, salinization of soil cover becomes salinized (1.6-4.5%[7]), fall in groundwater level by about 5 m (Kuz'mina & Treshkin, 1997), dead wood (40-50%), stag-headedness[8] (55%), water heart[9] (70%) occur, and there is no regeneration of wood and grass regeneration.

Monitoring of seasonal salinization and groundwater depth in 21 holes established that the ground water level is affected by the Kok-Darya tributary to a greater extent than by the Amu-Darya river. The influence extends about 1000m perpendicular to the river channel, whereas that of the Amu-Darya extends about 600 m perpendicularly from the channel. This can be explained by peculiar relief, notably a high natural levee near the bank of the Amu-Darya river.

Groundwater level dynamics studied in drainage holes indicates that landscapes in the central part of the nature reserve are now developed under automorphic conditions (GWT 5-5.5 m and more). Therefore ecotopes atypical of tugai ecosystems develop because the groundwater level does not meet optimum requirements for maintaining the tugai floodplain ecosystems (Kouzmina & Treshkin, 2003).

According to the main version of the project for SKMC construction either Biruni pumping station should be closed or the quantity of discharged waters should be partially decreased without improving the water quality. If so, the Kok-Darya tributary that replenishes ground waters through most of the Badai-Tugai nature reserve will entirely dry out or its flow will be reduced by two times. In different seasons of the year water salinization reaches 4.6-13 g/l in Kok-Darya. Low discharge in the Kok-Darya channel as well as complete drying can cause irreparable damage to the nature reserve. In the first case, the central and northern parts of the nature reserve would be gradually dried, resulting in disappearance of forest stands. Complete drying of the Kok-Darya channel could cause the same areas to be dried at once. The loss of major tugai ecosystems would differ only in time: either in 1/2 or in 10/15 years.

---

[7] Weighted average at a depth of 100 cm in percentage of salt.
[8] Stag-headedness – dry top of trunk trees.
[9] Water heart is disease (rotten stuff) of part trunk of tree.

The scientifically-based version of a project to change water impounded in the Kok-Darya tributary differs from that connected with the SKMC construction and financed by World Bank. According to this, Biruni pumping station would continue to convey fresh water from the Amu-Darya river through existing collectors into the Kok-Darya tributary to be subsequently pumped over the territory of the nature reserve for the next 20 years (maximum water discharge would be 15 m3/sec up to 30 m3/sec). This would improve the quality of water pumping by the Biruni station, from 4.6-13 g/l (at the present time) to 1-2 g/l (in future).

This alternative version of the project would not require additional money or construction but would not reduce overall costs of the SKMC scheme. The improved quality of discharged waters in the Kok-Darya tributary would be conducive to decreasing soil salinization and groundwater level near the channel as well as to creating a stable regime for wood tugai ecosystems. The food base for bukhara deer and pheasant would also be improved. It will be possible to preserve the hydrological regime over the whole territory of the Badai-Tugai nature reserve and the groundwater level under tugai forests at a depth of 2.0-3.0 m in the growing seasons. It would favour natural tugai regeneration over more than 2000 hectares of solonchaks, which would be dissolved thanks to this project. Dull landscapes of solonchaks could be transformed into high-productive tugai ecosystems with the whole associated fauna complex.

7. ASSESSMENT OF PERMISSIBLE ECOLOGICAL CONDITIONS FOR HYDROMORPHIC TUGAI AND SOLONCHAKOUS ECOSYSTEMS

The authors' long-term field and stationary investigations make it possible to evaluate the permissible groundwater level, the degree of salinization and the conditions favourable for different tugai and solonchakous ecosystems in the delta of Amu-Darya river including tugai complexes in zones of drainage water in the old course of Akcha-Darya, the Zhana-Darya and Badai-Tugai nature reserve.

- The permissible groundwater level, the degree of salinization of soils and ground waters reveal high fluctuations for various communities and different-aged stages of tugai ecosystems.
- Groundwater levels of 0.5-1.5 m are required for plant communities such as *Salix songarica, Calamagrostis dubia, Phragmites australis, Typha angustifolia, T. minima,* and *T. laxmannii.* For communities of *Salix*

songarica and *Calamagrostis dubia* the soil salinization and the groundwater level should be minimum within the first three meters – 0.1-0.3%[7] in soils and 0.8-1.2 g/l in the ground water.

- A groundwater level of 2-2.5 m. is required for *Populus ariana, Populus pruinosa, Elaeagnus turcomanica, Tamarix ramosissima, Halimodendron halodendron* and for grasses such as *Trachmithum scabrum,* and *Glyzyrrhiza glabra.* Taking into consideration the unfavorable conditions in the Badai-Tugai nature reserve a groundwater level fluctuating from 2.5 to 3.5 m. is recommended (depending on the water amount in the river, the season of the year and the relief) for all wood-shrubby communities. This is the maximum permissible value of the groundwater level for maintaining the tugai ecosystems. Soil salinization under such plant communities can be rather high only in the uppermost metre layer – 0.8-1.5%[7]. The salinization decreases with the depth reaching 0.1-0.7% in the second and 0.05-0.6% in the third metre layer. Wood-shrubby communities can become dried when the soil salinization makes up 0.9-1%[7] at a depth of more than 3 m. However, they can grow and develop in local salt-affected soil horizons (20-40 cm) at a depth of one metre (6-16%).
- The groundwater level in 3.5-4.5 m is tolerable for:
- solonchakous vegetation consisting of *Tamarix hispida, Lycium ruthenicum, Salsola dendroides, Halostachys belangerian, Limonium otolepis, Suaeda salsa,* and *Climacoptera lanata.* These are tolerant to soil salinization equall to 1.6-4.5%[7] in the first two metres of the soil layer and 9-23% in some surface horizons (10-20 cm); in the third metre layer the soil salinization falls to 0.5-0.7%[7] Plants grow when the salinization of ground waters fluctuates from 4 to 20 g/l.
- Grass communities, replacing former forests, comprising *Alhagi pseudalhagi, Karelinia caspia,* and *Aeluropus littoralis* are found to be tolerant to soil salinization of 1-2%[7] in the first two metres and the salinization of ground waters – 1-3 g/l.
- For seed regeneration of plant communities, surface flooding is required for 10-15 days in the period of fruiting, subsequent the maintaining the groundwater level at a depth of 0.3-0.5 m for a month. In this case, salinization should not exceed 0.05-0.07%[7] in the first two metres of the soil. For Lochmium the soil salinization degree can be somewhat increased to 0.1-0.15%[1] in the first two metres. The optimum degree of salinization of ground waters in the range of 0.7 - 0.9 g/l is found to be favourable for the activity of shrubby young growth in tugai wood. (Kuz'mina & Treshkin, 1997; Kouzmina & Treshkin, 2001; Novikova et al., 1998).

8.   CONCLUSIONS

Our studies have shown that the artificially formed lake Ayazkala is rather poor in flora and fauna. The lake habitat for fish and birds is now highly degraded. Today's ecological situation may be considered as critical. There are two possibilities for tackling this problem. The lake can be entirely drained for regeneration of the former sand desert vegetation and rehabilitation of the salt-affected bottom. Alternatively, the lake must become desalinized for regeneration of shrub and grass tugai, the latter being most important for increasing the lake productivity for fisheries and as a habitat for birds. It is inexpedient, from a scientific viewpoint to leave the lake in its present condition, because the lake and adjacent territories will continue to suffer from degradation.

As to the region of the Akcha-Darya old channel and delta, 25-40 m3/sec of water discharge, would lead to surface flooding of depressions between ridges in the desert. As a result, range lands would become degraded, and stagnant lakes would form and encourage numerous blood-sucking insects.

The least damage to ecosystems could be secured by deepening the collector channel without accompanying wetlands.

Aeolian sanding of collectors and seepage loss can be prevented by growing local tree and shrub species in wide strips along the collector channel.

When the South-Karakalpakian Collector is constructed the local population will be deprived not only of range lands located in the channel and near it, but drinkable artesian water in the area along the collector.

The problem of preserving the Badai-Tugai nature reserve after construction of the SKMC must be solved by regeneration of ecosystems through hydrotechnical measures designed to create environmental conditions similar to natural ones.

Chapter 13

IMPACT OF TECHNOGENIC DISASTERS ON ECOGEOLOGICAL PROCESSES

L.M. ROGACHEVSKAYA
Water Problems Institute of Russian Academy of Science

1. INTRODUCTION

Human society, as the most vigorously developing part of an ecosystem, essentially increases the speed of natural processes. Human industrial activity requires raw material, power and information exchange, all within the ecosystem. Thus quantitative changes in the life of society cause qualitative changes in bio- and geosystems. Therefore the consequences of the technological activity of a society are directly reflected in the ecogeological processes taking place within the Earth, its water and the atmosphere.

During the development of human civilization, new features have developed in interactions between the ecological and geological environments. As a greater variety of technical equipment was developed the influences on environmental components, including ecogeological processes, has grown. Since the start of the twentieth century the numbers of natural and technogenic accidents occurring in the world each decade, has increased by a factor of 9 (Profiles of Disasters in the World Summary of Statistics by Continent, 1994). A sharp rise took place in the second half of the century as a result of scientific and technical progress (Danilov-Daniliyan et al, 2001). The end of the twentieth century was characterized by increasing technogenic pressure upon all litho-and biosphere components of the environment. This caused more technogenic accidents. For example, in the USA, direct losses to gross national product from technogenic failures and accidents, pollution and the chronic diseases related to them was 4-6 %, and the increased premature death rate was 15-20 % of the premature death rate - 20-30 % for men and 10-20 % for women (Porfiriev, 1990).

An accident can be defined as a dangerous one-stage event that deeply affects a local population or an area, including loss of life, damage and destruction of the property, or damage to an environment (Cutter, 1993). Accidents caused by industrial activity, whether deliberate or unintentional, are referred to as technogenic accidents (techno (Greek.) - skill, genes - born). Scientific and technical progress can lead to such accidents and some may constitute powerful and dangerous risks to the ecosystem.

The analysis of serious accidents is often very difficult. It is sometimes impossible to distinguish fully between their natural and technogenic causes. Incidents that, at first sight, appear to be natural are often a result of human actions. Nevertheless, while the average number of natural accidents remains fairly constant, the proportion of technogenic ones grows and becomes more and more appreciable.

Technogenic accidents can be divided into local, regional and global events. Direct influences on a geosphere are exerted at a local level (e.g. separate settlements) but, depending on scale, accidents can also have regional or even global effects.

Most technogenic accidents are connected to household activity of people and industry, including extractive industries, but also with terrorism and wars (see Table 13-1).

Domestic technogenic accidents cause the most extensive background damage to environment because they constantly and steadily influence geological processes. Failures of transport systems, cumulative destructive effects of modern megacities, forest fires and deforestation, excessive regulation of drainage, emissions of polluting substances into the environment, and many other things, come into this category.

Transport systems cover the whole surface of the globe and are therefore a constant source of technogenic pressures on the environment. The systems extend through various landscape-climatic conditions and often cause activation of exogenic geological processes – landslips, landslides, karst, avalanches and mudflows, etc. Explosion of main gas pipelines can cause landslips within a radius of several kilometers. Increased pollution by heavy metals, especially lead, along routes used by cars also leads to technogenic damage.

Major urban concentrations such as megacities cover enormous areas and also influence on ecogeological processes. For large cities, processes caused by gravity (landslides, landslips), surface water (erosion), and ground water (flooding, subsiding, karst, subsidence) are commonplace. Rocks settle beneath cities both because of loading by constructions and vibration caused

Table 13-1. Ecological consequences of technogenic catastrophes

Accident	Consequences	Ecological processes
Failures of transport and municipal systems, explosions of major pipelines	Physical damage to the shallow geological environment	Landslips, collapses, subsidence, permafrost, avalanches, karst, permafrost degradation.
	Chemical and thermal pollution of water resources	
Influence of large technical constructions	Constant static and dynamic pressure upon a ground, vibration	Subsidings, collapses, landslips
	Change of underground water behavior, suffusion	Flooding, subsidence of karst, suffusion
Man-made forest fires	Destruction of turf and forest cover	Wind and water erosion, swamping
Excessive regulation of water	Changed balance of underground water	Flooding, suffusion, erosion, swamping, increased salinity, wind processes
Emissions of polluting substances in an environment	Change of substrate and ground water composition	Karst dissolution
Extractive industries	Artificial change of a relief	All types of gravitational displacement
	Change of mechanical pressure in rocks	Technogenic earthquakes
Nuclear tests	Powerful motions of an earth's crust	Technogenic earthquakes, permafrost degradation
Failures at the enterprises of the raised (high) risk	All mentioned above	All mentioned above
Terrorism and wars	Destruction of the economic infrastructure, and all mentioned above	All mentioned above

by transport, especially underground. For example, the base of the Palace of Fine Arts in Mexico has subsided 4,8 m. Aquifers become exhausted quickly due to extraction at a rate 250 percent above the norm. Such powerful technogenic pressures also result in serious pollution of the bio-and geospheres. According to the United Nations, Mexico City heads the list of ten cities of the world with serious water quality and supply problems.

Figure 13-1. The first stage of swamping in burnt outparts of woodland near Moscow (one year after a fire of 2002).

From the beginning of civilization man-made forest fires have been a serious problem both on local and regional levels. 97 % of forest fires are caused by artificial sources of ignition during human everyday life. Around settlements, along highways and main oil-and gas pipelines, and in large forests, often used for recreation, fires occur frequently. The natural phenomena that set the scene for major fires in central regions of Russia in 2002 (Fig. 13-1), extending over 1 million hectares, resulted from atmospheric conditions - an anticyclone collided with a warm front and remained above the affected region. Under these conditions, fires resulting from negligence of local residents and townspeople enjoying recreation became serious.

Woods are a major source of natural resources but are extensively damaged by fires worldwide. In the USA in 2002 more than two million hectares were damaged by fire. In areas that have lost their woodlands disasters may occur. Powerful rains on exposed areas around Durango (Colorado), where powerful forest fires occurred in the summer 2002, resulted in mudflows. According to experts, such phenomena will continue for at least five years until vegetation regenerates. Similarly large scale tree felling can lead to erosion of soil and turf and swamping of the ground.

Technogenic accidents often occur, when human activity changes the natural balance existing in geosystems. For example, the growing requirements of mankind for water have resulted in almost full regulation of surface water sources in areas with high population densities. Creation of water reservoirs causes flooding and swamping of area of up to millions of hectares. In seismically dangerous areas, filling of major water basins causes artificial earthquakes.

Between the 1970s and 1990s the Aral Sea gradually disappeared. The reduction of the Aral Sea area and the general degradation of environment connected to this process have resulted in the largest man-made ecological accident in the world (Novikova, 1999). Excessive withdrawals of water from the Syr-Darya for irrigation have intercepted flow into the lower reaches. At the confluence of the river with the Aral Sea, flow has been strongly reduced since the 1970s. As a result almost all mudflows, which were previously washed away from fields and settled in the sea before, are now deposited in river pools. On irrigated fields, salinization takes place. In the area of the former Aral Sea and coastal section of the Amu-Darya delta, wind blowing leads to deflation and formation of barchans. Surface water processes lead to planar washout, linear erosion etc. (Fig. 13-2). A higher annual inflow is necessary for preservation of a present sea level, but it is impossible to provide this without serious damage to agricultural production in the region.

Emissions of polluting substances to the environment are observed at all levels - local, regional and global. Pollution of the atmosphere leads, finally, to pollution of water resources, and pollution of underground and surface waters, in turn, results in chemical, thermal and radioactive pollution of geological media.

Water pollution is not a global ecological danger like atmospheric pollution because the World Ocean keeps its natural ecosystems pure in general. However, pollution of relatively still surface waters already reaches continental scale in places. Chemical and thermal pollutions of groundwater lead to a qualitative change in the rock mass. «Acid rain » influences dissolution of rocks in karstic areas, damages soil structure, causing subsidence and landslips.

Sources of oil pollution at sea are oil wells and platforms, and shipping; accidents to oil tankers; oil carried into the sea by rivers; and natural oil seeps from the seabed. An example of serious oil pollution was the catastrophic oil slick that covered large parts of the coasts of Spain and France in 2002 and severely damaged coastal ecosystems.

Environmental pollution is also caused by industrial enterprises involving high risks both during the normal operational cycle and, especially, as a result of accidents. Examples are nuclear complexes, and chemical enterprises. Other serious sources of environmental damage are military industries and activities. Chemical, bacteriological, nuclear weapons of mass destruction are especially dangerous during manufacture, storage and

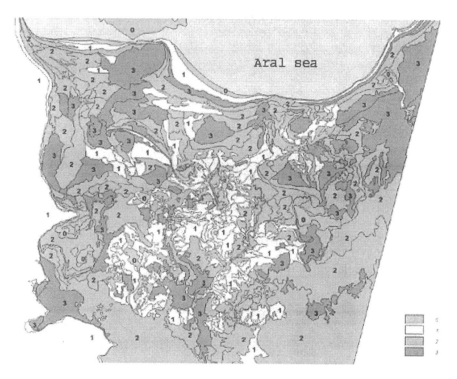

Figure 13-2. Barchans within a former channel of the Amu Darya River. Legend: Index 0 - Absence of hazardous level: in lakes there is a danger of accumulation of pesticides and heavy metals   from drainage waters; Index 1 - Low hazardous level, salinization and deflations in landscapes, from time to time full of river water; Index 2 - Middle hazardous level, accumulation of salts on a soil surface, sand dispersion on saline soils and suffusion in landscapes of the dried bottom of the sea, internal deltas and saline soils, Index 3 - High hazardous level, accumulation of sand and formation of mobile sand files, occurrence of the local centers - sources of salt.

transportation.

    In 1957, in the Chelyabinsk region (USSR) there was an explosion at a facility storing nuclear waste products. This resulted in strong radioactive contamination and the evacuation of population from a large area. On March 28 1979 there occurred the most serious failure in the USA, at the "Three Mile" reactor, Middletown, Pennsylvania. On February 11 1981, 400 thousand liters of a radioactive cooler landscaped at a factory "Sequoia -1" in

Tennessee (USA). On April 26 1986 there was the most terrible failure yet at the Chernobyl nuclear power plant (Ukraine, USSR). As a result of the fourth power unit explosion, some millions of cubic meters of radioactive gases were expelled into the atmosphere. This exceeded by many times the emissions from nuclear explosions above Hiroshima and Nagasaki. Winds carried radioactive substances across much of Europe. All inhabitants were evacuated from a zone of 30 kilometers radius from the blown up reactor. The basic sources of ground water pollution within the affected zone were the top layers of the aeration zone of soils and polluted sediment carried by water erosion into hollows and depressions. Radionuclides were later redistributed in the geological environment (Zektser and Rogachevskaya, 2002). Thus the basic processes of transportation were: infiltration of the atmospheric precipitation through the polluted soil layers and zone of aeration into the groundwater, filtration of ground waters though aquifers, and discharges of ground water into rivers and reservoirs. Thus, water erosion is at the root of pollution of zones of active water exchange.

Mankind increasingly uses the subsurface of the Earth for a variety of purposes – extraction of mineral resources, building of communications and other engineering constructions, and land filling of domestic, industrial and municipal wastes. Underground processes influence the formation and evolution of landscapes and strongly influence their biological efficiency and usefulness for functioning of living organisms, and aesthetic value. Intensive exploitation of the subsurface frequently causes damage with varied negative consequences for ecosystems.

Construction of underground communications, development of quarries and mines and other interventions by people into the subsurface of the Earth causes artificial change of relief. Development of mineral deposits breaks a natural balance. Exploitation of deposits is usually connected with extraction of rocks on the surface of the ground because opencast mining is economically attractive. Besides occupying significant areas of ground, quarries can lead to pollution of the environment by harmful components locally or even regionally. Wind processes may become especially active on exposed surfaces. From 1 hectare of the Kursk magnetic anomaly (Russia) – one of the largest deposits of iron ores in central - about 100 kg of dust is lifted by every dust storm.

Surface deformation of the ground results in activation of wind and water processes leading to subsidence of the ground surface, landslides, landslips and mudflows, thus agricultural areas and woodland may perish.

As a result of mining operations, the change of mechanical pressure in rocks promotes conditions for occurrence of sudden gas emissions, unexpected raisings in deep shafts, etc. Such explosions are often observed in mineral extraction areas. Large mining enterprises in mountainous areas may extract a weight of rock from the quarry and deposit mining wastes that may surpass the critical margin influence on the lithosphere by a factor of 10 or more, such that technogenic earthquakes become possible (Sashourin, 1996). The potential for technogenic earthquakes caused by extraction of minerals is greater in areas that already contain potential hazards. Tremors and earthquakes of different intensities are frequently observed after underground nuclear tests. Such powerful artificial influences on an earth's crust frequently create tremors, or trigger natural earthquakes. Some countries have claimed, over the years, that nuclear tests undertaken by the USA, Russia, China and France have caused earthquakes in their territories from the beginning of the 1950s at nuclear test sites used by the USA in the state of Nevada and on the Marshall Islands in Pacific Ocean, and by the USSR near Semipalatinsk (Russia) and on New Land, many nuclear weapon tests was carried out. Underground nuclear explosions were also made for the oil and gas industry, and for geological research in the former USSR. Near Semipalatinsk, until 1989, explosions with energy one thousand times greater than the bomb dropped on Hiroshima were carried out (Yakubovskaya, 1998). After the signing in 1963 of the treaty between the USSR, USA and Great Britain prohibiting tests in the atmosphere, space and under water, the USSR undertook only underground explosions. Since 1996 the National nuclear center of Kazakhstan and Agency on Nuclear Safety at the Ministry of Defense of the USA have started the removal of 186 tunnels and adits in which tests were carried out. In 2000 the last of the adits at the Semipalatinsk nuclear test site was blown up. Although testing has ended, a huge area remains totally unsuitable for living in or conducting economic activities. Due to high levels of background radiation cultivation is practically impossible.

It is often very difficult to distinguish between technogenic accidents and deliberate actions to cause events. Technogenic terrorism has become an additional threat for sociopolitical and criminal reasons. The technical achievements of civilization can be used to achieve strategic, political and other purposes at worst if weapons of mass destruction now existing are used. Nuclear weapons, for instance, can threaten the whole world if seized by terrorist groups.

This might occur in wars, including interethnic conflicts. Dangers arising from wars include not only global destruction and human victims, but strong influences on the environment. During military actions, water supplies and drainage systems, and industrial targets are the first to collapse, damaging the economic infrastructure. This also leads to ecological instability of the geological environment. Against this background, geological environmental conditions and ecogeological processes begin to change sharply.

For example, two decades of war in Afghanistan have caused such damage to the environment that this has became the main obstacle to restoration of the country's economy (Post-Conflict Environmental Assessment – Afghanistan., 2003). As a result of the conflict, all national programs for environment management have been suspended. Also, all structure of local and national self-management and the infrastructure have been completely destroyed. Agricultural activity is suspended, and the population has been compelled to move to cities. The water supply system in Kabul loses 60 % of water as a result of outflows from pipes and illegal use because of the damage sustained during the military conflict in the country, and lack of regular maintenance. A recent drought has aggravated the large-scale and serious degradation of natural resources. The low levels of subsoil waters and drying of the swampy ground are strengthening Aeolian, bringing soil erosion.

During the Iraq war of 1990-91 bombardments and moving of heavy military equipment caused major harm to Iraq's desert ecosystem (Desk Study on the Environment in Iraq, 2003). The thin layer of ground protecting the surface from erosion has been destroyed in many places. At present sand is exposed to winds and the formation of desert landforms is in progress. Also, due to failure of irrigation systems, agricultural lands have become flooded and salt crusts are forming.

Interventions by people in the geological environment create instability of natural geological processes, which usually are in exchange circulation. Accidents and failures have sudden and rectilinear character, provoking occurring of new ecogeological processes. Very often it is necessary to deal only with the center of a location where an accident has occurred, and then overcome the wider ecological consequences during a long period of time.

Chapter 14

EXOGENIC GEOLOGICAL PROCESSES AS A LANDFORM-SHAPING FACTOR

MAREK GRANICZNY
Polish Geological Institute, Centre of Geological Spatial Information, Warszawa, Poland

1.INTRODUCTION

Exogenic processes include geological phenomena and processes that originate externally to the Earth's surface. They are genetically related to the atmosphere, hydrosphere and biosphere, and therefore to processes of weathering, erosion, transportation, deposition, denudation etc. Exogenic factors and processes could also have sources outside the Earth, for instance under the influence of the Sun, Moon etc. The above mentioned processes constitute essential landform-shaping factors – Figure 14-1. Their rate and activity very often depends on local conditions, and can also be accelerated by human action. It is also true that combined functioning of exogenic and endogenic factors influences the present complicated picture of the Earth's surface.

Mountains, valleys and plains seem to change little, if at all, when left to nature (Adams and Wyckoff, 1971), but they do change continuously. The features of the Earth's surface temporary forms in a long sequence of change that began when the planet originated billions of years ago, and is continuing today. The processes that shaped the crust in the past are shaping it now. By understanding them, it is possible to imagine, in a general way, how the land looked in the distant past and how it may look in the distant future.

Landforms are limitless in variety. Some have been shaped primarily by streams of water, others by glacial ice, still others by waves and currents, and many more by slow movements of the Earth's crust or by volcanic eruptions. There are landscapes typical of deserts and others characteristic of humid regions. The arctic makes its special mark on rock scenery, as do the tropics. Because geological conditions from locality to locality are never quite the same, every landscape is unique. Rock at or near the surface of the continents breaks up and decomposes because of exposure. The processes involved are called weathering.

Weathering is the decomposition and disintegration of rocks and minerals at the Earth's surface. Weathering itself involves little or no movement of the decomposed rocks and minerals. The resulting material accumulates where it forms and overlies unweathered bedrock.

Erosion is the removal of weathered rocks and minerals by moving water, wind, glaciers and gravity. After a rock fragment has been eroded from its place of origin, it may be transported large distances by those same agents. When the wind or water slows down and loses energy, transport stops and sediment is deposited.

These four processes – weathering, erosion, transportation and deposition work together to modify the Earth's surface (Thompson and Turk, 1998).

2.   THE WORK OF WEATHERING

Weathering produces some landforms directly, but is more effective in preparing rocks for removal by mass wasting and erosion. Weathering influences relief in every landscape.

2.1   Freezing and thawing

Water expands when it freezes. If water accumulates in a crack and then freezes, its expansion pushes the rock apart in a process called frost wedging. In a temperate climate, water may freeze at night and thaw during the day. Ice cements the rock temporarily, but when it melts, the rock fragments may tumble from a steep cliff. Large piles of loose angular rocks, called talus slopes, lie beneath many cliffs. These rocks fell from the cliffs mainly as a result of frost wedging.

2.2   Temperature changes

Sudden cooling of a rock surface may cause it to contract so rapidly over warmer rock beneath that it flakes or grains break off. This happens mostly in deserts, where intense daytime heat is followed by rapid cooling after sunset or chilling by sudden rain. Ordinary temperature changes are, however, less destructive than once supposed.

2.3   Organic action

Plant roots expand as they grow, breaking up soil and moving small rock masses apart. Animals such as moles break up soil by burrowing. This phenomenon can also be seen in city sidewalks, where tree roots push up from below, raising the concrete and frequently cracking it.

2.4   Differential weathering

If the minerals in an angular block have about the same resistance to weathering, or differ in resistance but are evenly distributed, prolonged weathering will round off the block because exposure is greater along the angular edges. Rock containing minerals of varying resistance, segregated into layers or pockets, weathers unevenly as these become exposed. Edges of the more resistant layers stand out as ridges and pockets of resistant minerals as humps or knobs. This aspect of weathering is called differential. Along with erosion, which has a similar character, it accounts for much landscape relief.

2.5   Unloading

Rock formed at depth under high pressure expands as erosion removes, or "unloads" overlying rock. Expansion may cause fracturing parallel to the topographic surface, producing parallel "plates" or "sheets" from a few centimetres to meters thick. This process is called pressure-release fracturing. Many igneous and metamorphic rocks that formed at depth, but now lie at the Earth's surface, have been fractured in this manner.

2.6   Chemical action

Over longer expanses of geologic time rocks decompose chemically at the Earth's surface. The most important processes of chemical weathering are dissolution, hydrolysis and oxidation. Water, carbon dioxide, acids and bases, and oxygen are common substances that take part in these processes to decompose rocks. Decomposition, which is most rapid in warm, humid regions, produces new minerals that are usually mechanically weaker and of larger volume than the original minerals. The rock either crumbles or exfoliates (flakes off).

3.   FEATURES OF MASS WASTING

Rock material loosened by weathering is continuously pulled by gravity. If support is removed, the material tilts, falls, slides, flows or sinks. These movements from gravity alone are called "mass wasting". Some are dramatic and destructive. Every year small landslides destroy homes and farmland. Occasionally a big landslide buries a town or city killing thousands of people. Landslides cause billions of dollars in damage every year. In many instances, losses occur because people do not recognize dangers that are obvious to a geologist. But mass wasting, like weathering, is mostly slow and usually it does not alone produce large, spectacular landforms. Naturally, mass wasting is most effective in highlands, where slopes are steep.

Rockfalls are falls of individual rock fragments from a cliff. Fragments range from small flakes to blocks weighing hundreds of tons, which tend to break on the way down. An accumulation of fallen material at the foot of a cliff is known as a talus. Falls of mantle rock often mixed with vegetation are debris falls.

Landslides are movement of coherent masses of material along shears. A slide is usually faster than a flow, but it still may take several seconds for a block to slide down. The cause may be lubrication or weakening of the mass by heavy rains or undermining. Loosened material may accelerate to several tens of kilometres per hour, descend into a valley, break up and climb the opposite wall. Slides can wipe out towns; some block streams to form lakes. Slides of mantle rock mixed with some vegetation are termed debris slides. Landslides are often divided into slumps – downward slipping of a block of Earth material, usually with a backward rotation on a concave surface - and rockslides – usually rapid movement of a newly detached segment of bedrock.

Earthflow: during earthflow, loose, unconsolidated soil or sediment moves as a fluid. Some slopes flow slowly – at a speed of 1 centimetre per year or less. On the other hand, mud with a high water content can flow almost as rapidly as water. Earthflow can be divided into several subcategories.

Creep is the very slow, downslope movement of soil fragments at rates that diminish with depth. Creep tends to be fastest at the surface, where disturbance is more likely and there is least stabilizing friction. Evidence of creep includes downslope tilting of fences, telephone poles or breached retaining walls and trees that curve out from the slope before turning up.

In a debris flow more than half the particles are larger than sand size and the rate of movement varies from less than 1 m/ year to 100 km/hr or more.

A mudflow is the movement of generally fine-grained particles with large amounts of contained water. Mudflows – some fast, some slow – typically develop during heavy rains on fine, dry rock waste along steep slopes lacking vegetation.

Solifluction is the movement, usually very slow, of a mass of rock waste over frozen ground. The slope may be only one or two degrees. Flows occur on alpine and arctic terrains where rock waste gets saturated with meltwater that the frozen ground cannot absorb.

Landslides commonly occur on the same slopes as earlier landslides, because the geologic conditions that cause mass wasting tend to be constant over large areas and remain constant for long periods of time (Thompson and Turk, 1998). Thus, if a hillside has slumped, nearby hills may also be vulnerable to mass wasting. In addition, landslides and mudflows commonly follow the path of previous slides and flows. If an old mudflow lies in a stream valley, future flows may follow the same valley. Many cities and settlements were founded decades or centuries ago, before geologic disasters were understood. Often the choice of a town site was not dictated by geologic considerations but by factors related to agriculture, commerce, or industry, such as proximity to rivers and ocean harbours and the quality of farmland. Furthermore, geologists' warnings that a disaster might occur are often ignored. Geologists construct maps of slope and soil stability by combining data on soil and bedrock stability, slope angle and history of slope failure in the area. Awareness and avoidance are the most effective defences against mass wasting.

4.   SCENERY SHAPED BY STREAMS

Among all the exogenic land-shaping factors running water is by far the most effective. Streams, mainly, have cut the systems of valleys that drain the continents. Streams are the architects of flood plains, deltas and other common deposit features. Not only full-sized rivers, but also the countless small, temporary rivulets that result from every thaw and rain, work to shape the crust. With some notable exceptions the relief of landscapes generally, has been shaped largely by streams.

Valleys change in stages that suggest a life cycle. In youth a valley has a swift stream that cuts down vigorously, making a deep ravine or gorge with waterfalls, plunge pools, rapids and potholes. In maturity down-cutting has considerably lowered the valley bottom and the stream is less vigorous.

Down-cutting has waned relative to side-cutting and grading has eliminated waterfalls and other irregularities. The stream swings from side to side on a widened valley floor, depositing sediments to form small floodplains. In old age the valley bottom is near base level – i.e. the limiting level of erosion, such as a large river on a broad plain. Down-cutting has practically ceased, but side-cutting has made the valley still wider - sometimes kilometres wide. Changes affecting a stream's development, such as uplift, valley drowning or climate change may interrupt a valley's evolution at any stage.

Erosional features of valleys include:

- Potholes - deep, smoothly circular or elliptical depressions in bedrock made by abrasive sediment in eddies.
- Waterfalls - typical of young valleys. Cliff with falls may develop at the edge of rock masses more resistant than the rock below the falls. Cliff also may result from faulting, glaciation or valley-blocking by landslides or lava flows. Close series of falls form cascades.
- Rapids are areas of swift, dashing water around rocks projecting from the stream-bed. The rocks may be blocks weathered off valley walls or the edges of resistant strata, e.g. the last vestiges of waterfall. Rapids are common, as are waterfalls, in areas of recent volcanic activity or faulting.
- V-shaped valley cross-sections form where mass wasting on falls keeps pace with stream erosion. Loose mass-wasted material makes a slope of 35^0 (angle of repose) or less.
- Undercut slopes, steep slopes or cliffs may form on the outside of river bend. Side-cutting by streams as they round curves tends to produce unstable overhangs, which eventually fall, leaving steep walls.
- Vertical walls develop where down-cutting occurs without much mass wasting on the valley walls or side-cutting by the stream. Relatively rapid down-cutting may be favoured by vertical joints, faults or other weak zones.
- Walls of unequal slope developed where the effects of weathering, mass wasting and stream erosion differ on opposite sides of a valley.
- Ribs are the edges of relatively resistant, nearly horizontal strata projecting from valley walls. These generally alternate with the recessed edges of weaker strata.
- Benches are rock platforms extending from the base of a scarp or cliff on a valley wall towards the stream channel..

Depositional features of valleys include:

- Alluvial cones and fans are masses of sediment deposited at the foot of mountain ravines where streams reach a valley floor abruptly. Sediments are sorted, grading from coarse to fine at the bottom of the slope, according to the change of gradient. Alluvial cones, seen at the bottom of steeper slopes, differ from talus cones, in which material is not sorted. Alluvial fans, formed at the foot of gentler slopes, may grow much larger than cones, becoming plains.

- Braiding is the division of a stream into channels separated by elongated islands of sediment. This occurs when a stream has an overload of coarse material, slows and spreads laterally when the channel becomes choked with sediment.
- Alluvial terraces are level surfaces on deposits extending out from the valley wall. Terraces represent former levels of stream deposition. Changes in the conditions of flow may cause a stream to regard itself as a lower level, thus leaving the former deposition surface considerably above the stream. Alluvial terraces are subject to dissection by tributaries and by the main stream.
- Floodplains are nearly horizontal surfaces on fine-grained deposits left by floods. The flood-waters overflow from a curved channel that is no longer cutting down. The elevation of the channel floor varies according to flood conditions. The development of a flood-plain starts with the deposition of sediment as point bars at the base of slip-off slopes while the channel is shifting laterally around bends.
- Meanders are smooth, loop-like curves developed by continued cutting and filling in the channel. Meanders migrate downstream, and in the process valley side spurs are trimmed and eliminated.
- Cut-offs are short channels across the necks of meander curves. As the curve nears full circle, the neck narrows and floods across the neck may erode a new, shorter channel.
- Oxbow lakes often form in cut-off meander loops.
- Natural levees are accurate or curved, round-backed ridges flanking meanders, several meters above the floodplain. They form as floodwaters, overflowing the channel, and deposit most of their sediment load close to it.
- Deltas are accumulations of sediment deposited where streams empty into lakes, seas or oceans.

When stream discharge exceeds the volume of a stream channel, water overflows onto the floodplain creating a flood. Although flooding is a natural event, human activity can increase flood frequency and severity. Floods are costly because many people choose to live in floodplains. Many streams flood regularly, some every year. In any stream, small floods are more common than large ones, and the size of floods can vary greatly from year to year. For example, a stream may rise two meters above its banks during a once in ten-years flood, but 10 meters or more during a 100-year flood. Thus, a 100-year flood is higher, larger and much more dangerous for people living in floodplains.

5.   LANDFORMS IN LIMESTONE

Extensive landscapes have been developed on carbonate rocks in areas of the crust that were sea bottoms in the ancient geological past. Lime and other sediments accumulated on the sea bed as chemical precipitates or organic remains, gradually becoming rock. The solubility of this rock in water containing carbon dioxide gives rise to scenery characterised by caverns and sinkholes, dry valleys and "lost" rivers. Such features make up karst topography.

Surface landforms of karst topography include:
- Pillars (Pinnacles); mostly upland features, favoured by vertical jointing.
- Collapse sinks; irregular surface depressions formed by the collapse of cave or tunnel roofs. Upland features are less common than sinkholes.
- Sinkholes; deep, funnel shaped holes of all sizes, made by runoff entering joints, usually at intersections. They may contain ponds or lakes.
- Karst valleys; common upland valleys with prominent features due primarily to solution. The valley bottom is often poorly graded because of multi-level drainage; it may be dry or may contain a segment of a lost river.
- Valley sinks (Uvala); large depressions formed by the merging of sinkholes or collapse sinks.

6.   DUNE LANDFORMS

A dune is a mound or ridge of winddeposited sand. Approximately 20 percent of the world's desert area is covered by dunes. Dunes also form where glaciers have recently melted and along sandy coastlines.

Barchans are dunes of crescent form. Grains move up the windward side by saltation, then slide down the lee or slip side, forming a slope of 30 to 35 degrees. The slip face is concave upwards due to the action of wind eddies. Barchans usually develop in swarms near the edges of sand plains where sand is thinning.

Parabolic dunes are similar to barchans, but with longer horns pointing to windward.

Transverse dunes are long irregular features at right angles to the prevailing wind.

Seifs are long, sharp-ridged dunes, parallel to the prevailing wind direction. They may extend for hundreds of kilometres and reach several hundred meters in height.

Loess: wind can carry silt for hundreds or even thousands of kilometres and then deposit it as loess. As a result, even though the loess is not cemented, it typically forms vertical cliffs and bluffs. Large loess deposits accumulated during the Pleistocene Ice Age, when continental ice sheets ground bedrock into silt.

7.   COASTAL FEATURES

Where sea meets land, geologic agents work visibly, and landforms may change rapidly. A coast is a strip of land bordering the sea, within reach of marine processes and influences. Sandy coastlines occur where sediment from any source is abundant; rocky coastlines occur where sediment is scarce.

7.1   Sandy coastlines

Most of the sand along coasts accumulates in shallow water offshore from the beach. If a coastline rises or sea level falls, this vast supply of sand is exposed. Thus, sandy beaches are abundant on emergent coastlines. A long ridge of sand or gravel extending out from a beach is called a spit. A spit may block the entrance to a bay, forming a bay mouth bar. A spit may also extend outward into the sea, creating a trap for other moving sediment. A barrier island is a long, low-lying island that extends parallel to the shoreline. It looks like a beach or spit and is separated from the mainland by a sheltered body of water called a lagoon.

7.2   Rocky coastlines

In contrast to sandy beaches typical of emergent coastlines, submerged coastlines are commonly sediment poor and are characterized by steep rocky shores. A wave-cut cliff forms when waves erode a rocky headland into a steep profile. As the cliff erodes, it leaves a flat or gently sloping wave-cut platform. If the sea floods a long, narrow, steep-sided coastal valley, a sinuous bay called a fjord is formed. An estuary forms where rising sea level or a sinking coastline submerges a broad river valley or other basin.

8.   GEODYNAMIC RISK MAP

"The Map of Geodynamic Risks in Poland" was compiled at the scale of 1:500 000 using GIS technology (Graniczny, 1992). This map consists of three main tables; A - endogenic, B - hydrogeologic and C - exogenic factors. Table C lists the following exogenic agents:

- Loess (subsidence, cliffs, deep valleys).
- Karst (developed on limestones, marls, dolomites and gypsum).
- Mass movements (landslides, rockfalls – active, extinct etc.).
- Dunes (coastal and continental).
- Floodplains (high potential of flood occurrence).
- Organic sediments (weak substratum, shallow groundwater).
- Soil erosion.
- Wave-cut cliffs.
- Erosional escarpments.
- Areas of the potential mining risks.
- Devastated areas due to human activity (industrial areas etc.).
- Areas of the low exogenic risk.

The above mentioned map shows the importance different exogenic geological processes for human activities such as: territorial planning, urbanization, agriculture, environmental protection, sustainable development, recreation etc. This map, compiled at a regional or even review scale, is intended for use by non-geologists. It is aimed at decision makers and territorial planners at the national planning level, and gives primary information on geodynamic risks. The map is also expected to be widely applied for educational purposes and promotion of the Earth sciences; it will be followed by much more detailed maps and data bases illustrating different kinds of exogenic phenomena and processes. These maps are usually produced for special orders from government, local administration, industry and private investors.

In the last few years, in Central Europe, the effects of several exogenic geological processes have been seen. These are mainly related to mass movements (Poland, Slovakia, Czech Republic, Ukraine in the Carpathians Mts.), catastrophic floods (Poland, Czech Republic, Germany, Hungary, – Vistula, Odra, Danube) and coastal zone processes (Poland, Lithuania – Baltic Sea). These processes and their effects should be the subject of research and monitoring and be given special consideration by the administrative authorities.

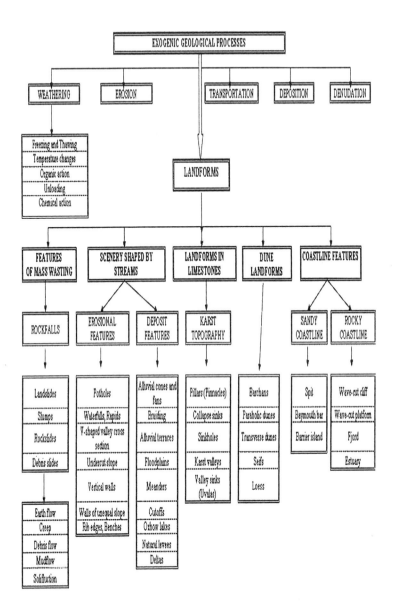

Figure 14-1. Exogenic geological processes

Chapter 15

ENVIRONMENTAL ASPECTS OF GROUNDWATER POLLUTION

ROALD G. DZHAMALOV
Water Problems Institute, Russian Academy of Science

The term groundwater pollution refers to adverse changes in groundwater quality (i.e., in its physical, chemical, and biological properties) caused by economic activities. These changes impair groundwater quality to the extent it does not meet water quality standards for use, thus partially or completely preventing specific types of consumption.

Near-surface aquifers, close to potential sources of surface pollution, are the most liable to groundwater pollution. These aquifers usually contain fresh groundwater, i.e., water with TDS total dissolved solids not exceeding 1 g/l. The term "groundwater pollution" is primarily applicable to this type of groundwater.

In order to estimate the degree of groundwater pollution, it is necessary to use water quality indices recorded under natural conditions (the background groundwater quality) and the accepted (established) standards of groundwater chemical composition and properties controlling the possibility of specific groundwater use. The change in groundwater quality may include increased salinity, rise or fall in the concentrations of certain groundwater constituents (chlorine, sulfates, bicarbonates, calcium, magnesium, iron, fluorine, and others), the occurrence of man-made substances foreign to groundwater (surfactants, pesticides, and petroleum products), variations in temperature and pH, the occurrence of odor and coloration, and other changes.

In addition to the background indices, the degree of groundwater pollution may be also characterized by the maximum permissible concentrations (MPC) of the substances that both naturally occur in natural waters and enter them as a result of the industrial, agricultural, municipal, radioactive and other types of pollution, as well as water treatment intended for various types of water use.

MPCs identify the permissible concentrations of certain chemical substances in natural waters and are a water-use-dependent system of reference indices of water chemical composition. Requirements of the MPC for different substances may vary from country to country. In the majority of countries, however, the potable water standards include the MPC for dry residue, chlorides, sulfates, total water hardness, pH, nitrates, fluorine, iron, manganese, copper, zinc, lead, selenium, strontium, uranium, radium, beryllium, molybdenum, arsenic, and a number of other elements. If water contains several limited substances (except for fluorine, nitrates, and radioactive substances), the sum of their relative concentrations C^* (the ratio between the actual substance concentration and its MPC) should not exceed unity.

Groundwater pollution may include several dozens of pollutants. In this case, this pollution may be characterized by concentrations of only two or three of these pollutants; namely, those that are most specific to this pollution and those with maximum ratios to the MPC. On the other hand, in order to generally characterize the groundwater pollution area in order to compare it with other pollution areas, this may also be described by some general indices. General indices of pollution usually include dry residue, total water hardness, and concentrations of chlorides, nitrates, heavy metals, and total organic carbon, as well as oxidizability and bacteriological and organoleptic properties. Thus, the comprehensive description of groundwater pollution should be based on the general and specific indices.

In order to determine the extent of multiple-source groundwater pollution, it is usually necessary to determine the area of pollution, the level of pollution, and the flow velocity of the polluted groundwater in the aquifer for each source of pollution (Hydrogeological Base..., 1984)..

The pollution area boundary is usually defined as the contour line of the MPC of the specific or most toxic pollutant or (for freshwater) as the contour line of the total groundwater salinity of 1 g/l. The area of groundwater pollution by several specific components is determined from both the MPC contour lines of these components and the contour line of the total salinity of 1 g/l. Thus, the groundwater pollution area includes the entire area limited by the contour line of the total salinity of 1 g/l and those of the MPC of separate pollutants. On the basis of available actual data, groundwater pollution areas are classified into 10 groups according to their areal extent: from those with an area less than 1 km^2 to those with an area greater than 200 km^2. For the maximum extent of this area one usually takes the length of the polluted area along the direction of groundwater flow.

The level of groundwater pollution within the determined polluted area is described in terms of the ratios of groundwater salinity to the relevant average and maximum permissible salinity, as well as in terms of the ratios of concentrations of separate pollutants within their MPC contours to the relevant average and maximum permissible concentrations. There are several levels of pollution, some of which correspond to the initial phase of pollution (below MPC) and others to supernormal groundwater pollution (from 1 to 100 MPC and higher). These levels are: moderate pollution (up to 1 MPC), considerable pollution (from 1 to 10 MPC), high pollution (from 10 to 100 MPC), and extreme pollution (higher than 100 MPC).

Seepage velocity of polluted groundwater within the aquifer is determined from the actual displacement of the TDS contour line of 1 g/l or the boundaries of selected MPC values during the estimated period of time. Estimation of seepage velocity should be performed along the direction of groundwater flow to the proximate water-pumped structures or drains.

The areal extent and the level of groundwater pollution primarily depend on human environmental impacts or the total amount of pollutants within the area under study, as well as on the natural geological and hydrogeological conditions controlling groundwater vulnerability. In other words, the liability of groundwater to pollution (the possibility of groundwater pollution) is in direct proportion to the human impact on the environment (the study area) and to the natural groundwater vulnerability within this area.

The natural groundwater vulnerability to pollution depends on the thickness of beds (particularly those that are impermeable and relatively impermeable) overlying the aquifer and preventing pollutants from entering groundwater from the earth's surface. The level of groundwater vulnerability should be assessed prior to designing and constructing projects affecting groundwater, as well as before developing and approving measures for preventing groundwater and water-supplystructures from pollution. The level of groundwater vulnerability depends on three groups of factors, i.e., natural, human-induced, and physicochemical.

The natural factors include:
- groundwater depth;
- thickness, lithology, and permeability of rocks overlying groundwater;
- presence of relatively impermeable beds, i.e., rocks with hydraulic conductivity less than 0.01 m/day;
- sorption properties of rocks; and
- relationship between groundwater levels in the aquifers.

The human-induced factors include the distribution of pollutants over the earth's surface and the nature of their infiltration into the groundwater. Physicochemical factors are characteristic properties of pollutants that control their migration capacity, sorption capacity, decay time, and the character and rate of interaction with rocks and water (Goldberg, 1987).

It is obvious that the deeper the groundwater and the thicker the relatively impermeable beds overlying it, the lower will be the groundwater vulnerability to any types of pollutants. It is good practice to assess the natural groundwater vulnerability in two steps. The first step primarily concentrates on the study of natural factors protecting groundwater on a regional scale. The second step includes detailed studies in order to account for the human impact and specific physicochemical properties of pollutants. The assessment of the degree of groundwater vulnerability may be both qualitative, i.e., based on the effect of natural factors, and quantitative, i.e., based on the effect of natural and human-induced factors.

The quantitative assessment of groundwater vulnerability may be based on determining the time "t" required for pollutants to reach groundwater from the earth's surface. This time depends both on natural factors and human-induced and physicochemical conditions. The degree of groundwater vulnerability may be assessed by the ratio of the calculated time to the decay time of the pollutant. For different schemes of wastewater infiltration, "t" is usually determined from the analytic relationships (formulae) known in hydrogeodynamics or from numerical simulations. The shorter the time of infiltration, the higher the degree of vulnerability.

It is worth noting that both the qualitative and quantitative estimates of the degree of groundwater vulnerability should be primarily carried out for unconfined aquifers, which are most liable to pollution from different sources (Bear, et al, "Flow and Contaminant Transport in Fractured Rock", 1993).

Groundwater pollution is closely related to general environmental pollution. Thus, it is virtually impossible to prevent groundwater pollution under conditions of considerable and progressive pollution of the atmosphere, surface waters, and soils and rocks of unsaturated zones. The relationship between groundwater pollution and environmental pollution may be classified according to the index of groundwater liability to pollution "P". This index is defined as the ratio of the modulus of human impact (M_{hi}) on the pollution area to the rank of natural groundwater vulnerability. The modulus of human impact is the ratio of the annual amount of all pollutants entering the area under study to the magnitude of this area.

Thus, the higher the value of the index P, the higher the potential liability of groundwater to pollution. Values of the index P may be divided into several ranges, which may serve as acceptable criteria for the hydrogeological areal subdivision of the pollution area according to the degree of groundwater liability to pollution (Goldberg, 1987).

Liquid wastewater, which is capable of infiltrating through the rocks of the unsaturated zone, is the main source of groundwater pollution. Large areas of groundwater pollution are usually confined to the industrial and agricultural enterprises with large wastewater discharge. Wastewater containing chemicals and radioactive substances causes the most persistent and toxic pollution.

The maximum groundwater pollution results from:
- surface waste liquid storage (sedimentation basins, receivers, evaporation basins, slurry tanks, tailing dumps, and others);
- solid waste storage;
- filtration beds;
- disposal fields for wastewater;
- storage for petroleum products, toxic chemicals, fertilizers, and other chemicals;
- filling-stations, oil pipelines, and highways; and
- polluted and substandard surface waters.
- The main pathways of pollutants entering aquifers include infiltration of pollutants from:
- the storage and landfill areas,
- agricultural fields,
- polluted and substandard surface waters, and (4) polluted precipitation.
  Among numerous groundwater polluters, the following are noteworthy:
- point sources, which include above all various types of waste storage, landfills, storage areas, and others;
- diffuse or nonpoint sources, which include precipitation, agricultural activity, industrial and urban agglomerations, highways, etc.

Point sources usually result in intense but local groundwater pollution, whereas diffuse sources usually cause less intense regional environmental pollution, which involves groundwater. Regional groundwater pollution caused by polluted precipitation and other sources is characterized both by a large affected area and spatial inhomogeneity of pollution: against the background of relatively low concentrations of pollutants (below MPC), there are localised areas with high concentrations of pollutants (above MPC). Local groundwater pollution is frequently caused by petroleum products, strong chloride solutions from industrial enterprises, nitrogen compounds in livestock waste, and bacterial pollution from various sources (Freeze & Cherry, 1979).

Groundwater pollution by oil and petroleum products, which are a mixture of various hydrocarbons, has specific features. The majority of petroleum hydrocarbons show low solubility in water and a density less than that of water. Therefore most hydrocarbons usually form lenses of single-phase liquid, which occupy the upper part of the aquifer section and vary greatly in size. These lenses float on a zone of two-phase liquid, i.e., the emulsified mixture of hydrocarbons and water. Beneath this zone, there is the zone of hydrocarbons, mainly aromatics, dissolved in water. Areal extent of these zones varies considerably. Areas occupied by the emulsified hydrocarbons and hydrocarbons dissolved in water are dozens of times greater than the areas occupied by the lens-shaped zone of single-phase hydrocarbons. The migration of petroleum products through the porous medium strongly depends on the ratio between oil and water in the two-phase mixture. If the oil content of rock exceeds 80-85%, the rock will be virtually impermeable to water and permeable only to oil. Conversely, after reduction in the oil content down to 20-15%, the rock will be permeable only to water.

Strong, predominantly chloride solutions are denser than freshwater. Since denser chloride solutions gravitate to the aquifer bottom, chloride pollution occupies predominantly the lower part of the aquifer. Chlorides are readily soluble in water. These persistent, undecomposable, and nonabsorbent substances show high migration capacity. Thus, chloride solutions may spread through the aquifer over long distances and form pollution areas large both in spatial extent and in area.

Groundwater pollution by nitrogen compounds is related predominantly to agricultural pollution sources. Thus, this type of pollution commonly occurs within areas with intensive agriculture and livestock waste. Due to the processes of successive oxidation of nitrogen compounds, groundwater contains three main forms of nitrogen, i.e., ammonium, nitrite, and nitrate. Ammonium nitrogen and nitrite nitrogen are unstable forms, which gradually pass into the most stable form, nitrate nitrogen which is the major pollutant accumulated by groundwater. The process of nitrification or conversion of nitrogen forms to nitrates lasts for approximately 1-1.5 month and strongly depends on the temperature and redox conditions in the unsaturated zone and in the aquifer. Nitrates are readily soluble in water, relatively weakly adsorbable by rocks, and capable of migrating considerable distances through the aquifer.

Under reducing conditions in the aquifer, the rate of nitrification slows down to zero. Under these conditions, the processes of denitrification begin to dominate and reduce nitrate nitrogen to nitrite nitrogen and ammonium nitrogen. Under these settings, ammonia nitrogen may persist over a long period of time, not converting into nitrate nitrogen. Thus, the succession of direct and reverse conversions of ammonium nitrogen into nitrate nitrogen and vice versa is a specific feature of the migration of nitrogen species in groundwater.

The main feature of bacterial groundwater pollution is its containment or local spreading within the aquifer. This is due to the relatively short lifetime of bacteria in groundwater. This time varies from 30 to 400 days and depends on a variety of factors: the species composition and abundance of bacteria in water, seepage velocity, geochemical environment, temperature, and the presence of other pollutants in water. The migration distance of microorganisms is controlled by, in addition to lifetime, the adsorption of microorganisms by water-bearing rocks during groundwater seepage. Thus, a 40-cm-thick soil layer retains about 90% of bacteria.

It is worth noting that increase in water temperature affects the toxicity of pollutants. Thus, a rise in water temperature from 15 to 25°C results in a nearly threefold increase in the toxicity of zinc. Increase in temperature may also augment the toxicity of cyanides, certain pesticides (HCCH and others), and a number of other toxic substances. Since the temperature of polluted wastewater is as a rule elevated, this may result in higher toxicity of some groundwater constituents.

The spatial extent and level of radioactive groundwater pollution are estimated with allowance made for the sorption capacity of water-bearing rocks and the half-life of pollutants. These characteristics control the concentrations of radioactive elements in groundwater. Of radioactive elements, those poorly adsorbed by rocks (iodine-131, sulfur-35, ruthenium-106, uranium, cesium-137, and strontium-90) constitute the principal hazard to groundwater. Of the latter elements, strontium-90, cesium-137, and uranium are especially dangerous because of their long half-life and high migration capacity in groundwater.

Acid rain is one of the major sources of diffuse regional groundwater pollution. Major factors in the processes of groundwater acidification and groundwater pollution caused by acid precipitation are as follows: the level of carbonate content of water-bearing rocks and concentrations of bicarbonate ions in the initial groundwater composition; conducting properties of the aquifer; depth to groundwater; water exchange time between different components of the aquifer system; type of groundwater hydrodynamic regime; and the proportion of forest and swamp areas within the catchments areas.

The effect of acid rain is increased migration capacity of major and minor groundwater constituents and pronounced disturbance of the alkaline-acid and redox conditions in the water-rock system. Especially abrupt changes in groundwater chemical composition occur under the effect of so-called "pH-shocks", which are usually due to spring melt water and strong rains. In this case, the initial pH value of precipitation may drop to 3-4, and precipitation begins to dominate in the water budget of groundwater recharge areas. Regional groundwater pollution caused by the effect of acid precipitation involves mainly sulfates and heavy metals. In the aquifer, the process of acidification may persist for decades and usually results in changed groundwater hydrochemistry, i.e., in replacement of the bicarbonate type of water by the sulfate type.

There are three major phases in the process of groundwater acidification and groundwater pollution: the first phase (when pH is higher than 6) is reversible, since the natural buffer capacity of the aquifer system may restore its initial hydrogeochemical conditions; the second phase (when pH drops to 5) is irreversible, since the natural buffer capacity of the aquifer system is incapable of neutralizing the long-term effect of acid precipitation, and concentrations of sulfates become as high as 40-100 mg/l and even higher; and the third phase (when the aquifer system is virtually depleted of bicarbonate ions and quite low in carbonate minerals) includes progressive groundwater acidification and pollution. In the latter case (when pH is below 5), groundwater becomes enriched in Al^{3+}, Fe^{3+}, and SO_4^{2-}. Thus, there are three phases in the history of groundwater acidification and pollution under the effect of acid precipitation: reversible, irreversible, and progressive. The second and third phases of acidification of unconfined groundwater are common in the north and north-west of Europe, whereas in the central and southern regions of Europe, the first phase is common and the second phase is under way.

Essentially, in the regions of intensive economic activity, especially in the industrial and agricultural regions, the formation of anthropogenic hydrochemical groundwater regimes is under way. The bulk of pollution due to components entering groundwater from the earth's surface occurs in unconfined aquifers, which serve at once as an accumulator of pollutants, protective layer, and pollution sources for deeper aquifers. Since the rate of the vertical water exchange between groundwater and the earth's surface is many times higher than the horizontal migration of pollutants along the aquifer, the main accumulation of pollutants occurs in the near-surface aquifers and especially in the unconfined aquifer. Thus, the emphasis in the study of groundwater pollution should be given to unconfined aquifers.

Predictive estimates of the area extent and level of groundwater pollution by individual components usually include determinations of the migration velocities of pollutants in the aquifer, and their travel times and concentrations at a given point. In this case, groundwater seepage should be considered in conjunction with the physicochemical interaction of pollutants with groundwater and rocks. Among the processes involved in this interaction, the most important are physical and chemical sorption, interstitial retention of emulsified and suspended pollutants, molecular diffusion, hydrodynamic dispersion, dissolution of rocks, and release of gases, heat transfer, and a number of other processes. It is expected that the ultimate result of this comprehensive consideration will include an estimate of the changes in the concentration of pollutants both along their migration paths and in time.

The practice of groundwater pollution forecasts is based on the following general trends in the migration of pollutants in water-bearing rocks:

- the velocity and range of the migration of pollutants is strongly controlled by convective transfer;
- at the interface between the polluted and clean groundwater, the transition zone of mixing (dispersion zone) is formed under the effect of sorption, dispersion, and other physicochemical processes.

The area extent of this zone is controlled by the duration and seepage velocity of pollutants, as well as by the parameters of relevant physicochemical processes.

In aquifers, the migration of pollutants is mainly due to the convective transfer. Pollutants may also migrate under the effect of molecular diffusion governed by the concentration gradient. However, this slow process may be ignored in forecasting the migration of pollutants for distances exceeding several meters or time intervals less than 100-200 years. Field studies indicate that there is a zone of groundwater with intermediate chemical composition rather than a well-defined front. Formation of mixing zones results from hydrodynamic dispersion of pollutants, as well as from sorption and chemical reactions. A distinction is made between microdispersion and macrodispersion of pollutants in seepage flow (Fried, 1975).

In aquifers, microdispersion and macrodispersion is controlled by heterogeneity in hydraulic conductivity of water-bearing rocks, which results in the formation of "fingers" differing in concentration from other portions of groundwater flow. The general pattern of dispersion and its effect on the migration of pollutants dissolved in groundwater are usually described by substitution of the coefficients of dispersion D, Dx, and Dy. In order to improve the reliability of forecasting the migration of pollutants, this forecast should be based on the maximum values of the seepage velocities and dispersion coefficients.

The migration of pollutants in aquifers is rather strongly affected by absorption of pollutants dissolved in groundwater to rocks. This process is favored by large contact areas between the surface of rock grains and water in porous media. Particulate pollutants and colloidal solutions, as well as some microorganisms, are retained mechanically in pores of the rock. Otherwise, pollutants are retained by either physical or chemical sorption. The rate of sorption is in direct proportion to the rate of change in the concentration of pollutants in the solution. Usually, the rate of sorption is sufficiently high for this process to be considered as being virtually at equilibrium. Thus, forecasts of the migration of pollutants may allow for the processes of equilibrium sorption using the sorption parameter b, which is equal to the ratio between the respective ultimate equilibrium concentrations of pollutants in the solution C_0 and in the sorbent N_0. Since the parameter of sorption depends on individual properties of rocks, it should be determined experimentally.

The main targets for groundwater protection include creation of a system of measures intended for preventing pollution, remediation of its consequences, and preservation and improvement of groundwater quality to provide safe water use. It is worth noting that problems of groundwater protection against pollution should be settled in conjunction with the protection of the environment as a whole. Experience shows that preventing pollutants from spread in aquifer and their subsequent extraction and removal is a complex technological problem, involving expensive and difficult measures. Thus, groundwater protection should concentrate on preventive measures controlling the possibility of groundwater pollution.

The first step in a system of preventive measures is the organization of a monitoring system to control groundwater quality and pollution, as well as the existing and potential polluters. The monitoring of groundwater pollution is usually carried out within areas where pollution has been already recorded or is highly likely to occur. The objectives of monitoring include:

- studying the reasons for groundwater pollution with identification and localization of existing polluters and determination of their main characteristics;
- identification of the pollution area and determination of its dimensions and depths to groundwater;
- analyzing the chemical composition of pollutants and distribution of their concentrations over the area;
- development and accomplishment of measures for remediation of the consequences of groundwater pollution; and
- development of monitoring systems or refinement of the objectives for existing systems.

In order to develop a monitoring system to control groundwater pollution, the top-priority objective should be organization of an observation network that covers the major sources of groundwater pollution, as well as major protected units and, above all, centralized water-supply structures that are liable to groundwater pollution. The main objectives of the observation network are:

- timely revealing of groundwater pollution;
- studying the dynamics of pollution in time and space;
- studying the migration of pollutants in aquifers with allowance made for the physicochemical processes occurring during the interaction of pollutants with groundwater and water-bearing rocks; and

- correction of the forecasts of the spread of pollutants in the aquifer and improvement of the methods of forecasting.

For each specific case, the number of observation stations (usually, observation wells) and their distance from the boundary of groundwater pollution are determined with allowance made for the actual conditions and the results of predictive estimates of the residence time and migration velocity of pollutants. The need to enlarge the observation network is controlled by the nature and migration velocity of pollutants. Enlargement should be implemented along the direction from the source of pollution to pumped water facilities. In order to control the location of the boundary of pollution, some observation stations should be placed within and some outside the pollution zone. It is worth noting that the special observation network for controlling the groundwater pollution should be also used as a component of any general monitoring system for the environment as a whole.

In addition to the monitoring system to control the groundwater pollution, the protection of water- supply structures may be achieved by establishing special groundwater protection areas. The establishment of groundwater protection areas implies the protection of the groundwater use against the bacterial and chemical pollution, consideration of specific hydrogeological conditions, and regulation of economic activities within these areas.

If the environmental pollution affects groundwater quality, the polluted groundwater in turn adversely affects the environment. This effect shows itself in the transport of pollutants by groundwater to surface water bodies, in the formation of geochemical and thermal anomalies on the earth's surface, in the pollution of rocks of the unsaturated zone as a result of fluctuations in the level of polluted groundwater, etc. Thus, measures to protect groundwater should be taken both in the context of general water protection and environmental protection. Otherwise, all of the special measures intended to protect groundwater will prove to be inefficient.

PART IV:
MEDICAL PROBLEMS RELATED TO GEOLOGY AND ECOSYSTEM INTERACTION

Chapter 16

HUMAN HEALTH AND ECOSYSTEMS

OLLE SELINUS[1], ROBERT B. FINKELMAN[2], JOSE A. CENTENO[3]
[1.]*Geological Survey of Sweden*, [2] *U. S. Geological Survey, USA*, [3] *U. S. Armed Forces Institute of Pathology, Washington*

1.   INTRODUCTION

Emerging diseases can present the medical community with many difficult problems. However, emerging disciplines may offer the medical community new opportunities to address a range of health problems including emerging diseases. One such emerging discipline is Medical Geology. Medical geology is the science dealing with the influence of natural environmental factors on the geographical distribution of health in humans and animals. Medical Geology is a rapidly growing discipline that has the potential of helping medical and public health communities all over the world pursue a wide range of environmental and naturally induced health issues. In this article we provide an overview of some of these health problems being addressed by practitioners of this emerging discipline.

2.   BACKGROUND

Medical Geology is concerned with the relationship between natural geological factors and human and animal health, as well as with improving our understanding   of   the   influence   of   environmental   factors   on   the geographical distribution of health problems. Medical Geology brings together geoscientists and medical/public health researchers to address health problems caused, or exacerbated by geologic materials (rocks, minerals and water) and geologic processes including volcanic eruptions, earthquakes and atmospheric dust (Selinus and Frank, 1999).

"Whoever wishes to investigate medicine properly, should proceed thus: in the first place .. must also consider the qualities of the waters ... they differ much in their qualities" (Hippocrates 460 – 377 BC). These words are often quoted to show the antiquity of the belief that human health owes much to environmental and geological factors. Medical geology is not strictly an emerging discipline but rather a re-emerging discipline. The relationship between geological such as rocks and minerals and human health has been known for centuries. Ancient Chinese, Egyptian, Islamic and Greek texts describe the many therapeutic applications of various rocks and minerals and many health problems that they may cause. More than 2,000 years ago Chinese texts describe 46 different minerals that were used for medicinal purposes. Arsenic minerals, for example orpiment (As2S2) and realgar (As2S3), were extensively featured in the materia medica of ancient cultures. Health effects associated with the use of these minerals were described by Hippocrates (460-377B.C.) as "... as corrosive, burning of the skin, with severe pain..."

Among the environmental health problems that geoscentists are working on in collaboration with the medical and public health community are: exposure to toxic levels of trace essential and non-essential elements such as arsenic and mercury, trace element deficiencies, exposure to natural dusts and to radioactivity, naturally occurring organic compounds in drinking water; volcanic emissions, etc. Geoscientists have also developed an array of tools and databases that can be used by the environmental health community to address vector-borne diseases and to model pollution dispersion in surface and ground water They can be applied also to some aspects of industrial pollution and occupational health problems (Thornton, 1983).

3.   TRACE ELEMENT EXPOSURE

Geological materials are the ultimate source of all elements on our planet. Elements are ubiquitous in the lithosphere where they are inhomogenously distributed and occur in different chemical forms. Ore deposits are thus merely natural concentrations of elements which are commercially exploitable. While such anomalous accumulations are the focus of mineral exploration, the background concentrations of metals which occur in common rocks, sediments and soils are of greater significance to the total metal loading in the environment. All known elements are present at some level of concentration throughout the natural environment, in humans, animals, vegetables and minerals, and their beneficial and harmful effects have been present since evolution began (Garrett, 2000; Kabata-Pendias, 2001; Låg, 1990).

Table 16-1 Average abundance of some elements in the earth's crust and rocks (all values in ppm)

Element	Earth's crust	Ultrabasic	Basalt	Granite	Shale	Limestone
As	1.8	1	2	1.5	15	2.5
Cd	0.2	-	0.2	0.2	0.2	0.1
Co	25	150	50	1	20	4
Cr	100	2000	200	4	100	10
Cu	55	10	100	10	50	15
Pb	12.5	0.1	15	20	20	8
Se	0.05	-	0.05	0.05	0.6	0.08
U	2.7	0.001	0.6	4.8	4	2
W	1.5	0.5	1	2	2	0.5
Zn	70	50	100	40	100	25

Geology may appear far removed from human health. However, rocks are the fundamental building blocks of the planet's surface and different rock mineral assemblages contain all of the naturally occurring chemical elements found on Earth. Many elements are essential to plant, animal and human health in small doses. Most of these elements are taken into the human body via food and water in the diet and in the air we breathe. Through physical and chemical weathering processes, rocks break down to form the soils on which the crops and animals that constitute the food supply are raised. Drinking water perclates through rocks and soils as part of the hydrological cycle and much of the dust and gases in the atmosphere are of geological origin. Hence, through the food chain and through the inhalation of atmospheric dusts and gases, there are direct links between earth materials and health.

Understanding the nature and magnitude of these geological sources is a key requirement if we are to be able to develop efficient approaches to assess the risk posed by toxic elements and minerals in the environment. It is important to be able to distinguish between natural and anthropogenic contributions to element loadings. Table 16-1 shows the significant differences between different rock types and their contents of selected elements. Concentrations of these elements can range over orders of magnitude among different types of rocks. For example the concentrations of elements such as nickel and chromium are much higher in basalts than in granites, whereas the reverse is true for lead. These types of bedrock are also easily weathered and the elements will be mobilized into the environment. In sediments, the heavy metals tend to be concentrated in the fractions with the finest grain size and the highest content of organic matter. Black shales tend, for example to be enriched in these elements.

Figure 16-1. Volcanic eruption at Krafla, Iceland 1980. Photo Olle Selinus

The volcanic eruption of Mount Pinatubo is a compelling example of the dramatic effects of geology, environment and human health. Volcanism and related activities are the principal processes that bring elements to the surface from deep within the Earth (Figure 16-1) . During just two days in June 1991 Pinatubo ejected 10 billion metric tonnes of magma and 20 million tonnes of SO_2; the resulting aerosols influenced global climate for three years. This single event introduced an estimated 800,000 tonnes of zinc, 600,000 tonnes of copper, 550,000 tonnes of chromium, 100,000 tonnes of lead, 1000 tonnes of cadmium, 10,000 tonnes of arsenic, 800 tonnes of mercury and 300.000 tonnes of nickel to the surface environment. (Garrett, 2000). Volcanic eruptions redistribute many harmful elements such as arsenic, beryllium, cadmium, mercury, lead, radon, and uranium. Many other redistributed elements have undetermined biological effects. At any given time, on average, 60 volcanoes are erupting on the land surface of the Earth, releasing metals into the environment. Submarine volcanism is even more significant than terrestrial volcanism, and it has been conservatively estimated that at least 3000 vent fields are currently active along the mid-ocean ridges. One interesting fact is that about 50% of the deposition of SO_2 is natural, mainly from volcanoes, and only 50% is from anthropogenic sources.

It is also necessary to determine the preexisting natural background. This question of natural background levels has important economic implications for decisionmakers and politicians but also for environmentalists. Human activities of all kinds have led to metals being redistributed from sites where they are fairly harmless to places where they affect humans and animals in a negative way. This is particularly important since acid rain and associated acidification accelerates this process so as to make some heavy metals, e.g. mercury, easily accessible and thus absorbed in the nutritional chain while essential trace elements, such as selenium may become unavailable to living organisms.

4.   DEFICIENCY AND TOXICITY

The naturally occurring elements are not distributed evenly across the surface of the earth and problems can arise when element abundances are too low (deficiency) or too high (toxicity). The irregular distribution of essential elements in the natural environment can lead to serious health problems particularly for subsistence populations who are heavily dependent on the local environment for their food supply. Approximately 25 of the naturally occurring elements are known to be essential to plant and animal life in trace amounts, these include calcium, magnesium, iron, cobalt, copper, zinc, phosphorus, nitrogen, selenium, iodine and molybdenum. On the other hand, an over-abundance of these elements can cause toxicity problems. Some elements such as arsenic, cadmium, lead, mercury and aluminium have no or limited biological function and are generally toxic to humans (Table 16-2).

Table 16-2 Diseases at state of deficiency, respectively toxicity, caused by the same element

ELEMENT	DEFICIENCY	TOXICITY
Iron	Anaemia	Haemochromatosis
Copper	Anaemia,"Sway back"	Chronic copper poisoning, Wilson-, Bedlington-disease
Zinc	Dwarf growth, Retarded development of gonads, Akrodermatitis entero pathica	Metallic fever, Diarrhoea
Cobalt	Anaemia, "White liver disease"	Heart failure, Polycythaemia
Magnesium	Dysfunction of gonads, Convulsions, Malformations of the skeleton, Urolithiasis	Ataxia
Chromium	Disturbances in the glucose	Kidney damage(Nephritis)

	metabolism	
Selenium	Liver nechrosis, Muscular dystrophy("White muscle disease")	"Alkali disease", "Blind staggers"

Most elements are known as trace elements because their natural abundances on Earth are generally very low (mg/kg concentrations in most soils). Trace element deficiencies in crops and animals are therefore commonplace over large areas of the world and mineral supplementation programmes are widely practised in agriculture and in animal husbandry. Trace element deficiencies generally lead to poor crop and animal growth, poor yields, and to reproductive disorders in animals. These problems often have greatest impact on poor populations who can least afford mineral interventions.

Paracelsus (1493-1541) defined the basic law of toxicology *"All substances are poisons; there is none which is not a poison. The right dose differentiates a poison and a remedy."* This relation between the dose and effect for any substance is shown by the curve starting at zero in Figure 16-3. Increase of the amount/concentration (on the horizontal axis) causes increasingly negative biological effects (on the vertical axis), which may lead to inhibition of biological functions and eventually to death. Evidently, minimizing concentrations of *non-essential* elements/substances would be beneficial. The situation for *essential* elements is different. Negative biological effects increase both for increasing and decreasing concentrations, illustrated by the continuous curve in the form of a trough in Figure 16-2. Thus both too much and too little are equally harmful, sometimes fatal.

In discussing medical geology it is necessary to know which elements are essential for humans and animals. Essential major elements include: calcium, chlorine, magnesium, phosphorus, potassium, sodium and sulphur. Essential trace elements are chromium, cobalt, copper, fluorine, iron, manganese, molybdenum, zinc and selenium. Elements with no recognized biological role are called non-essential elements. Over-exposure to some of these elements such as cadmium, arsenic, mercury and lead may have harmful consequences (Table 16-3).

Most of the heavy metals in different amounts are essential for biological functions of organisms (e.g., cobalt, copper, manganese, molybdenum, zinc, nickel, and vanadium). They are called micronutrients. At high concentrations, however, all metals may negatively influence physiological function. Cadmium, mercury, lead, copper, and other metals have all been linked to various toxic effects in living organisms. Of these, mercury and lead do not seem to serve any biological functions in living organisms (this is, however a matter of discussion). Examples of health effects of elements in living organisms can be seen in Table 16-2.

Table 16-3 Heavy metals overview

Metal	Symbol	Biological role	Toxicity	Presence
Iron	Fe	Essential for all organisms. Deficiencies widespread	To humans at drinking water level >200 mg/l	Common in many minerals
Copper	Cu	Essential for all organisms	Toxic at high doses	In many sulphide deposits
Zinc	Zn	Essential for all organisms	Toxicity low	In many sulphide deposits
Nickel	Ni	Essential for some organisms	Some compounds extremely toxic	In many sulphide and laterite deposits. In deep-sea nodules
Manganese	Mn	Essential for all organisms.	Mainly non-toxic	In many minerals and in deep-sea nodules
Molybdenum	Mo	Essential for all organisms except some bacteria	Considered toxic. More toxic to ruminants than to humans	In pegmaitites, some copper deposits and U-deposits
Cobalt	Co	Essential	Toxic in higher doses	In sulphide deposits, deep-sea nodules and U-deposits
Vanadium	V	Essential for some organisms	Toxic	In many types of mineral deposits and oil deposits
Lead	Pb	Considered non-essential	Toxic	In many types of mineral deposits
Mercury	Hg	Considered non-essential	Very toxic	In some gold deposits and other deposits
Cadmium	Cd	Seems to be essential to some animals	Toxic	In Zn-deposits
Chromium	Cr	Essential to some organisms	$Cr(6+)$ highly toxic	In some sulphide and platinum group deposits
Arsenic	As	Essential for some organisms (e.g. humans)	Toxic	In many types of sulphide deposits

5.   INTERACTIONS, SPECIATION AND BIOAVAILABILITY

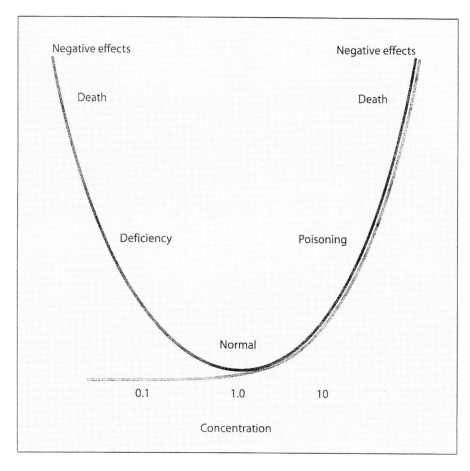

Figure 16-2. Dose-effect curve showing the relationship between concentrations and biological effects of essential (dark) and of non-essential (light) elements

Pathways through which trace elements enter the body

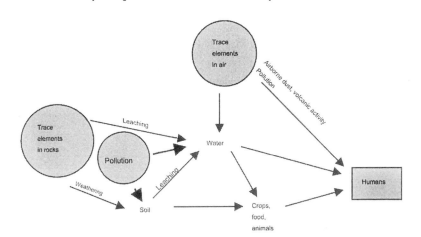

Figure 16-3. Pathways through which trace elements enter the body

In addition to understanding both natural and anthropogenic sources of harmful substances in the environment, it is also important to consider exposure and bioavailability. Exposure is the qualitative and/or quantitative description of total exposure to a given chemical substance via a range of pathways such as ingestion, inhalation or dermal contact. Bioavailability is the proportion of a chemical available for uptake into the systemic circulation of a given target organism following a given mode of exposure. Bioavailability directly influences exposure and therefore the effect and risk of health. Large quantities of a potentially harmful substance may be present in the environment, but if it is in a non-bioavailable form, the risk to health may be minimal. Bioavailability depends not only on the physical and chemical forms in which the element is present, but also on local factors in the environment, for example pH. The bioavailability and mobility of metals such as zinc, lead and cadmium is greatest under acidic conditions, while increased pH reduces bioavailability. Also soil type for instance clay and sand content, and its physical properties, affects the migration of metals through soils. The organisms present in soils also affect metal solubility, transport, bioavailability, and bioaccessibility

Pathways of exposure to environmental/natural contaminants may include diet (food, water, deliberate/inadvertent soil ingestion), dermal absorption and inhalation (Figure 16-3). In terms of ingestion, much emphasis has been placed on water (Combs, 2005). However, soils and food stuffs are likely to be far more important dietary contributors because the concentrations of potentially harmful substances in soils are much greater (parts per million) than in water (parts per billion). Whether soil ingestion is inadvertent or via the deliberate eating of soil known as geophagia, this exposure route should not be under-estimated. For example, studies in Kenya have shown that 60 – 90% of children between 5 – 14 years of age practice geophagia and consume 28 g of soil per day (Abrahams, 2005).

6.   SELENIUM DEFICIENCY IN CHINA

Selenium is an interesting element because the concentration range between deficiency (< 11 µg/day) and toxicity (> 900 µg/day) is very narrow (Fordyce, 2004). In the 1930s, China reported on a heart muscle disease that was to be called the Keshan disease. The first record of this disease is from the beginning of this century. In 1935 the disease became prevalent in Keshan County. It has since been recorded in 309 counties in China and appears in a large area from southwestern to northeastern China (Fordyce, 2005). In the 1960s scientists started to suspect that the disease was of natural origin and in the 1970s the probable solution was found - selenium. Two major findings in the study of the relationship between selenium and the Keshan disease have drawn the interest of researchers. These findings are: (1) The disease was always located in low selenium areas which geographically form a low selenium belt (2) Selenium use in prevention and treatment of the disease was highly successful It has been proposed that the Keshan disease may be caused by very low contents of selenium in bedrock, soils and natural waters.

One other area of active research in China is the Kaschin-Beck disease. Kaschin-Beck disease (KBD) has occurred in China for centuries but, despite extensive research, there is still much to learn about its epidemiology, etiology, biochemistry and prevention. Initial symptoms of KBD are joint swelling, pain, and general malaise, often affecting children. Skeletal remains indicate that KBD goes back to at least the 16th century. KBD is known from the scientific literature since the beginning of the 19th century and is mainly limited to a southwest - northeast belt of China. The disease was later found to be fairly common in China and has been studied for more than 40 years. The size of the affected population is not known, but estimates are about one to three million in China. The probable cause of KBD is the Coxsackie virus, but only in individuals who are very selenium deficient, caused by low natural contents of selenium in bedrock. Numerous field trials in China, (involving adding selenium to fertilizers, spraying it on crops and adding it to chicken feed) have shown selenium addition to the diet greatly reduces the incidence of Kashin-Beck disease (Fordyce, 2004).

7. GLOBAL IMPLICATIONS AND MEDICAL GEOLOGY EXAMPLES OF CHRONIC ARSENIC AND FLUORINE POISONING

In Bangladesh, India, China, Taiwan, Vietnam, Mexico and elsewhere, high levels of arsenic in drinking water have caused serious health problems for many millions of people (Kinniburgh & Smedley, 2001). Inorganic arsenic is one of the oldest known poisons to man. Chronic arsenic exposure particularly affects the skin, mucous membranes, nervous system, bone-marrow, liver and heart. Arsenic toxicity is strongly dependent on chemical form. Essentiality of arsenic has been shown in studies with animals but not in humans. Worldwide natural emissions of arsenic into the atmosphere in the 1980s totall 1.1 to 23.5 tonnes per year, derived mostly from volcanoes, wind born soil particles, sea spray and biogenic processes. Coal combustion alone accounts for 20 percent of the atmospheric emission, and arsenic from coal ash may be leached into soils and waters.

There is growing concern about the toxicity of arsenic and the health effects caused by exposure to elevated concentrations of arsenic in the geochemical environment. The danger to human health due to arsenic poisoning has now been recognised by WHO and the provisional guideline value for arsenic in drinking water has been lowered from 50 μg/l to 10 μg/l. The United States has adopted the same levels. Among the countries that have well documented case studies of arsenic poisoning are Bangladesh, India (West Bengal), Taiwan, China, Mexico, Chile and Argentina (Figure 16-4). The common symptoms of chronic arsenic poisoning are conjunctivitis, melanosis, depigmentation, keratosis and hyperkeratosis. The source of arsenic is geological, the element being present in many rock-forming minerals, including iron-oxides and clays, but mostly in sulphides minerals, most commonly in iron sulfides.

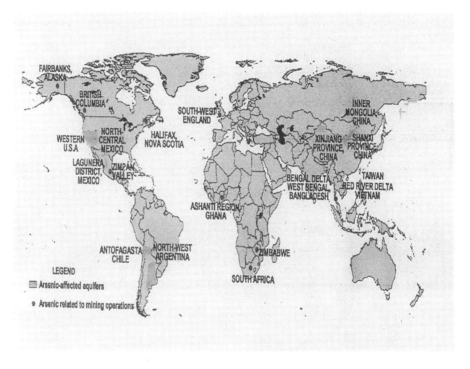

Figure 16-4. Examples of arsenic contamination

The worst arsenic problem in the world so far is in India and Bangladesh. It is estimated that 200 000 people have developed As poisoning including skin and internal organ cancers. It is also estimated that about 38 million people in West Bengal and 40 million in Bangladesh are at risk of arsenic poisoning, many of them consuming water with more than 100 ppb arsenic, five times more than the WHO recommended limit. Some estimates show that more than a million may be drinking water with arsenic more than 100 times this limit. This incidence of arsenic groundwater contamination in West Bengal and Bangladesh is considered to be the biggest arsenic poisoning case in the world so far (Smedley & Kinniburgh, 2005). The contamination occurs in groundwater from the alluvial and deltaic sediments that make up much of the country. The arsenic is of geological origin and has probably been present in the groundwater for thousands of years. Problems are only apparent now because it is just within the last 20-30 years that groundwater has been extensively used for drinking water and irrigation in the rural areas. Geoscientists from several countries are working with public health officials to seek solutions to these problems. By studying the geological and hydrological environment, geoscientists are trying to determine the source rocks from which the arsenic is being leached into the ground water. They are also trying to determine the conditions under which the arsenic is being mobilized. The arsenic is desorbed and dissolved from iron oxide minerals by anerobic (oxygen-deficient) groundwater. These questions will allow the public health communities around the world to identify aquifers with similar characteristics and more accurately determine which populations may be at risk from arsenic exposure.

International incidents of arsenic contamination in groundwater and the consequent ill health of people have been widely reported also from other places (Figure 16-4). The arsenic contamination incident in the well-water of Taiwan (1961-85) is well known. The population of the endemic area was about 100 000. Similar problems were reported from Antofagasta in Chile where almost 100 000 people out of a total population of the city of 130 000 were drinking water with high arsenic content for 12 years between 1959 and 1979. Chronic arsenic poisoning is also reported in some parts of Region Lagunera, Mexico where the arsenic concentration in groundwater exposed to the population was 0.41 microgrammes/l. Similar incidents were reported in Argentina in 1955 (Smedley & Kinniburgh, 2004).

Contamination of the environment with arsenic both from natural and anthropogenic sources is wide-spread and may be regarded as a global issue. We will continue to find many more situations where As contamination of surface and sub-surface waters and soils will increase. New occurrences will be found particularly in Central and Eastern Europe and the developing world. The bioaccessibility and exposure will be influenced by the nature of the contamination, including the chemical and mineral forms of arsenic and their solubility (Centeno *et al.* 2002a, 2002b, Tchounwou *et al.* 2003).

In China, geoscientists are working with the medical community to seek solutions to arsenic and also fluorine poisoning caused by residential burning of mineralized coal (Belkin *et al.* 1997). Chronic arsenic poisoning affects several thousand people in Guizhou Province, P.R. China. Those affected exhibit typical symptoms of arsenic poisoning including hyperpigmentation (flushed appearance, freckles), hyperkeratosis (scaly lesions on the skin, generally concentrated on the hands and feet), Bowen's disease (dark, horny, precancerous lesions of the skin). Chillie peppers dried over open coal-burning stoves may be a principal source for the arsenic poisoning. Fresh chillie peppers have less than one part-per-million (ppm) arsenic. In contrast, chilie peppers dried over high-arsenic coal fires can have more than 500 ppm arsenic. Significant amounts of arsenic may also come from other tainted foods, ingestion of dust (samples of kitchen dust contained as much as 3,000 ppm arsenic), and from inhalation of indoor air polluted by arsenic derived from coal combustion. The arsenic content of drinking water samples does not appear to be an important factor (Wang *et al.* 2004).

Detailed chemical and mineralogical characterization of the arsenic-bearing coal samples from this region (Belkin *et.al.* 1997) indicate arsenic concentrations as high as 35,000 ppm. Typically coals have less than 20 ppm arsenic (USGS unpublished data). Although there are a wide variety of arsenic-bearing mineral phases in the coal samples, much of the arsenic is bound to the organic component of the coal. This observation is important for two reasons. Firstly, because the arsenic is in the organic matrix, traditional methods of reducing arsenic, such as physical removal of heavy minerals, primarily arsenic-bearing pyrite, would not be effective. Secondly, because the visually observable pyrite in the coal is not a reliable indicator of the arsenic content, the villagers had no way of predicting the arsenic content of the coals that they mined or purchased. To overcome these problems a field test kit for arsenic was developed (Belkin *et.al.* 2003). This kit gives the villagers the opportunity to analyze the coal in the field and identify the dangerous high-arsenic samples as well as the safer low-arsenic coals.

8.   RADON – COMMON IN MANY PLACES

Radon is a naturally occurring radioactive gas that you can not see, smell or taste and can only be detected with special equipment (Appleton, "Radon in air and water. In Selinus", 2005). It is produced by the radioactive decay of radium, which, in turn, is derived from the radioactive decay of uranium. Uranium is found in small quantities in soils and rocks, although the amount varies from place to place. Radon decays to form radioactive particles that can enter the body by inhalation. Inhalation of the short-lived decay products of radon has been linked to an increase in the risk of developing cancers of the respiratory tract, especially of the lungs. Breathing radon in the indoor air of homes contributes to about 20,000 lung cancer deaths each year in the United States and 2,000-3,000 in the UK

Geology is the most important factor controlling the source and distribution of radon. Relatively high levels of radon emissions are associated with particular types of bedrock and unconsolidated deposits, for example some, but not all, granites, phosphatic rocks, and shales rich in organic materials. The release of radon from rocks and soils is controlled largely by the types of minerals in which uranium and radium occur. Once radon gas is released from minerals, its migration to the surface is controlled by the permeability of the bedrock and soil, the nature of the carrier fluids, (including carbon dioxide gas and groundwater), and meteorological factors such as barometric pressure and rainfall.

Radon released from rocks and soils is quickly diluted in the atmosphere. Concentrations in the open air are normally very low and probably do not present a hazard. Radon that enters poorly ventilated buildings, caves, mines and tunnels can reach high concentrations under certain circumstances. The construction method and the degree of ventilation can influence radon levels in buildings. Exposure to radon will also vary according to how particular buildings and spaces are used.

The concentration of radon in a building primarily reflects (1) the detailed geological characteristics of the ground beneath the building, which determines the potential for radon emissions, and (2) the structural detail of the building and its mode of use, which determines whether the potential for radon accumulation is fulfilled. Geological radon potential maps show that the relative hazard from radon may be derived from a range of data including radon measurements in soil and dwellings; soil and rock permeability; uranium concentrations in rocks and soils; and gamma spectrometric data. Radon potential maps have important applications, particularly in the control of radon through environmental health and building control legislation. They can be used

- to assess whether radon protective measures may be required in new buildings;
- for the cost-effective targeting of radon monitoring in existing dwellings and workplaces;
- to provide a radon potential assessment for homebuyers and sellers; and
- to provide exposure data for epidemiological studies of the links between radon and cancer.

Whereas a geological radon potential map can indicate the relative radon hazard, it cannot predict the radon risk for an individual building. This can only be established by having the building tested.

Radon dissolved in groundwater migrates over long distances along fractures and caverns depending on the velocity of fluid flow. Radon is soluble in water and may thus be transported for distances of up to 5 km in streams flowing underground in limestone. Radon remains in solution in the water until a gas phase is introduced (e.g. by turbulence or by pressure release). If emitted directly into the gas phase, as may happen above the water table, the presence of a carrier gas, such as carbon dioxide, would tend to induce migration of the radon. This appears to be particularly the case in certain limestone formations, where underground caves and fissures enable the rapid transfer of the gas phase. Radon in water supplies can result in radiation exposure of people in two ways: by ingestion of the water or by release of the radon into the air during showering or bathing, allowing radon and its decay products to be inhaled. With household drinking water supplies the main pathway is by ingestion. Radon in soil under homes is the biggest source of radon in indoor air, and presents a greater risk of lung cancer than radon in drinking water. However radon in drinking water could be a problem for children.

9.   IODINE AND FLUORINE

Even if the relationships are somewhat more complicated than previously believed, the connection between goitre and iodine deficiency is nevertheless a classic example of Medical Geology. The connection between geology-water-food chain-diseases can clearly be shown for iodine (Fuge, 2005). Goitre was common in ancient China, Greece, Egypt and amongst the Inca but was successfully treated by sea weed, which contains high levels of iodine. Iodine has long been known as an essential element for humans, and mammals in general, being a component of the thyroid hormone thyroxine. Deprivation of iodine results in a series of Iodine Deficiency Disorders (IDD) the most common of which is endemic goitre. The areas where IDD are concentrated tend to be geographically defined. Thus the most severe occurrences of endemic goitre and cretinism have been found to occur in high mountain ranges, rain shadow areas and central continental regions (Earthwise, 2001).

Goitre is the enlargement of the thyroid gland as it attempts to compensate for insufficient iodine. It has, for example, been estimated that out of a population of 17 million in Sri Lanka, nearly 10 million people are at risk of goitre. Iodine deficiency in pregnant mothers can also lead to cretinism and impaired brain function in children. These are serious and debilitating consequences, as the capability of children is severely restricted and they become a burden on the family. The WHO currently estimates that over 1.6 billion people are at risk from iodine deficiency and that it is the single largest cause of mental retardation in the world today. A dreadful form of IDD is cretinism. A cretin is physically and mentally retarded and deaf-mute. China alone has 425 million people who are at a risk in regard to IDD.

The link between environmental iodine and IDD has been known for the last 80 years. During this time, many aid agencies and governments have attempted to solve the problem by increasing dietary intakes of iodine via the introduction of iodinised salt and iodinised oil programmes. Despite these interventions, IDD remain a major problem globally. It is likely that IDD are multi-causal diseases involving factors such as trace element deficiencies, goitre-inducing substances in foodstuffs (goitrogens) and genetics. Geochemists have an important role to play in determining the environmental cycling of iodine and its uptake into the food chain if levels of dietary iodine are to be enhanced successfully.

Another important element is fluorine (Dissanayake & Chandrajith, 1999). The geochemistry of fluorides in ground water and the dental health of communities, particularly those depending on groundwater for their drinking water supplies, is one of the best known relationships between geochemistry and health. Many water supply schemes, particularly in developing countries where dug wells and deep bore holes form the major water sources, contain excess fluoride and as such are harmful to dental health. For most trace elements required by man, food is the principle source. In the case of fluoride, however, much of the input into the human body is from drinking water and the geochemistry of fluoride in groundwater is therefore of particular significance in the etiology of dental diseases.

Fluorine is an essential element with a recommended daily intake of 1.5 - 4.0 mg/day. Health problems may arise from deficiency (caries) or excess (dental mottling and skeletal fluorosis). Unlike other essential elements, for which food is the principal source to the extent of about 80%, the principal source of fluoride is water (Edmunds & Smedley, 2005). Fluorine in surface and ground waters is derived from the following natural sources:

- Leaching of the rocks rich in fluorine.
- Dissolution of fluorides from volcanic gases by percolating groundwater along faults and joints of great depth and discharging as fresh and mineral springs.
- Rainwater, which may acquire a small amount of fluoride from marine aerosols and continental dust.
- Industrial emissions and effluents
- Run-off from farms using phosphatic fertilisers extensively.

After, for example, the eruptions of the volcano Hekla on Iceland in 1693, 1766 and 1845, detailed descriptions of fluorosis were presented. Acute poisoning was described. Since world war II Hekla has had eruptions in 1947, 1970 and 1980 and a number of analyses of fluorine have been performed. The volcano delivered huge amounts of fluorine and concentrations of 4300 mg/kg in grass have been found (Weinstein & Cook, 2005).

In the beginning of the 20th century it was known that high contents of fluorine could cause fluorosis. The natural content of fluorine in drinking water is normally 0.1-1 ppm. In many places around the world, for example India, China, Africa etc., the contents could however be about 40 ppm which leads to serious fluorosis. The picture is rather complicated because there are also antagonistic effects. Molybdenum and selenium can reduce the effects of high contents of fluorine. One of the major benefits of fluorine seems to stem from its antagonism to aluminium. In Ontario, there is less Alzheimers disease where the drinking water contains more fluorine.

10.   GEOPHAGIA – DELIBERATE EATING OF SOILS

One quite different aspect of Medical Geology is geophagia, the deliberate eating of soil. In many ancient and rural societies and amongst a wide variety of animals, exposure of chemical elements occurs principally through the deliberate ingestion of soil, or soil-derived "medical" preparations (often associated with immigrant communities). Such behaviour is medically known as either pica or more specifically as geophagia (Abrahams, 2005). Other forms of pica commonly reported since the sixteenth century include the eating of coal, cinders, plaster, dung, ash, snow, and ice (pagophagia). Whilst increasingly uncommon in modern societies, geophagia is common among traditional societies and has been recognised since the time of Aristotle as a cure-all for health problems. Soil may be eaten from the ground as a paste, but in many situations there is a cultural preference for soil from special sources, such as termitaria.

Geophagia is considered by many human and animal nutritionists to be either a learned habitual response, in which clays and soil minerals are specifically ingested to reduce the toxicity of various dietary components common to the local environment, or an attempt to reduce their in-built response to nutritional deficiencies resulting from a poor diet, often rich in fibre but deficient in magnesium, iron, and zinc. Such diets are common in tropical countries, particularly where the diet is dominated by starchy fibre-rich foods such as sweet potatoes and cassava

From a historical perspective, geophagia has also been commonly associated with various mental disorders and afflictions with a wide variety of rather unpleasant cures. Even today the theory of geophagia as a subconscious response to dietary toxins or stress must be balanced against the habitual eating of soil that has been reported to develop into extreme, often obsessive, cravings. These cravings are often reported to occur immediately after rain. Typical quantities of soil eaten by geophagics in Kenya have been reported to be 20 grams per day. This is almost 400 times more than typical quantities of soil thought to be ingested as a result of inadvertent ingestion through hand-to mouth contact (e.g. 50 milligrams per day). Whilst eating such large quantities of soil increases exposure to essential trace nutrients, it also significantly increases exposure to biological pathogens and to potentially toxic trace elements, especially in areas associated with mineral extraction, or in polluted urban environments.

Geophagia is also increasing in developing countries. Because of immigration the tradition of geophagia is brought into western societies and specially imported soils can often be found in local immigrant food stores for sale to immigrants.

11. NATURALLY OCCURRING ORGANIC COMPOUNDS IN DRINKING WATER

Balkan endemic nephropathy (BEN) is an irreversible kidney disease of unknown origin, geographically confined to several rural regions of Bosnia, Bulgaria, Croatia, Romania, and Serbia. The disease occurs only in rural areas, in villages located in alluvial valleys of tributaries of the lower Danube River. It is estimated that several thousand people in the affected countries are currently suffering from BEN and that thousands more will be diagnosed with BEN in the next few years (Feder *et.al.* 1991; Orem *et.al.* 1999; Tatu *et.al.* 1998).

Many factors have been proposed as etiological agents for BEN, including bacteria and viruses, heavy metals, radioactive compounds, trace element imbalances in the soil, chromosomal aberrations, mycotoxins, plant toxins, and industrial pollution . Recent field and laboratory investigations support an environmental etiology for the disease, with a prime role played by the geological background of the endemic settlements. In this regard, there is a growing body of evidence suggesting the involvement of toxic organic compounds present in the drinking water of the endemic areas. These compounds are believed to be leached by groundwater from low rank Pliocene lignite deposits, and transported into shallow household wells or village springs. Analysis of well and spring water samples collected from BEN endemic areas contain a greater number of aliphatic and aromatic compounds, and in much higher abundance (>10x) compared with water samples from nonendemic sites. Many of the organic compounds found in the endemic area water samples were also observed in water extracts of Pliocene lignites, suggesting a possible connection between leachable organics from the coal and organics in the water samples.

The population of villages in the endemic areas uses almost exclusively well/spring water for drinking and cooking, and are therefore potentially exposed to any toxic organic compounds in the water. The presumably low levels of toxic organic compounds present would likely favour relatively slow development of the disease over a time interval of 10 to 30 years or more. The frequent association of BEN with upper urinary tract (urothelial) tumours suggests the action of both nephrotoxic and carcinogenic factors, possibly representing different classes of toxic organic substances derived from the Pliocene lignites. Pliocene lignites are some of the youngest coals in the Balkans and are relatively unmetamorphosed in the endemic areas. They retain many of the complex organic compounds contained in the decaying plant precursors and many kinds of potentially toxic organic compounds may be leached from them.

In the Pliocene lignite hypothesis for BEN etiology, however, other factors besides the presence of low rank coals must also be in play. The hypothesis also implies many or all of the following circumstances:

- the right hydrologic conditions for leaching and transport of the toxic organic compounds from the coal to the wells,
- a rural population largely dependent on untreated well water,
- a population with a relatively long life span (BEN commonly becomes manifest in people in their 40s and 50s),
- a relatively settled population for long exposure to the source of nephrotoxic/carcinogenic substances,
- and a competent and established medical network for recognition of the problem and proper, systematic, diagnosis.

It may be that BEN is a multifactorial disease, with toxic organics from coal being one necessary factor in the disease etiology. The challenge to researchers is to integrate studies among disparate scientific disciplines (medicine, epidemiology, geology, hydrology, geochemistry) in order to develop a reasoned conceptual model of the disease etiology of BEN.

12.   GLOBAL DUST

Dust is a global phenomenon. Dust storms from Africa regularly reach the European Alps, and the Western hemisphere and Asian dust outbreaks can reach California in less than a week, some ultimately crossing the Atlantic and reaching Europe. The ways in which mineral dust impacts upon life and health are wide-ranging (Derbyshire, 2005; Plumlee & Ziegler, 2004). These include:

- changes in the planet's radiative balance
- transport of disease bacteria to densely populated regions
- dumping of wind-blown sediment on pristine coral reefs
- general reduction of air quality
- provision of essential nutrients to tropical rainforests.
- transport of toxic substances

Mobilization of dust is both a natural and a humanly triggered process. We mobilize dust when we disturb the land surface or strip it of its vegetation. Changing climatic conditions play a key role as natural changes occur in available moisture and wind speeds. Although vegetation exerts a critical control on dust mobility, vegetation itself is influenced by climate, human activity and other factors. A better understanding of dust, including the processes that control its sources and transport as well as its impacts, is needed if its negative consequences are to be mitigated; identifying and controlling at least those due to human activities would be a good start.

Neither the precise nature nor the epidemiology of the impact on human health of natural atmospheric dusts (non-occupational lung disease) is known in any detail. Very fine particles may penetrate deeply into the lungs to cause silicosis, asbestosis and other lung conditions. The denser the dust concentration, the higher the rates of chronic respiratory disease and associated death rate.

Natural (non-occupational) silicosis was first reported in some Bedouin people in the Sahara Desert in the middle of the 20th century and has since been found in Pakistani farmers, Californian farm workers, Ladakh villagers, and Thar Desert residents (NW India), as well as in northern China. Whilst little quantitative information on natural silicosis is available, studies showing an incidence of over 22 % of the population of some Ladakh villages and over 21 % in people over 40 years of age in settlements in North China make it likely that the population affected in Asia is probably numbered in millions.

Both land surface and atmospheric dusts will have to be more closely studied if this question is to be answered satisfactorily. On land, dust source and deposition regions must be identified and the ways in which movement of dust (dust fluxes) have varied in the recent past and under various climatic conditions must be determined. There is still much to learn about dust transport mechanisms and pathways, as well as the influence of atmospheric dust on the Earth's radiation balance. Such work requires expertise in land surface processes, geochemical/isotopic 'fingerprinting', analysis of past climates, remote sensing, and close investigation of atmospheric radiation and dynamics. Incorporating dust into climate models (from source to sink) will improve understanding and enable us to provide predictions on several time scales (weeks to centuries).

13.   CONCLUSIONS

Medical geology is an emerging discipline and a science which will grow rapidly. Several geological surveys are integrating medical geology in their work and medical geology is now taught in some university courses for medical students. In the future it will be important to continue improving communication amongst the various disciplines concerned with diseases caused by geological factors which influence the well being of humans and animals. It will also be important to develop information material for the use of schools, public and private organizations interested in medical geology problems to show the impact of geologic factors on well being of humans and animals as well as arranging joint technical meetings to address issues of mutual concern amongst geoscientists and other disciplines concerned with medical geology. Geological surveys, universities and geological and medical societies should take a more active role in providing useful information on geologic conditions in medical geology and encourage the development of local working groups of multi-disciplinary medical geology experts. It would also be useful to encourage research in the area of producing more effective methodologies for the study of geological factors in environmental medicine and formulate recommendations for mitigation of effects of natural and man-induced hazardous geochemical conditions. Medical geology is thus an interdisciplinary science which will be heard of increasingly in the future.

Chapter 17

MEDICAL AND ECOLOGICAL SIGNIFICANCE OF THE WATER FACTOR

LEONID I. ELPINER
Water Problems Institute of Russian Academy of Sciences

A worsening ecological situation in many regions of the world calls for consideration of medical aspects of human living conditions. Special attention should be paid to ecological causes of alterations and changes in physiology, adaptation, morbidity, and reproduction that give concern to medical services and authorities. It is vital to increase knowledge in this area. Clarifying cause-and-effect relationships between life support mechanisms and human health, on one hand, and the environment, on the other, should perfect the theoretical basis. Qualitative characteristics of water have a major significance for environmental safety.

Human health is affected by external pathogens and biological peculiarities of the population – the so-called "complex of medical ecological factors". The theoretical and methodical basis of medical ecology consists of environmental hygiene, including sanitary toxicology, physiology, microbiology, epidemiology of infectious and non-infectious diseases, and medical geography. The knowledge accumulated within these disciplines includes evidence of real and possible influences of water on the character and scale of infectious and non-infectious diseases, genetic alterations, and peculiarities of human development. A positive medical ecological significance of water is determined by its prevailing role in optimal and safe human living conditions at the population and individual level.

In assessing the role of the water factor, its influence on human living conditions should be considered according to the hydrogeological situation in the particular territory concerned. Assessment includes the quality, quantity and regime of waters. Each component may directly and/or indirectly impact upon human living conditions and health. However, qualitative characteristics are of dominating significance for a safe human environment.

Significant increases in man-made pollution of natural waters in the last few decades has led to intensive medical ecological assessments of water quality changes under strong anthropogenic impacts such as industrial and sewage water discharge, surface run-off from urban and agricultural territories, atmospheric pollution, etc. The problem of drinking water quality is now of primary importance. Water quality directly depends on the quality of ground and surface water sources, intensity of water protecting measures, effectiveness of water treatment and disinfection technologies, and safety of water pipes and water supply networks (Yakovlev, 1998).

Nowadays the broad ranges of human pathologies related mainly to drinking water quality are known. These include infectious, parasitic, non-infectious and genetic diseases.

The role of disease transmission via water of a number of infectious intestinal diseases (enteric fever, paratyphoid, dysentery, cholera, salmonellas, viral hepatitis, and others less commonly occurred diseases) has been proven by thorough long-term epidemiological investigations started as early as the nineteenth century. Infectious diseases caused by pathogenic bacteria, viruses, protozoa or parasitic agents are the most typical and widespread health risk factor related to water (Onishchenko, 2002).

Drinking water contamination by sewage water at its source or in the water supply network is the established cause of a large number of intestinal infection outbreaks (Koshkina, et al, 1996). Modern techniques of epidemiological analysis applied to determine the routes of spreading of intestinal infections are sufficiently informative. These are based on the identification of pathogenic organisms discovered in drinking water and egesta from those who are affected. The pattern of spreading of disease is characteristic. On a water supply network, cases are grouped along the infected water flow. In case of non-centralized water supply, the relationship with infected water source, for instance, a well, is usually established.

Existing data on the possible role of water in the rise of infectious, particularly, gastro-intestinal, diseases can be combined with new data to show that the spreading of infections is not the same in different territories, even within borders of one geographical region. Regularities of epidemiological findings are not always clear enough, however, to establish obvious causes (for instance, infected water source). In addition, inter-specific relationships between macro- and microorganisms may possibly be driven by, as yet, unestablished biological laws (Cherkassky, 1981). There is a need for epidemiological investigations to better determine the influence on human health of levels of microbial water pollution in specific geographical regions.

Many parasitic diseases and those related to natural habitats (malaria, opistorkhosis, tularemia, leptospirosis, encephalitis, etc.) are also closely related to water. Lyambliosis, which may cause intestinal and liver alterations among humans, is closely associated with drinking water. Modern epidemiological data demonstrate that water is a major route of the disease transmission. The presence of people infected with such diseases is usually considered as an indication of the presence of an infectious agent in water (Be'er, 1996). The range of non-infectious diseases related to the accumulation in humans of high concentrations of harmful chemical substances – heavy metals, microelements, toxic organic compounds, and radioactive substances - due to drinking water consumption is broad. It includes cardio-vascular, gastro-intestinal, and immune diseases, as well as diseases of the nervous system, locomotory apparatus, allergies, genetic alterations, etc. (WHO, 1993).

Researchers have linked man-made non-organic and organic pollution of industrial, agricultural, household origin with the possibility of specific or non-specific influences of water on human health. The latter expresses itself as a decrease in an organism's resistance to the impact of physical and biological factors, allergies, etc. Modern toxicological investigations have significantly broadened views about relationships of a number of non-infectious diseases, including oncological ones, to water pollution by obvious toxicants and carcinogens as well as by the substances the presence of which was previously not considered to be direct evidence of pollution. An example is nitrogen-containing compounds turning within the human organism into N-nitrosamines. (WHO, 1993).

The problem of transformation of chemical substances in water has become very important in the assessment of water treated by strong oxidants. It has been established that carcinogenic haloid-substituted hydrocarbons (trihalomethans) form during the chlorination of water containing natural and anthropogenic organic compounds make, leading to modifications to, and new technologies for, water treatment.

It should be noted that the problem of the overall increase in oncological diseases is undoubtedly related to the increased pollution of water bodies by organic synthetic substances, primarily pesticides. A list of organic substances of anthropogenic origin that are capable of polluting water sources, including groundwater, is large – there are hundreds of such compounds. These include alkanes, ethylenes, benzols, aromatic hydrocarbons, pesticides, water chlorination by-products as well as a range of other organic components such as products of organic synthesis, the oil-chemical industry, solvents, detergents, dye-stuffs, etc.

Many substances capable of contaminating water may cause one or more toxic effects: carcinogenic, nephrotoxic (kidney alterations), hepatotoxic (liver alterations), genotoxic, or mutagenic. Large-scale laboratory investigations on animals have been undertaken to establish concentrations of organic and inorganic substances in drinking water that can be tolerated as a basis for drinking water quality standards.

However, in recent years, ecological epidemiological investigations establishing relationships between human pathologies and natural or anthropogenic components in drinking water have taken on a special significance (Onischenko, 2002). For a number of substances which may be found in drinking water, cause-and-effect relationships of several non-infectious pathologies are obvious. Examples linked to groundwater uses are presented in special chapter of the present book. More examples can be added from WHO Guidelines for drinking water quality control (WHO, 1993).

Toxicological characteristics of more than 130 in organic and organic substances and compounds are registered in the Guidelines. Most of them are capable of exerting adverse influence on human organism when occurring in sufficient concentrations in contaminated drinking water. For instance, high Mn concentrations cause alterations in the central nervous system and thyroid gland, Se – caries in children, malignant tumors, Mo – cardio-vascular diseases, podagra, goiter, alterations in the ovarian menstrual cycle, Ba – alterations in cardio-vascular, haematogenic (leucosis) systems, B – alterations in carbohydrate metabolism, decrease in enzyme activity, irritation in the gastrointestinal tract, decrease in reproductive capacity among men, alterations in the ovarian menstrual cycle among women, As – nervous system, organs of vision, Ni – heart, liver, organs of vision, Hg – strong alterations in kidney functions, nervous system. This list may become larger. More complete information is available in a number of recent publications including IRPTS, 1996.

Information on positive influences of chemical substances, depending on the character of their biological action and concentration, is also available. Mainly, these are biogenic trace elements – Cu, F, Fe, Zn, Ca, Mg, and a number of other substances (WHO, 1993). It should be remembered that part of the biologically necessary elements has to be secured by humans from water in non-bound state. On the other hand, concentrations of such biologically important substances must not exceed established standards to avoid negative impacts. In most cases, adverse impacts result from the long-term influence of excessive concentrations.

As far as the role of water in the development of a number of diseases is concerned, special attention should be paid to natural mineral water composition. National and international data provide evidence of a relationship between several pathologies and consumption of water that is "too soft" or "too hard". The differences between these types of water are Ca and Mg carbonate concentrations. Soft water is reportedly associated with cardio-vascular pathology, and hard water with urolithiasis, nephrosis, gastrointestinal tract's diseases (Elpiner & Vasiliev, 1983). Water with excessive mineralization (2-2,5 mg/l) may exert adverse influence on childbearing or cause menstrual irregularities (Ob'edkova, 1983).

More and more evidence appears in the scientific literature that drinking water with moderate mineralization (200-500 mg/l) and the presence of Ca and Mg ions may reduce adverse effects of some other substances. These ions are considered to have protective effects against toxic microelements (for instance, Ca) and macroelements (for instance, Na) that provoke high blood pressure (Revis et al., 1980).

The idea of physiological soundness of drinking water led to consideration of problems associated with the salt ratio, hygienic significance of hydrochemical classes, and the role of macro- and microelements ratio. That is why in the 1970th the basics of optimal salt composition were set out, along with water safety criteria for optimal reactions at the cell, organ, and organism levels (Elpiner, 1975). Therefore, new optimization approaches to drinking water microelement composition assessment, and investigation of the capacity of microelements to contribute to the organisms biological needs are now of interest (WHO, 1993).

The shift, from approaches that establish only highest and lowest acceptable concentrations in drinking water according to organoleptic and toxicological characteristics, to an optimization approach in which the physiological (or - more precisely – biological) soundness of water is assessed, undoubtedly reflects progressive tendencies in drinking water doctrine. Until recently, this approach covered only optimal concentrations of fluorine – a microelement, deficiency of which causes caries, and excess of which leads to fluorosis of enamel or, in highest concentrations, bone damage).

All these data led researchers to assess in new ways the natural processes by which drinking water attains its chemical composition and the implications for water quality.

Information on water induced non-infectious diseases has been added to investigations of hydrobiological processes such as eutrophication of water bodies caused by intense development of, mainly blue-green, algae as a result of the discharge of a number of toxic organic that accumulate in water and in hycrobionts and causing pathogenic effects if consumed (Elpiner, 1986). It should be recalled that the mechanism of eutrophication is related to excessive incoming of biogenic elements (N and P) of anthropogenic origin. Data on poor quality in eutrophic water bodies were limited, until recently, to water poor organoleptic characteristics (smell) appearing during water treatment.

Better information on toxicity of eutrophic water bodies gives special significance to hydrobiological prognoses and their new medical ecological interpretation.

Complicated accumulative processes of harmful substances in food chains of aquatic ecosystems are generally not taken into account during the definition of water quality criteria for sources of drinking water. However, there are sufficient data to show that at each food (trophic) level (for example, organisms   protozoa   mollusk   fish) concentrations of these substances increase ten- or hundredfold (Krassovsky et al., 1982).

Extinction of biological communities is known in some cases where accumulated substances that have lost or changed structure to become more toxic are discharged into water (Krassovsky et al., 1992). These processes help the identification, from the medical biological point of view, of which ecosystems are capable of water self-purification.

It is also possible to assess land ecosystems and their relationships to water resources.

Flora and fauna, including wild animals and pets, determine the level and character of biogenic nutrition and raw material productivity as well as influencing human health. Along with well-known positive influences of much vegetation on gas and ionic air composition, relationships between a number of human zoonoses and natural living conditions of their agents, especially arthropoda, exist (Pavlovsky, 1964). The influences of water are both direct and indirect. In some cases water bodies are the place where insects are grown and propagate. They are the habitat of several heminth bridging species (fish, mollusks). In some water bodies, shore vegetation influences the habitats of inoculable agents of disease. Changes to water volume changes can change vegetation, improving or worsening these conditions. These changes are also effected by formation of new land, often linked to hydroeconomic development. For instance, tick habitats are enlarging (ticks are agents of encephalitis) along new roads under construction (Cherkassky, 1981).

Today many scientific investigations are available on the relationships between the quality of water used for irrigation, stock watering, swimming birds, breeding and biological soundness and quality of agricultural production. From the medical ecological point of view these issues should be considered taking into account data on toxic substances circulating in agro-ecosystems (Elpiner & Bezdnina, 1986; Bezdnina, 1997). This includes transfer from water into agricultural plants and animals (during watering or feeding) and then to humans through consumption of vegetables, meat and dairy products.

Data referred to above illustrates the significance of water and land ecosystems for pathogenic influences on water. Abiotic causes of worsening chemical composition of water used for human purposes are important, too. However, anthropogenic influence is the main factor affecting the functioning of the ecosystems, and the composition of water used by humans.

Therefore, investigations about water and the human environment take on a special significance in the light of increasing anthropogenic impact on hydrological situations. Here is where hydrotechnical construction particularly attracts the attention of ecologists, environmental specialists and communities.

Nowadays, there are no schemes for complex use of water resources, protection of large water bodies, or projects to create large water bodies that do not involve predictions of changes in ecological conditions. However, such investigations commonly do not have sufficient depth and range of study of phenomena. At the same time, in recent decades, an acute water supply situation in a number of large world regions necessitates broadening of knowledge of water resources management and, linked to that, important medical biological aspects of social ecological problems.

Studies of aspects of the theory of water resources management to protect nature, carried out by the Institute of Water Problems, Russian Academy of Sciences, highlighted the fact that life support and human health protection related to hydrotechnical construction needs a fundamentally new approach to medical ecological predictions. The aim of the predictions is to determine acceptable levels of human interference in natural hydrological processes and pathways, and improved techniques of medical ecological assessment, to prevent negative changes resulting.

Up-to-date data on direct and indirect influences of water on human health, possibilities for rational recreational use of water bodies and surrounding territories including areas with curative microclimates, bathing sources, etc. are required. The close links between social and ecological processes accompanying hydrotechnical construction make it necessary to take these aspects into account, including such phenomena as provision of established standards of the population nutrition, changes in the conditions of employment, migration, and human adaptation to new modified climatic conditions.

Accomplished works provide the basis for researchers to set out scientific grounds and principal methods for medical ecological forecasting of consequences of hydrotechnical construction consequences (Scientific council ..., 1990).

Without going into details of the techniques, it should be noted that resulting predictions help to substantiate and positively plan social economic developments while minimizing negative influences on the medical biological situation. These measures include the provision of new water bodies with a regime that eliminates or decreases the possibility of adverse medical ecological effects.

Experience accumulated in Russia testifies to the necessity and reality of the solution of medical ecological problems related to hydrotechnical construction. However, the need for the development of a medical ecological dimension to water resource management is only part of the human ecology problems emerging in connection with modern water supply.

One of the main tasks of scientific and technological progress is to ensure an adequate and safe water supply.

To achieve this goal demands major development of medical ecological investigations in two main interconnected directions. The former concerns the acceleration of the shift into intensive water resources use. The latter concerns the amelioration of water resources management in order to minimize qualitative and quantitative deficits of water in vulnerable areas using hydrotechnical construction and water protection measures.

Intensive use of water resources use requires the development of water saving technologies in industry and agriculture, and introduction of techniques and economic stimuli for water saving in the household sector, as well as the use of new innovative sources. This requires a more thorough approach to water use, and scientific substantiation of practical measures to prevent adverse effects on human health and for amelioration of sanitary problems.

Firstly, innovative sources (salty brackish ground and surface water, treated sewage water of different origin) may serve for household water supply and recreational water use (Avakyan et al., 1987). Successful use depends on the soundness of the investigations aimed at the perfection of a legislative base of requirements for water treatment, neutralization, and disinfection of natural and sewage water, including recycling systems.

Modern approaches to household and drinking water quality standardization need to be broadened. There is a strong need to investigate transformation products of chemical compounds and long-term consequences of consumption of adverse substances. However, the most important step is to shift to the consumption of water, which is not only organoleptically good and chemically and epidemiologically harmless but also valuable in terms of trace and macro-component content. This is of utmost importance in the light of scientific and technological progress in treatment of natural and sewage water, especially that with high salt content, as well as because of the prospective use of new water sources from different biogeochemical regions.

Amongst the problems of water quality standardization, agricultural water standardization has special medical ecological aspects. Work in this field is related to the delivery of sound (in the light of health-giving substances) and harmless (with no harmful admixtures) agricultural products taking account of increased crop capacity and productivity of livestock. As mentioned above, such medical ecological criteria relate to the basis of circulation of chemical elements in which humans are the final link.

Medical biological criteria are nowadays set to ensure the effectiveness of water protective measures by establishment of maximum acceptable levels of sewage discharge. In Russia their calculation is based on maximum acceptable concentrations of hazardous substances established by hygienists for water intakes. However, such modern standards do not allow specialists to solve problems on ecological safety of water bodies as a whole. Reverse connections determining the influence of water ecosystems on water quality are also revealed. Areas of water bodies with previously sound water quality also need to be taken into account. Problems in ecological hygienic standardization of water bodies in terms of water quality are emerging (Krassovsky et al., 1982)

There is a need for a system of measures that provides for reduction of divergences of water quality from modern standards. The most rational economic option could be the setting of priorities for water protective measures over a period of time in terms of the extent of pollution and form of water supply (Ministry ..., 1989).

In order to develop this approach it is necessary to further explore methods and predictive techniques for possible changes to the medical ecological situation in parallel with changes to the hydrological-hydrogeological situation.

The use of a sound theoretical and methodical basis will allow specialists to plan medical ecological investigations in the light of predictive problems and to develop appropriate guidelines for use in specific geographical regions where changes to land water regimes are expected. Specific predications of such hydrological and hydrogeological changes are necessary to determine the medical ecological investigations on water regime changes that need to be undertaken. These issues take on added significance because of possible global hydroclimatic changes (Elpiner, 2003)

Such investigations are necessary to forecast environmental (hydrochemical, hydrobiological, zoological, and botanic) changes in connection with planned economic activity and expected climate changes.

PART V:
PREDICTION OF THE GEOENVIRONMENTAL EVOLUTION OF ECOSYSTEMS

Chapter 18

PREDICTION OF EXOGENIC GEOLOGICAL PROCESSES

ARKADY I.SHEKO, & VLADIMIR S.KRUPODEROV
All-Russian Research Institute for Hydrogeology & Engineering Geolog,y, Moscow region, Russia

1. INTRODUCTION

Prediction of exogenic geological processes (EGP) involves scientifically substantiated forecasting of events in space and time under the action of natural and anthropogenically induced factors. The goal of the EGP prediction is to give the answers to the three basic questions – where, when and of which activity (size) one or another type of exogenic geological process can happen, and also to address some specific questions such as: to what distance and with what velocity collapsed rocks will move; how large will be the area affected by the process, etc.

At present, the problem of predicting events is generally poorly investigated, and especially in respect of the prediction of exogenic geological processes in time.

In 1975 the All-Russian Research Institute for Hydrogeology and Engineering Geology (VSEGINGEO) developed a systematic approach to the theoretical principles and methods of long-term predictions of EGP-activity, and, on that basis has compiled, for the first time in the world, special engineering-geological maps for the prediction of landslides, mudflows, abrasion and erosion to the year of 2000 (scale: 1:200 000) for northern coast of the Black Sea - from the Danube to the Chorok River (The authors: A.I. Sheko, V.S.Krupoderov, V.I. Diakonova, I.V. Malneva, P.A.Dvortsova).

Later, similar maps were compiled for the territories of North Caucasus, the Baikal-Amurskaya Railway Magistral, Sakhalin, Kurils and other regions of Russia.

The accuracy of the predicted results was quite satisfactory especially in respect of tendencies in EGP-activity. Thus, the catastrophic landslides of 1989 and 1998 in the Chechen Republic were forecast, respectively, in 1980 and 1994. Catastrophic events that took place over the entire territory of North Caucasus in 2002 and the mudflow that destroyed the town of Myrnyauz (Kabardino-Balkaria) in 2000 had been also forecast in 1994.

Theoretical principles and modern methods for the prediction of exogenic geological processes are described in many publications (Osipov, 1999; Sheko, 1980; Sheko & Krupuderov 1984; Sheko & Grechishchev, 1988).

Exogenic geological processes are considered to be complicated open multi-component equifinal geological systems where the evolution and manifestations of any process is the result of the interaction of all the formative factors.

Because of the diversity of these factors, they are conventionally subdivided into three groups: constant (unvarying), slowly varying and rapidly varying.

Constant factors are those which during the prediction period can be considered as unchanged. These are the geological structure and relief. Although the action of weathering, lithological composition and other characteristics of rocks slowly change, causing changes in the relief, these are so slight during the period of prediction that they may be neglected. However this group of factors does determine the genetic features and intensity of exogenic geological processes.

Slowly varying factors can be subdivided into two subgroups: basic and derivative. Basic (independent) factors include current tectonic movements and climatic conditions. Derivative factors are isostatic and eustatic changes in water levels of seas and oceans; and changes to hydrogeological and geocryological conditions, vegetation and soils. The factors of this group determine a general tendency towards EGP-activity.

The third group of rapidly varying factors can also be subdivided into the same two subgroups: basic and derivative. The basic (independent) factors include the meteorological (precipitation, temperature, etc.), hydrological (river water loss, water levels in seas, rivers and lakes), and seismic conditions, as well as human activity (anthropogenic factors). These factors determine the EGP-activity regime in time, but act via derivative factors including surface run-off, and moisture content, strength and deformation properties of rocks.

It is possible to define a hierarchy of the factors, within such a complicated multi-factor system, determine the role of each component in the evolution of one or another type of EGP, and to work out the theoretical principles and methods for the prediction of these processes.

One can distinguish three basic steps in the prediction process:

- prediction in space;
- in time, and
- assessment of a hazard, risk and damage.

These three can also be considered as independent types of the prediction.

2.   SPATIAL PREDICTIONS

Spatial predictions aim to estimate the probability of EGP in a given area. Such predictions are based on the special engineering-geological zoning of territories made according to the scheme of N.V.Popov. In accordance to this scheme, territory under study is subdivided into 1-order regions (according to structural characteristics), areas (according to geomorphological characteristics), and 2-order regions (according to stratigraphic-genetic characteristics). Mostly methods are based on quantitative characteristics of exogenic geological processes themselves or the conditions in which these occur.

One of the most widely used current methods of prediction is territorial zoning by EGP intensity. There are alternative quantitative techniques, such as territorial zoning by value of geodynamic potential, ordered classification of constituent elements etc. However, these are difficult to use and insufficiently objective. More detailed information about prediction methods can be found in the book "Landslides and Mudflows" prepared under the aegis of UNESCO in 1998 (Kozlovsky, 1988).

The zoning of territory by EGP intensity is the most objective method estimating the probability of exogenic geological processes for a specific area.

The intensity and activity of exogenic geological processes are the most important concepts in engineering geodynamics. These were originally proposed by E.P.Emelianova (Sheko & Grechishchev, 1988).

The intensity of one or another EGP genetic type is estimated by the coefficient of territory affected, which, in turn, is determined as a relationship between the area (distance, number) of all forms of a given EGP genetic type, not depending on age, and the area (distance) of the entire territory affected. The resulting parameter characterizes the intensity of the given genetic type for the area under study for the entire continental history of its geological evolution as well as its vulnerability to this EGP type in future.

The coefficient of territory affected is not only a quantitative parameter of EGP intensity. It also reflects, in an integral form, the interaction of all the influencing factors and the degree of those influences upon the given exogenic geological process.

The coefficient of territory affected is determined (calculated) for areas with similar geological and geomorphological conditions. As a rule, these are regions composed of rocks of a certain stratigraphic-lithological type coincident with a specific range of geomorphological elements. If the region is unevenly affected then sub-regions are distinguished.

The engineering-geological map of the territorial zoning according to EGP-intensity shows the following taxonomic units: regions distinguished by the geostructural parameter; areas – by geomorphological conditions; districts – by stratigraphic-lithological characteristics of rocks; and sites – by composition and intensity of exogenic geological processes. A formula indicates the degree to which each site is affected by all the types of EGP spread within it.

An extract from the above-mentioned map, describing one of the sites on the Black Sea coast, is shown in Figure 18-1.

The classification of EGP intensity on the map depends on the map scale. Thus, for the present map at a scale of 1:200 000 – 1:500 000 the following classification is used:

1. rather low (<0.01);
2. low (0,01-0,1);
3. mean (0.1-0.3);
4. high (0.3-0.5);
5. very high (0.5-0.7);
6. rather high (>0.7).

The given figures are obtained according to Equation:

$$K_{intensity} = S_{affected\ area}/S_{total\ territory}$$

This approach to the EGP-intensity assessment is widely used. Maps of this type have been compiled for the entire territory of Russia, based on the results of engineering-geological examination.

3.   TEMPORAL PREDICTIONS

The basic parameter used in the preparation of temporal predictions of exogenic geological processes is the intensity of these activities. Activity is the quantitative relationship between the area (distance, number) of all the active forms of a given type of EGP in a given site and the summed area

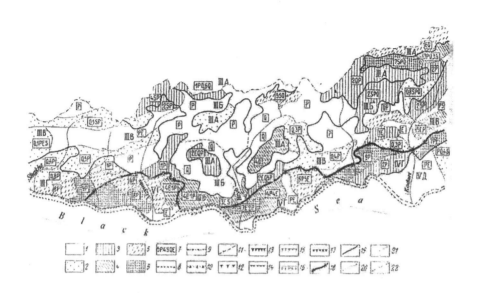

Figure 18-1. The map of the territory zoning by the exogenic geological processes intensity for the Black Sea coast (Shakhe River – Kodori River). Regions: III – meganticlinorium of the Main Caucasian Ridge; IV - Georgian median massif. Areas: A - nival alpine (>2000 m); Б – highland (1300-2000 m); B – medium-height land (600-1300 m); Г – lowland; Д – hilly piedmonts. Intensity of exogenic geological processes, expressed as a territory affectedness (for landslides, erosion and collapsed debris – in shares of a unit of water stream length: 1 – less than 0.01; 2 – 0.01 to 01; 3 – from 0.1 to 0.3; 4 – from 03 to 0.5; 5 – from 0.5 to 0.7; 6 – over 0.7; 7 – formula of territory affectedness by exogenic geological processes (P – landslides, S – mudflows, Q – collapses-debris, E – erosion; a number before a letter means the affectedness coefficient in the tenth shares). The composition of beaches: 8 – pebble; 9 – sand-clayey; 10 – pebble-clumpy. Genetic types of shores: 11- rugged with shore swells; 12 – steep, abrading in loose and coherent rocks; 13 – the same in hard and semi-hard rocks; 14 – sliding, abrading. Shore-reinforcing constructions: 15 – buns; 16 – wave pushing-off wells with bun and breakwaters; 17 – discontinuous wave pushing-off wells. Boundaries between: 18 – engineering-geological zones; 19 – engineering-geological areas; 20 – groups of sites by affectedness; 21 – combination of processes genetic types; 22 – regions of investigations.

(distance, number) occupied by this process, independent of its age, in that site.

Predictions of exogenic geological processes in time can be subdivided into those based on :
- analysis of the geological history of a territory;

- analysis of large climatic changes;
- long-term and very long-term and
- short-term predictions.

The theoretical basis for temporal predictions of exogenic geological processes are the regular features shown by the development of these processes over time.

It has been established that exogenic geological and other natural processes have a cyclic evolution caused by quasi-periodic changes of rapidly varying factors. The durations of cycles vary widely - from parts of a second to tens, hundreds and millions of years and longer. Short overviews of cyclicity of natural processes' evolution can be found in a number of publications (Osipov, 1999; Sheko, 1980; Sheko & Krupuderov 1984; Sheko & Grechishchev, 1988).

The main influences on the development of many natural processes, including EGPs, are phenomena occurring on the Sun. This has been confirmed by many researchers working in different scientific spheres (Vytinsky, 1973; and others). Therefore more attention should be paid to solar activity.

One can observe in the processes occurring on the Sun, basic cycles with average durations of 11, 22 and 80-90 years. It is also suspected that there are longer cycles but these have not yet been sufficiently verified. Amongst these are cycles of 55, 180 and 400-600 years.

We can consider the characteristics of some solar activity cycles that make significant contributions to the development of exogenic geological processes.

4.   THE 11-YEAR CYCLE

According to the data of Wolfdemeier, the duration of this cycle varies within 7 to 17 years (Osipov, 1999; Kyuntsel, 1980; Kozlovsky, 1988; Pletnev, 1972).

Among the solar activity parameters are "Wolf's numbers" meaning a quantity of spots on the Sun. These are used widely in investigations. A.P.Reznikov has analyzed the entire series of observations of Wolf's numbers and has confirmed earlier statements of many researchers that the activity of solar processes is highly influenced by tidal forces of planets in the Solar System.

The duration of an upward branch of the 11-year cycle of Sun activity is in inverse proportion to its amplitude, i.e. the bigger the amplitude, the shorter the period of rise, whereas the duration of the downward branch is in direct proportion , i.e. the greater the amplitude, the longer the fall. Yu. I.

Vitinsky (1973) reported that the strongest active formations of spots on the Sun more often take place not during the maximum of the 11-year cycle, but in the 1-2 years after it or, sometimes, prior to it. In different hemispheres of the Sun this cycle may develop in different ways. Thus one of the hemispheres (northern or southern) leads 11-year cycle development. The northern and southern hemispheres do not always have synchronous maxima. The start of these periods in different hemispheres may differ by 1-2 years. A.M. Ol' established through the analysis of the index of recurrent geomagnetic activity that the following 11-year solar cycle originates during the falling segment of the previous one about four years prior to a minimum extremum.

Yu. I. Vitinsky (1973) concluded that the real 11-year cycle begins in the middle of the falling period with the appearance and strengthening of polar magnetic areas. This first phase of the development ends by the beginning of the observed 11-year cycle at which time a second phase begins with the development of bipolar magnetic areas. This continues until the middle of the declining branch of the 11-year cycle, when a new cycle originates.

Another important feature in the development of 11-year cycles is that odd cycles are characterized by higher extremes than even ones.

Long-term predictions of exogenic geological processes can be subdivided into very long-term (within a century-long solar cycle to hundreds of years) and long-term (within a 11-year solar cycle –10-15 years) ones. The methods for these predictions differ only by the greater or lesser detail of initial data. Long-term predictions relate to the general tendencies in the development of one or another EGP genetic type, determined through the analysis of geological evolution of the territory and large climatic changes. The general technological scheme of long-term regional predictions is shown in Figure 18-2.

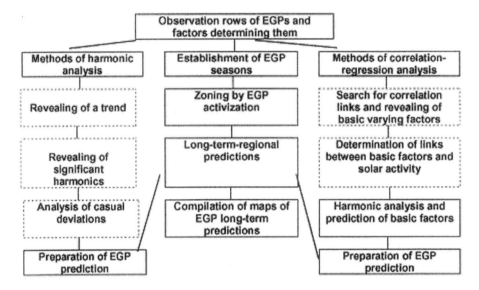

Figure 18-2. The technological scheme for long-term predictions of exogenic geological processes

More detailed information on the methods of long-term prediction can be found in the above mentioned publications. The principles behind these methods rest in the following. By analyzing the time series of exogenic geological processes and rapidly varying factors that control activity of these processes, one can determine regularities in development of the processes through time, as well as correlations between basic rapidly varying factors and processes. Furthermore, territorial zoning according to EGP activity is linked to temporal zones. In respect of a temporal zone, area has a similar composition and regime of rapidly varying factors that determine the development of one or another EGP genetic type. The results of zoning can be reflected in an engineering-geological map of EGP activity predictions on a scale of 1:200 000 – 1:500 000.

5.   PREDICTION BASED ON ANALYSIS OF INTEGRAL CURVES OF MODULAR

5.1   Coefficients of anomalies

Such predictions are made using the following techniques. For the time series of exogenic geological processes and factors determining them, the modular coefficients of anomalies are calculated by the equation:

$$K_\Delta = \frac{S_i - S_0}{S_0} \qquad (18\text{-}1)$$

where S_i is the value of a row term (for example, the number of mudflows in a given region in a given year); S_0 – arithmetical mean of EGP manifestation in this row.

The integral curve of anomalies is plotted by the results of algebraic summing:

$$K_i = \sum_1^i K_\Delta \qquad (18\text{-}2)$$

where K_i is the algebraic sum of modular coefficients of anomalies for the i-th year.

Then the resulting integral curves are compared with the rapidly varying factors. Figure 18-3 shows the comparison of mudflow events in different regions with solar activity (Wolf's number) and macro-forms of atmospheric circulation observed by Wangenheim. As can be seen from the Figure, the general forms of integral curves for all of the regions under study (Caucasus as a whole, Armenia, Caucasus without Armenia, Middle Asia as a whole, the right-hand bank of the Syr Daria and the Naryn River basin, Zailiskij Alatau, Kopet-Dag) is the same and coincides with the general course of the integral curve of the solar activity anomalies. It follows from this analysis that increase of mudflow processes is observed throughout the epoch of century-long solar activity minimum. The maximum of mudflow activity is seen especially distinctly for Middle Asia, except Zailiskij Alatau.

The maximum on the integral curve for mudflow activity in Middle Asia fell in the year of 1934, i.e. it coincided with the century-long minimum of solar activity anomalies. For Turkmenia this maximum fell in 1940-1941, i.e. offset by a half a phase from the 11-year solar cycle, and associated with the maximum of the 17-years solar cycle. These maxima coincide with an increase of anomalies of the number of days with the E-macroform of atmospheric circulation (meridional circulation) and weakened W-macroform (after Wangenheim). The maximum of modulus coefficients of mudflow activity in Turkmenia in 1940-1941 coincides with the maximum of anomalies of the C-macroform of atmospheric circulation (meridional circulation). The minimum is shown on integral curves of mudflow activity anomalies for Middle Asia in 1924-1925. It is associated with the next to last minimum of the 11-year cycle, relative to the century-long minimum and

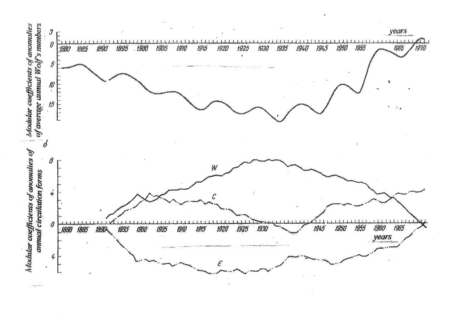

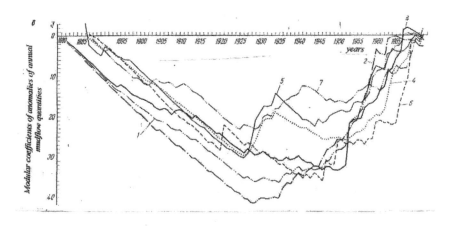

Figure 18-3. The integral curves of solar activity in the Wolf's numbers (a); macroforms of atmospheric circulation after Wangenheim (b), and quantity of mudflows per year in different regions (c). Caucasus (as a whole); 2 – Armenia; 3 – Caucasus (without Armenia); 4 – Middle Asia (as a whole); 5 – the right-hand shore of the Amu-Daria and Naryn River basin; 6 – Zailiskij Alatau; 7 – Kopet-Dag.

coinciding with the turning-point on the integral curve for anomalies of the E-macroform. A sharp activation of mudflow processes in this region in 1920-1921 coincided with the turning-point on the integral curve of anomalies of the W-macroform (zonal circulation characterizing westward transport of air mass) and a lesser peak on the integral curve of anomalies of the E-macroform.

The character of the integral curves of mudflow activity anomalies in the Caucasus differs from those described above. The absolute minimum for the two curves (for Caucasus as a whole and the Caucasus without Armenia) coincides with that for the integral curves for Middle Asia (1924-1925). However, for the Caucasus without Armenia this extreme is expressed more weakly. The absolute minimum of this curve is observed for 1948-1949.

As for Middle Asia, the epoch, corresponding approximately to the century-long minimum (1935-1940) of the integral curve of solar activity, is characterized by some increase in mudflow activity, though, graphically, this is not expressed as distinctly.

By comparing the integral curves of mudflow anomalies with those for anomalies of solar activity and macroforms of the atmospheric circulation, a clear relationship can be identified between anomalies of these three factors, which relate to the character of weather. At present, solar activity anomalies are on the rising branch of the integral curve. Therefore, the general background of anomalies in the mudflow activity will be positive and generally, in the next decade mudflow activity in Caucasus and Middle Asia is expected to increase. Events in the Caucasus during recent years substantiate this supposition.

6.   PREDICTION BASED ON THE ANALYSIS OF PARTS OF THE 11-YEAR SOLAR CYCLE

A more obvious relationship between EGP-activity and solar activity can be established by analyzing the activity of exogenic geological processes within particular parts of the 11-year solar cycle. For this purpose, the 11-year solar cycle is subdivided into: the branches of rise and fall (PC), quarters (ЧА), and epochs of maximum and minimum (MM). Quarters I and II lie within the rising branch; quarters III and IV in the falling branch. The epoch of the maximum includes quarters II and III; the epoch of the minimum – quarter IV of the previous cycle and quarter I of the next 11-year solar cycle.

Analysis of curves for average annual frequency of mudflows in terms of the ЧА, MM and PC branches has helped to establish the cycles of different orders of mudflow activity:

- cycles of the 1-st order within the 11-year cycle of solar activity; these include also shorter cycles, the maxima of which are in neighbouring 11-year cycles and offset by only one quarter of the 11-year cycle;
- cycles of the 2-nd order, the maxima of which are, in neighbouring 11-year solar cycles separated from each other by more than one quarter of the solar cycle;
- 3-rd order cycles with maxima separated from each other by the entire 11-year solar cycle;
- cycles of the 4-th order with maxima separated from each other by two complete 11-year solar cycles;
- cycles of the 5-th order, the maxima of which are separated from each other by three complete 11-year solar cycles.

The established cycles of EGP activity, the extremes of which are associated with different parts of the 11-year solar cycle, can be extrapolated for future years. If EGP events is short then, by comparing the curves of EGP events and rapidly varying factors that determine them, plotted for the parts of the 11-year solar cycle, correlations between them can be established. Further, through extrapolating curves of rapidly-varying factors predictions of the types of EGPs can be made.

Figure 18-4 illustrates mudflow activity in the parts of the 11-year solar cycle on the right-hand bank of the Syr Daria and in the Naryn River Basin. The graphs distinctly show cycles of the 4-th order presumably for the quarters (ЧА) of 29-30 years; maxima and minima (MM) of 31 years; and periods of rise and fall (PC) of 29-30 years. The extremes on the ЧА-graphs are associated with the third quarters of the solar cycle; on the MM-graphs with the epochs of maxima; and on the PC-graphs with the falling branches. The first extreme is in the 18th and the second in the 19th, 11-year solar cycles. Both extremes lie on the rising branch of the century-long solar cycle: the first at the very beginning, and the second with the century-long maximum. As can be seen from the graphs, a notable extremum is observed, probably, in the 13th or 12th 11-year solar cycles.

If to suppose that the mudflow activity will be similar to that which has been observed in the past, then the highest mudflow activity should be expected in the epochs of maximums in the 22nd or 23rd solar 11-year cycles.

7.   PREDICTIONS BASED ON CORRELATION-REGRESSION ANALYSIS

Correlation-regression analysis is used to:

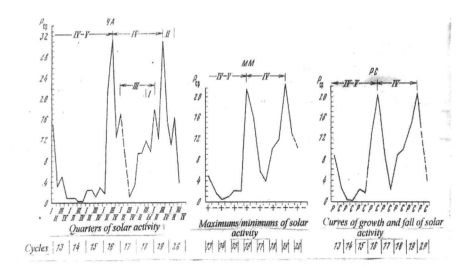

Figure 18-4. Curves of mudflow occurrences for parts of the 11-year solar cycle on the right-hand shore of the Amu-Daria and Naryn River basin. ЧА - by quarters; MM – by maximums/minimums, and PC – by rise and fall branches. Roman numbers indicate the orders of cycles.

- reveal links between EGP activity and fast-varying factors determining it;
- extrapolate the EGP activity, basing on the revealed regularities, for a specified period;
- reveal inertia (lagging) of processes relative to fast-varying factors; reveal links between factors in territory spatial zoning by the EGP activity regime;
- determine the closeness of links between time rows of exogenic geological processes, and others.

The use of the correlation matrix in the analysis of fast-varying factors aims at determination of an influence degree upon activity, and provide a basis for predictive calculations. Thus, regression analysis makes it possible to detect a link between mudflow activity and the W and E atmospheric circulation with a high degree of confidence (95 %), and macroform C with a lower confidence (95 %). The regression line relating the quantity of mudflows in a year to the number of days of the E-macroform obviously demonstrates a direct relationship, whereas an inverse relationship is observed between the quantity of mudflows in a year and the number of days with the W macroform. No clear correlation is observed between the quantity of mudflows in a year and the number of days with the C macroform.

Coefficients of correlation between mudflows and macroforms of atmospheric circulation are not high for different regions of the former USSR. This indicates a weak link between these parameters. For some regions, however, coefficients are high. For example, for Armenia at a zero level of significance the correlation coefficient with macroform W is 0.5 and with macroform E 0.4 to a very high degree of confidence (the probability is actually equal to 1).

Practically all the exogenic geological processes are characterized by inertia (lagging) relative to rapidly varying factors. This inertia takes into account the previous history of a process.

According to A.A.Pletnev (1972), a mathematical model includes temporal and spatial correlation links. The numerical values of the "memory" implicit in modeling is determined by time intervals where these links are significant.

A.A.Pletnev further suggests that, to establish the memory and, hence, the most significant correlation coefficient, one should undertake analysis of the mutual correlation function:

$$R_{xy}(\tau) = \frac{M[x(t)-\bar{x}][y(t-\bar{e})-\bar{y}]}{\delta_x - \delta_y} \tag{18-3}$$

where M is the symbol of mathematical probability;
δ_x, δ_y – average quadratic deviations of x(t) and y(t);
x , y – mean values of correlated phenomena;
τ - time when the correlation coefficient has the highest significance.
In predictions the highest coefficient of correlation is taken.

8.   PREDICTIONS BASED ON HARMONIC ANALYSIS

The advantage of harmonic analysis is that prediction of rapidly varying factors is not needed, as it is supposed that the influence of all the rapidly varying factors is already taken into account in the fragment of the EGP time row for which the modelling is being done.

The mathematical model most suitable for analysis and prediction is:

$$X(t) = \eta(t) + z(t) + \varepsilon(t) \tag{18-4}$$

where η is the trend; z and ε are the cyclic and casual components, respectively.

The presence of the trend, cyclic and causal components in the time series can be explained by regularities in the multi-year activity variability of processes and rapidly varying factors.

In the majority of cases the trend is represented by an algebraic polynomial not higher than third degree (Pyrkin, 1977). Otherwise, the polynomial would derive cyclic components. The time series can be with or without a trend. Without a trend, the series can be considered to be stationary, which facilitates extrapolation and earlier warning.

Figure 18-5 presents the mudflow epignosis and prediction for the right-hand bank of the Syr Daria, made on the basis of harmonic analysis.

9. ENGINEERING-GEOLOGICAL MAPS OF EXOGENIC GEOLOGICAL PROCESSES

9.1 Long-term predictions

The final step in long-term regional predictions is the compilation of special engineering-geological maps. These usually depict a prediction of one EGP genetic type. They are compiled on the basis of zoning the area in terms of intensity of the exogenic geological processes on a scale of 1:200 000 – 1:500 000. The following units are distinguished: regions, areas, temporal zones, districts and sites.

Regions are defined by major geostructural features; areas by geomorphological conditions; temporal zones by the types and regimes of fast-changing factors that determine EGP events; districts by stratigraphic-lithological features; and sites by the intensities of processes. Within a temporal zone, exogenic geological processes occur simultaneously but their intensities depend on the susceptibility of specific areas to these processes.

Figure 18-6 illustrates part of an engineering-geological map of landslide activity prediction for an area on the Black Sea coast. This map was compiled in 1985 with predictions to the year 2000. The legend describes expected landslide activity over different periods. The accuracy of the predictions shown on the map has proved to be quite satisfactory (about 75-80 %).

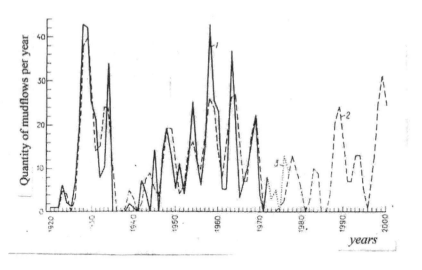

Figure 18-5. The mudflow epignosis and prediction on the right-hand shore of the Syr Daria, made on the basis of harmonic analysis. The values: 1 - factual; 2 – predicted; 3 – verified predicted.

9.2   Short-term predictions

Short-term predictions of exogenic geological processes are made to give warnings of a few hours to a year so that effective control can be exerted over EGP activity and actions for prevention or mitigation of adverse consequences can be planned.

The short-term prediction should be of an operative character and must be specified for the entire EGP-prone period.

Predictions of this type can be regional or local. Regional predictions are usually compiled for administrative regions/republics, and local ones for small areas (e.g. for a settlement or a particular site). In preparing short-term predictions, both probabilistic and deterministic methods are used as described above. The latter are used for local predictions.

Prediction of exogenic geological processes also includes the identification of EGP-hazardous seasons. These can be determined both for engineering-geological regions and areas, and for administrative territories (republics, regions).

The EGP-hazardous season is the period during the year (months) when there is an increased possibility of one or another EGP type of event occurring . It is reasonable to distinguish within an EGP-hazardous season a

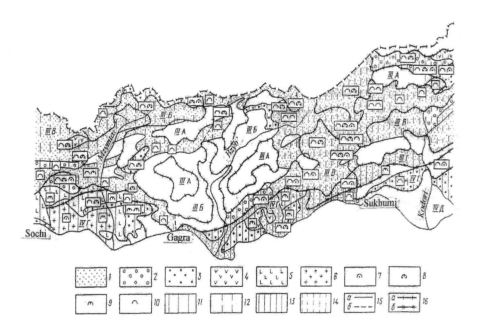

Figure 18-6. Part of a map of long-term landslide predictions. Landslide intensity determined as a ratio of the area of landslide forms to the entire territory affected: 1 - very low (<0.05); 2 - low (0,05-0,1); 3 - mean (0.1-0.3); 4 - high (0.3-0.5); 5 - very high (0.5-0.7); 6 - rather high (>0.7). The rest unshaded is the territory where landslides are not revealed. Genetic types of landslides: 7 – block-type landslides in bedrocks; 8 - block-type landslides in Quaternary sediments; 9 - landslide streams; 10 – mud avalanches. Prediction of landslide activization: 11 – the highest activization, caused by anomalous precipitation, is expected in 1997-2000, the mean one – in 1988, 1992-1993. The landslide activization in the coastal zone, caused by abrasion, is expected in 1995-1997; 12 – low activity is expected in 1992-1993 and 1996; 13 – the highest activization caused by anomalous precipitation is expected in 1995-2000; the high activization caused by abrasion, is expected in 1995-1997; 14 - the highest activization, caused by anomalous precipitation, is possible in 1995-1996, low one – in 1992-2000.

Boundaries of: 15 – sites with different landslide intensity (a) and temporal zones distinguished by landslide activization, not coinciding with the boundaries of regions or areas (b); 16 – areas (a) and regions (b). Regions: megantiklinorium of the Major Caucasian Ridge, IV – Georgian median massif; areas: IIIA – nival Aplian, IIIБ – highland, IIIB – mean-height land, IIIГ and IVГ – lowlands, IVД – hilly plain.

period with the highest probability of EGP events, in particular time (months) when the EGP-probability is over 70 %.

For example, in Georgia the mudflow-hazard season lasts seven months from April to October; in Uzbekistan, nine months from January to September. However, the period of the most probable mudflows in Georgia is in August when about 62% of these occur, and in July (14 % of cases), amounting to a total for these two months of 76 % of events. In Uzbekistan the period of the most possible mudflows are the three months from April to June.

In conclusion, we now have at our disposal sufficiently well developed methods for the prediction of exogenic geological processes to forecast with reasonable reliability, supported by appropriate observation (monitoring), potentially catastrophic exogenic geological processes to provide for safety of people living in dangerous zones.

Chapter 19

PREDICTION OF ENDOGENIC GEOLOGICAL PROCESSES

G. VARTANYAN
Russian National Research Institute for Hydrogeology and Engineering Geology, Moscow region, Russia

Endogenic geological processes include a wide spectrum of evolutionary changes in the state, structure and properties of rocks.

This chapter discusses the possibilities of predicting those abyssal processes which are able to exert a direct and decisive influence upon the state of the geological conditions in a particular region and, through their combined action, affect the degree of comfort or discomfort of the ecosystems formed there.

Other endogenic processes, such as metamorphism, degassing, dehydration of matter, etc., which do not present potentially catastrophic hazards for ecosystems, are not discussed here.

Amongst various processes of the Earth's geological evolution, endogeodynamic development is especially significant due to its major contribution to the formation of current geoenvironmental conditions.

One of the most typical and catastrophic consequences of endogeodynamics are multiple and intensive earthquakes, each demonstrating the completion of an evolutionary step in the development of the lithospheric cover.

Traces of past strong seismic catastrophes (such as elongated faults of different morphologies) indicate that earthquakes have accompanied the entire geological (and, perhaps, much earlier) history of the Earth and represent a regularly repeated and normal, though horrible in human terms, aspect of the Planet's geological life.

For many years, the problem of prediction of these destructive endogenic phenomena has attracted the attention of researchers.

However, in spite of more than one thousand years of study of these natural processes, there is still no uniform understanding of the processes leading to a basis for sound prediction of events and no established systems for forecasting seismic events amongst the seismological community.

This situation has arisen because geotectonic evolution of the Earth as a geophysical object is insufficiently developed despite many conceptual and hypothetical approaches to the problem. The mechanisms of "preparation" and occurrence of earthquakes are complex and varied.

However, thanks to the concentrated efforts of the international scientific community, especially scientists of the USA, Japan, China, Russia and Italy, during recent decades, the search for reliable precursors of active endogenic geodynamic processes has made progress. Some techniques have been developed for prediction of earthquakes, tsunami and volcanic eruptions (Mogi, 1985; Sholtz, 1967; Bredehoeft, 1967; Vartanyan, 1979; Vartanyan, 1999).

Cuurrent world practice in studying and predicting endogeodynamic phenomena is based on long-term monitoring of natural processes and is based on the principle of consequent approximations.

Thus, a system for prediction of strong earthquakes is constructed from a complex of the graphical solutions including such factors as: the degree of complexity of the geotectonic structure of a region under study; the degree stress in the subsurface; current motions of the earth's crust; and historical seismicity.

With the aid of such reconstructions, long-term (for time intervals of a few tens of years) estimates of the probability of strong seismic events are made.

Through analyzing earthquake precursors observed in different geophysical and biogeophysical fields, the time of appearance and amplitude of which can be linked using empirical relationships to the magnitude and location of a future earthquake, medium-term prediction (months to years) can be carried out.

Short-term predictions are made through analyzing sets of observations of short-living precursors but these have to be separated from background "noise" in the data. The number of the precursors and total area covered by them determines the probability, place and magnitude of a future earthquake.

In some cases, it becomes possible to make a so-called operative prediction, warning of an event a few hours before it takes place.

This ideal prediction scheme is not always possible. Normally consequent approximations must be applied to the solution of the complicated geological-geophysical problems.

At the same time, the practical need for protection of people and material property against natural cataclysms have led many countries that are susceptible to earthquakes, tsunami and volcanic eruptions, to take steps to set up monitoring systems of the state of the subsurface state, as a basis for assessing the risks from geodynamic hazards.

Chapter 20

MATHEMATICAL MODELS OF THE INTERACTION BETWEEN THE GEOLOGICAL AND ECOLOGICAL ENVIRONMENT

HE QINGCHENG
Department of Groundwater Monitoring, China Institute for Geo-Environmental Monitoring, Beijing, China

1.   INTRODUCTION

The earth is a huge, varying, and complete dynamic system consisting of the geosphere, atmosphere, hydrosphere, and biosphere. The human race is a significant component of the biosphere, which is intermedial to the other three spheres. Human activities extend to the range of the biosphere, from the surface of the crust to the surrounding atmosphere. The natural environment on the earth's surface is vital to the existence and development of mankind. Primarily, the environment includes climate, geology, water and ecology, which interrelate to, interdepend on, and interact with one another. In other words, the natural environment is a complex system.

For mankind, as the main place for activities, the geological environment is the most important element in the natural environment system, affecting both our existence and development. It interacts or interrestricts with the climatic, ecological, aquatic, and land environment. The geological environment interacts with the natural environment through many kinds of geologic processes of mass and energy exchange, such as erosion, denudation, corrosion, gravitation, deposition, geochemical action and geostress action. It constantly evolves and transforms, and gradually builds balance and stability in the long term.

On the other hand, however, human influence cannot be ignored, and can exceed that of natural processes. Urban, villages and towns, industry, agriculture, and communication and transportation form tightly interrelated social-economic-environmental systems. Social-economic activities, including urban construction, development of enterprises in villages or towns, constructions of seaports, airports, railways, and highways, population migrations, explorations of mineral resources or water resources, building water projects such as canals, channels, reservoirs, and sea embankments, and waste discharge. These directly or indirectly influence the geological and other natural environments. Thus the human race lives in two environments-the natural one and the social-economic one. The geological environment is dominant in nature. People can improve or ameliorate the natural environment to favor the social economy through science and technology. On the other hand, human activities, especially production and construction, can damage or destroy the natural environment, social economy and human life if they do not conform with the rules of nature.

The results of seeking superficial economical interests with no regard for negative environmental effects can be severe. The intensification of human activities increases the frequency of occurrence of many geo-hazards. Consequently, geo-environmental problems are becoming obvious and more often brought to public attention. Such problems involve landslide, subsidence of mines karst and foundations, water bursts and geofracture, seawater intrusion, water quality degradation, exhaustion of famous springs due to over-exploitation of ground water, soil and water contamination by pesticide and chemical fertilizer, and soil erosion following agricultural devegetation, deposition of silt in canals or seaports, flood aggravation resulting from artificial conversion of lakes or sea to agriculture land, and pollution caused by unreasonable discharges of wastes. All of these problems have contributed to deterioration of the ecological environment.

In brief, the geological environment provides resources for human existence and development, but the unreasonable exploration and use of geological resources can trigger serial environmental problems, especially eco-environmental problems, which do great harm and feedback to worsen the geo-environment. Therefore assessment of the adaptability of geo-environment and eco-systems should be done before exploration and use of the geological resources. A scientific and reasonable exploration plan can be made in order to realize sustainable development of fully used geological resources at the same time as good protection of the eco-environment.

The state and quality of the geo-environment can be assessed and modeled. Recent mathematical models for evaluation of geo-environmental quality are roughly of four types:
- Index Model of Geo-environment Quality,
- Mathematical Statistical Model of Geo-environment Quality,
- Classification Model of Geo-environment Quality, and
- Bio-index Model of Geo-environment Quality.

Table 20-1 Mathematical Models of the Geo-enviroment and Eco-system

Type	Sub-type	Logical conception	Characteristic of the evaluation element (factor)	Note
1 Index Model	1 typical index model 2 classification index model 3 sensitive factor index model	The environment quality can be ascertained and reasoned within certain time and space	1 physical and chemical indicators 2 objective and reasonable scores	The first three models interrelate and can be synthetically employed in the complex assessment of the geo-environment quality
2 Mathematical Statistics Model	1 qualitative analysis model 2 multi-index classification model 3 factor analysis model	The environment quality varies randomly within certain time and space	1 physical and chemical indicators 2 objective and reasonable scores	
3 Classification model	1 integral classification model 2 fuzzy classification model 3 neutral-color classification model 4 classification by the amount of information model 5 sensitive index classification model	The class limit and the environment quality change are both indefinable.		
4 Bio-index model	1 biology indicator method 2 bio-index method 3 others	Biology is accordant to living environment and sensitive to environment change	1 physiological reaction indicator 2 the variety of genus and group of biology in the environment	1 Bio-index can also be classified by the second and third models. 2 Bio-index is an environmental index

Though differences exist between these four types, they can be employed synthetically. Their differences and relationships are shown in Table 20-1. All except the fourth model are often applied nowadays.

2.   CONTENT OF GEO-ENVIRONMENTAL QUALITY ASSESSMENT

Geo-environmental quality assessment mainly includes the following aspects.

2.1   Natural Geological Conditions

Human activities are firstly influenced and restrained by geological conditions, so the natural geological conditions should be assessed in the first place.

2.1.1   Components of rock and soil

The material components of the geological environment - rock, soil and water - are generally considered at the outset.

The first is the abundance of geological resources which are, in fact, components of the geo-environment, in a specific structure and state. These mainly include mineral, energy, building material, land, water, geological landscape, and geological space resources. Their types and quantities decide the function of the geo-environmental system in some degree and thus become the main content of the evaluation.

The second is the original geochemical background of rock, soil and water in the geo-environment which directly affects human existence and development. All humans are subject to the effects of the geochemistry field to some extent and the average amount of various chemical elements in the human body adapts to that in the local crust. For instance, the amount of more than 60 chemical elements in human blood apparently relates to the abundance of them in rock. Specified amounts of some chemical elements are necessary to the growth and development of organisms. In the geological environment, excessively high or low levels of some chemical elements can harm organisms. In addition, the chemical characteristics of soil, original rock, and water determine the growth and harvest of crops. Consequently, geochemical background is an important indicator of geo-environmental quality.

The third one is the mineral composition of rock, soil and water. Particular minerals and structures in rock, soil and water can be detrimental to the geological environment. For instance, the erosion of concrete by ground water and of foundations in button mud are harmful for construction activities.

2.1.2 Geological Structure

The geological structure and state is one of important aspects in the geo-environment quality assessment. It is a major component with those of atmosphere or surface water quality. Even if the material components in the geo-environment did not vary, the change in the physical properties (formation or state) alone of some geo-environmental factors would trigger geo-environmental problems.

2.1.3 Dynamic Force

The geological dynamic forces that drive crustal movement and development control and transform the structure and morphology of the earth's surface. Therefore, the stability of regional crust and the extent of geological dynamic development are also significant during geo-environment evaluation, especially the impacts on the geo-environment of various types and sizes of geo-hazards.

2.2 Capability to Resist Interference of Human Activities

The geological environment changes with natural environment evolution, sometimes slowly and imperceptibly and sometimes catastrophically. As a whole, human activities accelerate environmental change, especially in the circumstances of rapid population increase and large-scale development of productivity, science and technology. Human activities have strong impacts on the earth's surface and geological dynamic forces cannot be ignored. These artificial geological forces are similar to natural processes in scale and rate, and can interfere strongly with the geo-environment. The geological environment has a capability to resist such interference but, since that capability is limited, assessment of resistance is another important aspect of geo-environment evaluation.

2.3   Extent of Damage and Contamination

There exist hardly any areas unaffected by human activities and the less natural the geological environment, the more the effects of human factors. These must be taken into account in the assessment. The principal factor is geo-environmental damage resulting from social and economic activities (especially introduced geo-hazards). Environmental contamination (mainly of surface water, ground water, and soils) produced by all kinds of wastes also change material components in the geo-environment.

The overall quality of the geo-environment is subject to the positive or negative effects of each constituent element. However, when we evaluate the geological environment we should not depend on the average state of each factor. We should evaluate the positive and negative elements, respectively, find out which factors are most sensitive to human activities, and base the whole analysis on these.

The following section briefly introduces some models for geo-environment evaluation.

2.3.1   Index Model of the Geo-environment Quality

Geo-environmental evaluation can begin with qualitative assessment by sensory perception indicators, such as smell, taste, color, transparency, and sediments of ground water, then quantitative assessment by uni-index indicator, and later developed to the complex assessment of single or multi geo-environmental elements in an area, or even a whole area using comprehensive quality indicators. There are many parameters affecting geo-environmental quality, but the chemical content of components in the geo-environment is usually employed because of extensive and mature research on the geo-environmental pollution problems (such as surface water, ground water and soil pollution). Given the nature of the geo-environment, changes of geo-environment quality are demonstrated not only in the variety of the chemical components, such as contamination by specific elements, but also, more importantly, in the transformation of physical properties such as structure and state that leads to problems of surface collapse, land surface subsidence, landslides, and mud-rock flow. Damage caused by the physical state is often more directed, rapid and immediately harmful than chemical pollution. Therefore when data about geo-environment are used to express the geo-environment quality, both physical and chemical indicators should be considered.

Institution of the index model

The process by which large geo-environmental data sets are transformed into a simplified and systematic index of the geo-environment, is that of information transmission. As shown in Figure 20-1, there are multiple steps by which geo-environment quality indexes are obtained.

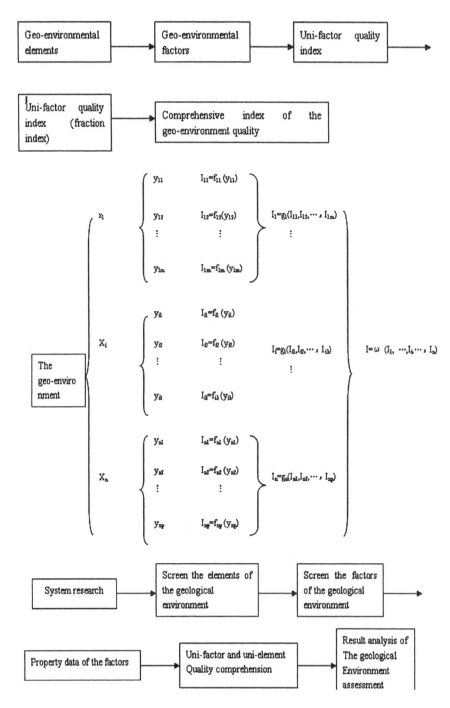

Figure 20-1. Information flow in the institution of the index model

Step 1: undertake systems research on the whole geological environment,

Step 2: screen the geo-environmental elements to determine which ones would be involved in the assessment,

Step 3: screen the factors of the involved geo-environmental elements to determine which ones would be used in the assessment. If these factors need further analysis, screen these again till the most basic ones are identified,

Step 4: transform the data or information of monitoring to the uni-factor quality index,

Step 5: transform the uni-factor quality indexes to the uni-element ones (factional indexes) by overlaying them as a certain mode,

Step 6: synthesize these factional indexes into a total index of the geo-environment and analyze and explain the total index.

From Figure 20-1, the geo-environment quality indexes are gradually attained through the uni-factor quality indexes, then the uni-element quality indexes, and last the comprehensive quality index. So the index model of the geo-environment generally has one of the following forms.

- Uni-factor Quality Index Model

The general formula of this model is $I_{ij}=f_{ij}(C_{ij})$, in which I_{ij} is the quality index of the jth factor of the ith geo-environmental element; C_{ij} is the physical and chemical indicator of this factor, which reflects the physical structure and state of the measured chemical concentration.

Usually there are two formats:

$$I_{ij}=C_{ij},$$

which uses the physical and chemical indicator to directly demonstrate the environment quality, and

$$I_{ij}=\frac{C_{ij}}{S_{ij}},$$

in which S_{ij} is the evaluation standard of the jth factor of the ith geo-environmental element. The second format employs the level of excess of the physical and chemical indicator over the standard, say, the index of standard excess, to express the quality index state. That I_{ij} equals 1 shows the environment quality is in the critical condition. This format is often used to express the mass of the uni-factor.

Because the value of I_{ij} is relative to some certain standard and the evaluation standard involves many factors, different areas have different standards, requirements and goals. A uniform national standard should be taken if it exists. In any case of parallel comparison of geo-environment quality, the same standard should be used.

The uni-factor geo-environment quality index only stands for the state of a single evaluation factor or physical and chemical indicator, and hardly reflects the whole geological environment. But it is the basis for all other indexes, classification, and comprehensive evaluation of geo-environmental quality.

- Fraction Index Model

  The general formula of this model is

  $I_i = g_i(I_{i1}, I_{i2}, \ldots, I_{im})$,

  in which I_i is the mass fraction index of the ith geo-environment element and m is the number of the evaluation factors of the geo-environment element.

  Generally it has following forms:

1. Linear function: $I_i + \alpha I_{ij} + \beta \alpha + \beta$ are constants,
2. Subsection linear function: $I_i + \alpha I_{ij} + \beta$, in which α and β have different values in the different defined areas of I_{ij} or C_{ij}. The physical indicators often have this feature.
3. Nonlinear function:

   Usually two equations: $I_i + \alpha I_{ij}^{\beta}$ or $I_i + \alpha \beta^{I_{ij}}$

- Comprehensive Quality Index Model

  When synthesizing geo-environmental indexes, we can model the actual value, average value, maximum value, or minimum value, or can take them all into account. Hence from different approaches we will get different comprehensive quality indexes. The most commonly employed are the following ones.

1 Representatives overlain: $I = \sum_{i=1}^{n} I_i$ (n is the number of the evaluation elements),

2 Averages overlain: $I = \dfrac{1}{n} \sum_{i=1}^{n} I_i$, and

3 Weighted averages overlain: $I = \dfrac{1}{n} \sum_{i=1}^{n} W_i \times I_i$ (W_i is the weight of the evaluation elements).

2.3.2   Mathematical Statistical Model of Geo-environment Quality

As more and more background information the geo-environment accumulates with the development of research on environmental geology, data for environmental monitoring becomes more complete and systematic. Particularly, computer technology has built many kinds of mathematical statistical models to analyze and address environmental problems, and to qualitatively analyse huge amounts of information and data. So far, all mathematical statistical approaches can be employed in environmental science to some degree. The applied models are mainly of three types: qualitatively discriminant analysis models, multi-index classification models, and factor analysis models. Qualitatively discriminant analysis models will be briefly described here.

Qualitatively Discriminant Analysis Model

Discriminant analysis, extensively used in many disciplines, is a branch of multivariate analysis. In discriminant analysis, the variable (discriminated object) is a quantitative index, and the dependent variable (discriminant standard) is the known, qualitative class.

If the unit of geo-environmental evaluation is n and the taken evaluation factor is p, available qualitative data of the jth evaluation unit can be symbolized as

The known evaluation standard has k classes, which is labeled as T_1, T_2

$$X_j = \begin{pmatrix} x_{j1} \\ x_{j2} \\ \vdots \\ x_{jp} \end{pmatrix} , \quad j=1, 2, \cdots n$$

$\dots T_k$, respectively. Taking Unit n_j from the evaluation units in Class T_i, and the qualitative data of all the evaluation factors in every unit gives

Discriminant analysis is undertaken to determine to which class

$$x_a^{(i)} = \begin{pmatrix} x_{a1}^{(i)} \\ x_{a2}^{(i)} \\ \vdots \\ x_{ap}^{(i)} \end{pmatrix} , \quad I=1, \cdots, R$$

Evaluation Unit j belongs, based on the class information of the known evaluation standard.

2.3.3   Classification Model of the Geo-environment Quality

During research on geo-environmental quality, it is always necessary to study multi environmental elements as well as serial ones. Because of differences in background, causes and processes to changed environmental quality, each evaluation factor can belong to different standard. As a result, a central problem is how to synthesize these evaluation factors to determine which class a specific environment state belongs to..

Currently classification method applied in geo-environmental evaluation are the Integral Method, Value M Method, Fuzzy Classification Method, and Neutral-color Classification Method. We mainly discuss the integral method because of the limitation of length of this article.

Based on the existing environmental standards, Integral Methods give each evaluation factor a certain score according to the data and qualitative characteristics and then sum the scores of all the factors. That is to say,

$$M = \sum_{i=1}^{n} a_i$$

in which M is the score of an evaluation unit, a_i is the score of Evaluation Factor I, and n is the number of the evaluation factors.

The environmental quality class can be defined according to the integral (Value M). This is a direct scoring method, which can straightforwardly relate to every class of the environment standards. The more integral the result the better the environmental quality. But the main difficulty lies in how to give each evaluation factor a reasonable score. Especially in geo-environmental evaluation where the physical indicators (state and structure) play a large part, it is much more difficult to score physical indicators than chemical indictors. Finding a basis for reasonable scores requires some overall, systematic and deep research on geo-environmental conditions.

In the application of this method, the number of evaluation factors and score standard should be determined as the specific geo-environmental condition. Generally, an environment quality standard is divided into five classes and the correspondent score to Class 1-5 is 100, 80, 60, 40, and 20, respectively. The classification of the environment quality is seen in Table 20-2.

Table 20-2 Classification Standard of the environment quality (Integral Method)

Integral (M)	M 96	96 M 76	76 M 60	60 M 40	M<40
Class	I	II	III	IV	V

As one of the approaches applied in the environment quality evaluation of recent years, the Integral Method can clearly and easily address multi-factor environment problems, but the simple addition used in the calculation cannot precisely reflect the relative consequence (weight) of the individual factors. However, for the quality of geological environment, the different weight of each individual evaluation factor must be indicated, therefore this method could be improved to meet the evaluation requirement.

Chapter 21

THE INFLUENCE OF CLIMATE CHANGE ON GEOLOGY AND ECOSYSTEMS INTERACTION

JONAS SATKUNAS[1], JULIUS TAMINSKAS[2], & NAUM G. OBERMAN[3]
[1] *Geological Survey of Lithuania, Ministry of Environment,* [2] *Institute of Geology and Geography, Vilnius, Lithuania,* [3] *Mining and Geological Company "Mireko", Komi Territorial Centre for State Monitoring of Geological Environments, Syktyvkar, Russia*

1. INTRODUCTION

Climate is the principal geographical factor influencing the main parts of the Earth's surface – it determines processes of landscape formation and of its specific elements, affects all exogenic geological processes, aquatic and terrestrial organic life and social conditions of human communities as well as economic development (Bukantis et al., 2001).

The surface of the continents is the result of interaction of the lithosphere and atmosphere. Climate change directly influences the well-being of the ecosystem and modifies it slowly or abruptly. Climatic conditions are directly and vitally important for variety and abundance of living organisms. However, climatic conditions are also very significant for physical processes and, in turn, determine the state of the "mineral foundation" occupied by living organisms. In addition, abrupt physical processes (floods, landslides, glacial advances) caused by can instantly destroy long established habitats and ecosystems.

Climate parameters determine the whole range of exogenic geological processes, that destroy rocks, form new sediments, and shape the landform. Climate influence is very clearly reflected by the intensity of weathering, formation of landslides, river erosion, aeolian processes, fluctuation of groundwater level, changes in cryosphere etc. (see chapter 6).

The present chapter discusses climate change as the most import agent of change of the physical environment and thus of ecosystems. The importance of climate change to landscape formation and physical environment is exemplified by the case studies of the development of karstic features in Lithuania and the investigations of changes of state of permafrost in Subarctic and Arctic Russia (Republic Komi)

2.   CLIMATE CHANGE AND KARST PROCESSES (THE EXAMPLE OF THE NORTH LITHUANIAN KARST REGION)

Active karst landscapes are highly vulnerable and complicate the regional economic development and the protection of nature. In the intensive karst zone the most serious hazards occur where ground collapse damages buildings, roads and crops. Another serious effect is seen in the deterioration of groundwater quality.

Upper Devonian gypsum and dolomites occur beneath the Quaternary sediments in North Lithuania. Sinkholes appear frequently where the Quaternary is particularly thin and underlain by gypsum. The geology and landscape of North Lithuania's karstic region is discussed more fully in the Chapter 6 on "Karst and ecosystem" of the present volume and this chapter only deals with the relation of climate change to the karstic process.

Monitoring of the karst process was undertaken in the zone of the most active sulphate karst – mainly in the Tatula River cachment and karstic lakes territory. Data on hydrology, climate, hydrogeology of the period 1962–2000 were collected and analysed. Also, on the basis of runnoff and hydrochemical measurements, the rate of chemical denudation of gypsum was calculated for the Tatula River basin. Comparative analysis of results tested the hypothesis that intensification of the karst process is related with climate change (Taminskas & Marcinkevičius, 2002).

During the last two decades of the 20th century an increase of near surface winter temperature was observed in Lithuania. This trend still persists. It is considered that this phenomena could be caused by the repeated very low temperatures in the stratosphere of the European Arctic sector, accompanied by increased zonic circulation intensity in the troposphere in Europe and the North East Atlantic. Changes of atmospheric circulation caused changes to all other climatic parameters. The amount of winter precipitation, duration and depth of the seasonal frozen ground especially shifted during this period. Changes of winter climatic parameters had important influences on the change of water balance and chemical denudation intensity.

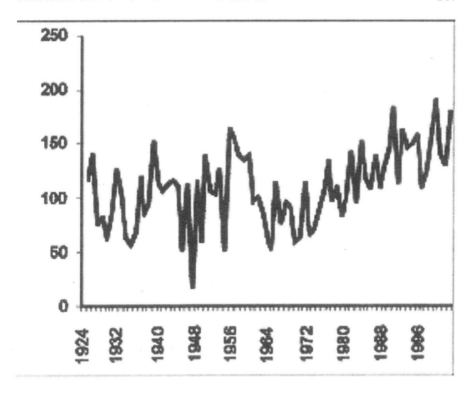

Figure 21-1. Winter precipitation (mm) fluctuation in Birzai

The annual precipitation in the study region ranges from 434–921 mm, with an average of 640 mm (1924–2000 years). From 1924 to 1999, a trend of slight decrease of annual precipitation was recorded in the karst region, though in general the 20th century, according to L. Lizuma's data (Lizuma 2000), is characterized by a trend of annual precipitation increase. The lowest annual precipitation was in period of 1961–1980 (Table 21-1).

We can distinguish a period between 1949–1996 when precipitation in warm seasons was decreasing, except for short intervals, whereas in cold

Table 21-1. Amount of annual and winter season precipitation during different seasons

		1924–1940	1941–1960	1961–1980	1981–2000
Annual	Average	690	656	576	660
	Min.	567	473	434	513
	Max.	921	852	733	846
Winter	Average	96	108	89	139
	Min.	56	16	53	95
	Max.	154	165	136	192

seasons it was decreasing only until 1969. Since 1964 the precipitation in cold seasons increased significantly, except over certain very short intervals (Figure 21-1).

These variations in precipitation during cold and warm seasons are characteristic of the whole Lithuanian territory (Bukantis, Rimkute, & Kazakevicius 1998). In year 1980 the sum of precipitation of three winter months exceeded 100 mm and until year 2002 it was not lower 95 mm. In the period 1981–2000, compared to 1961–1980, the amount of winter precipitation increased by 36% (Table 21-1) and constituted 21% of annual rate (Figure 21-1).

Since 1850 the annual mean and maximum temperatures have been constantly increasing (Lizuma, 2000). A particularly distinct warming of the air temperature has been observed since the 1960s. In the period 1961–1990, compared to 1931–1960, the mean temperature of January and February increased in the karst region by 0.5 ° C (Bukantis, Rimkute, & Kazakevicius 1998). An especially distinct increase of the annual and winter temperatures occured in the period 1994–2002 (Figure 21-2). Such climate warming during the cold and warm periods influenced other hydrometeorological parameters. Warmer winters (Table 21-2), in 1974–2000 may be related to a shift in the last 25 years of the NAO (North Atlantic Oscillation) northern atmosphere activity center from South-East Greenland to the Norwegian Sea (Corti et al. 1999; Hilmer et al. 1998, Hilmer and Jung 2000, Hurrell and van Loon, 1997).

Due to warmer winters in 1981–2000 compared to 1955–1980, the seasonal frozen ground period shortened from 110 to 96 days, and the mean depth of freezing decreased from 42 to 30 cm. More often thaw periods

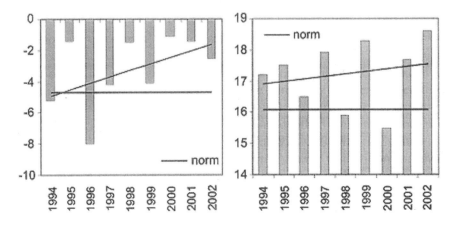

Figure 21-2. Mean winter and summer temperature (° C)

Table 21-2. Mean annual and winter air temperature during different periods

	1924–1940	1941–1960	1961–1980	1981–2000
Annual	6.0	5.6	5.6	6.5
Winter	-5.1	-4.8	-5.1	-3.3

increased the surficial runoff to karstic sinkholes and lakes.

There is direct dependence of runoff variations on precipitation, except in following long dry periods. The response of river runoff to an abrupt change of precipitation usually occurs a year later. This is, presumably, due to temporary storage in the karst landscape.

The winter runoff variations in the 1962s–1980s were negligible, but in the beginning of the 1980s winter runoff increased rapidly (Figure 21-3). This can be explained by a long period of warm years with higher winter precipitation. The higher precipitation and warm winters caused rapid increase of groundwater recharge and accordingly minimal winter runoff. In the 1960s and early 1970s the falling level of groundwater created good

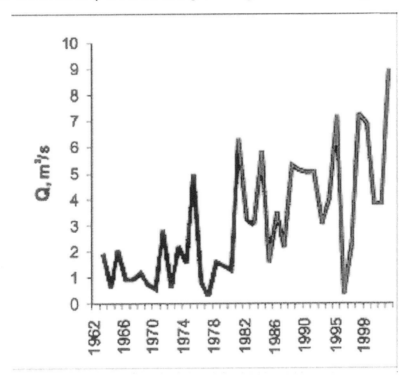

Figure 21-3. Mean winter runoff of Tatula River

conditions for a more intensive water circulation in the karst rocks and reduction of hydrostatic pressure supporting the arch of Quaternary deposits above the karst cavities. Besides that, in period 1970s – 1990s the annual amplitude of groundwater fluctuations increased rapidly, indicating more intensive annual water circulation, which is one very important factor in karst rock denudation.

In the 1980s, due to increased winter precipitation and higher air temperature, the level of groundwater (D_3t aquifer) increased. Such high groundwater levels remained until the end of the 20th century (Figure 21-4). Increasing groundwater resources of winter period enlarged the winter runoff almost by three times – from 26 mm in 1962–1980, up to 74 mm in 1981–2000. Minimum seven days winter runoff increased from 0,348 m³ s⁻¹ in 1962–1980, up to 0,718 m³ s⁻¹ in 1981–2000.

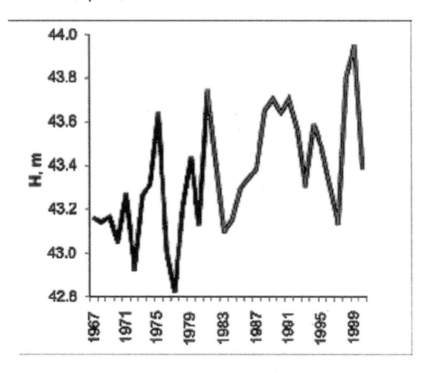

Figure 21-4. Mean winter groundwater level

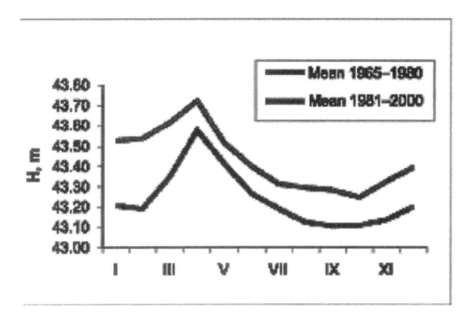

Figure 21-5. The intensity of gypsum denudation in Tatula River basin

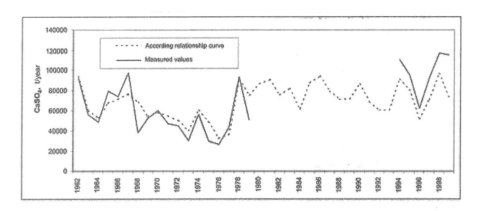

Figure 21-6. The chemical denudation in Tatula River basin

During winters of the period 1967–1980, compared to 1981–2000, the mean level of the D_3t aquifer was lower by 31 cm. Such significant increase of recharge of the D_3t aquifer in winters caused a higher level of this aquifer during the whole year (Figure 21-5). On the other hand, the increase of resources in the D_3t aquifer was the main reason for the increased gypsum chemical denudation by 30% (Figure 21-6). This caused formation of sinkholes that destroyed some buildings and communications.

So it can be concluded that during the last two decades of the 20th century sinkhole formation has become more active (Taminskas, & Marcinkevičius, 2002. The most likely explanation of intensification of sinkhole formation is the influence of the climate change on karstic processes. At present there is no reason to suppose that anthropogenic activities were not significantly involved in the intensity of karstic process. However the formation of sinkholes, that became more active during the last two decades of the 20th century, was mainly determined by the climate change.

3.   CLIMATE CHANGE AND PERMAFROST

Within the last century climatic fluctuations have caused changes in some of the characteristics of permafrost, with numerous studies showing the impacts of 20th century climatic warming that have been registered in in many regions and continents. These impacts include a rise of the active layer and permafrost temperatures, an increase in the depths of seasonal thaw and a reduction in the depths of seasonal freezing, losses of permafrost at both zonal and local scales, a decrease in permafrost thickness, local changes in the vertical structure of permafrost and a boost to some cryogenic geological processes. A summary of such studies is given in Chapter 28. Here, as an example, we discuss in more detail the European Russian Northeast. There are several reasons behind the choice of this region. Firstly, most of the region belongs to the fairly dynamic southern geocryological zone, which actually represents a transition from the zone of continuous and relatively stable low-temperature permafrost to the zone of annually thawed grounds. Secondly, the very high humidity, characteristic of the region under discussion, makes the permafrost even more dynamic. In such conditions, the effects of permafrost on ecological and urban systems are especially powerful. Thirdly, continuous and comprehensive monitoring of many characteristics of permafrost and adjacent natural media influencing it, has been conducted in the region for the last 35 years. The monitoring records are among the longest available from the world's permafrost regions and are, therefore, very informative.

The study area is a basin of the Usa River, the largest tributary of the Pechora River. The basin is more than 93 thousand km^2 in area and is located on a border of the Pechora lowland and the Ural Mountains. Most of the area is a taiga-tundra ecotone. The most complete permafrost data were obtained in the Vorkuta area, in particular at the Sub-Polar Tundra field station belonging to the Polyarnouralgeologia stock company. The station is located in the Pechora lowland in the zone of discontinuous permafrost.

Monitoring conducted at the station was headed by N. Oberman, V. Yakovlev, N. Kakunov.

The discussion presented below is mostly based on long-term monitoring records of cryolithozone temperatures, active/seasonally frozen layer depths, snow depths, seasonal and long-term frost heave and thermokarst settling of the ground surface, ground water levels and other characteristics of the environment. Temperature observations were conducted with the use of mercury thermometers with a scale factor from 0.05 to 0.1°C, in rare cases 0.2°C. The thermometers were put in cases filled with an inert (freezing by considerably lower temperature, compare to the temperature of permafrost) material, for example, grease. Five weather stations within the basin limits have sufficiently long records of air and seasonally frozen ground layer temperatures to be included in the analysis. Values of thermokarst settling and frost heave were determined by repeated instrumental levelling of special geodesic markers: from 36 to 48 such markers were established at each of 18 sites representing typical landforms and microrelief.

In the 1940s a change of phase in a century-long natural climatic cycle took place in the Arctic, warming giving way to cooling (Gruza & Rankova, 1980). In the European Russian Northeast, quarter-century periods of cooling and warming were overlaid on the background of this longer-term cooling trend (Oberman & Kakunov, 2002). An example from the Vorkuta weather station is given in Figure 21-7. Among these quarter-century periods, only the warming occurring in 1970-1995 can be correlated with continuous ground temperature records available for major physiographic divisions of the region. An increase in ground temperatures in the region had been previously reported (Oberman, 1996; Kakunov & Pavlov, 1997; Oberman & Yudina, 2000; Oberman & Mazhitova, 2001; Pavlov et al., 2002). The most detailed characterization of the "permafrost warming" is given in the 2001 paper, which is widely cited in further discussion.

In order to trace the dynamics of the active layer and seasonally frozen layer during a study of a 25-year warming period, Oberman & Mazhitova (2001) subdivided the period into three sub-periods and then compared mean values of the active layer depth for each of them. The comparison shows that the active layer depth increased during the warming period and the depth of seasonally frozen layer decreased. The main reason for this was a considerable increase in mean annual air temperature. However, it was not the only reason. Less powerful, yet significant controls over active layer characteristics are atmospheric precipitation and snow pattern, the latter depending on winds and topography. The effects of these controls on the active layer characteristics depends directly on the thermal properties of the ground and the hydrogeological features of an area. For example, in the same area, under practically equal long-term dynamics of maximum snow depths, the average increase in the active layer depth during the final sub-period (by comparison with the initial sub-period) was 24 cm in loams and only 8 cm in peats. The average decline in the seasonally frozen layer depth during the same final sub-period, as compared with the initial sub-period, varied from 21 to 51 cm depending on the controls mentioned above.

Data on recent temperature changes at the bottom of the active/seasonally frozen layer were also discussed in Oberman & Mazhitova (2001) by reference to an average depth of 2 m in loam, and 1 m in peat. The largest positive temperature increments were observed at the specified depths during the final third of the warming period. In pediment plains and piedmonts, where the seasonally frozen layer is underlain by fissures and karst-affected rocks, the largest temperature increments were 1.1 to 1.2°C. The increments in both frozen and annually thawed ground of glacial-marine plains were 0.7 to 1.0°C respectively. The minimal increments, from 0.2 to 0.7°C, were observed under permanent or temporary water bodies. Under lakes the increases were larger with decreasing lake depth, i.e., where the effects of the water body were weaker.

Warming penetrated below the active/seasonally frozen layer and encompassed the whole layer of annual temperature fluctuations. Analysis of Table 21-3 shows that temperature increments at a depth of zero annual amplitude during the period of warming vary greatly depending on physiographic divisions of the area and landforms within the limits of these divisions. This dependency manifests itself differently in permafrost and in thawed grounds. The lower the initial permafrost temperature, the larger the temperature increases during the period of warming. This is in agreement with a theoretical proposition made by Ershov et al. (1996) suggesting that the lower an initial permafrost temperature, the higher the speed of its increase under conditions of warming, as a result of reduced heat consumption by phase transitions of water. Degrees of warming of thawed ground are affected significantly by infiltrating atmospheric precipitation: a level pre-watershed part of a slope serving as a ground water alimentation supply area shows an increase in ground temperature that is twice that observed in the slope of a basin to a valley (Table 21-3; 5 and 4). Comparison of lower slopes composed of solid calcareous rocks and of loamy deposits (Table 21-3; 7 and 4) shows similar relationships: the former receive from 2.5 to 5 times more atmospheric heat due to water infiltration (according to our data). Thawed watersheds in the glacial-marine plain are warmed approximately 1.5 times more than frozen watersheds; in practice, the latter do not receive atmospheric heat (Table 21-3; 5 and 1).

Table 21-3. Changes in mean annual ground temperature in northeastern Europe during recent climatic warming (1970-1995)(according to Oberman and Mazhitova, 2001, with our supplements. Temperature at a depth of zero annual amplitude – geocriological term, meaning the temperature of occurrence of base of layer with annual temperature fluctuations)

Period of time over which changes were assessed	№	Physiographic divisions and landforms	Lithological composition of the layer of annual temperature fluctuations	Temperature at a depth of zero annual amplitude, °C	Period of time over which changes were assessed
				initial	increment by the end of period
1974-1995		Glacial-marine plain:			
	1	convex watershed	Loam	-1,15	0,7
	2	lower part of slope (residual-polygonal relief)	Loam (3,5-4,8 m – sand)	-1,75	1,0
	3	peatland	0-1,7 m, peat; deeper loam	-2,7	1,2
1970-1996	4	lower part of slope (local depression)	Loam	0,1	0,65
	5	concave watershed	Loam	0,5	1,15
1969-1996		Pediment plain:			
	6	watershed	0-3,0 m, loam, deeper terrigenous rocks	0,6	1,0
		Piedmont:			
	7	foot slope	0-3,0 m, loam, deeper calcareous rocks	0,25	1,15

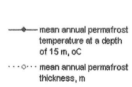

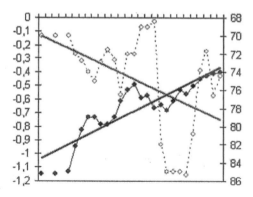

─◆─ mean annual permafrost
temperature at a depth
of 15 m, oC

···○··· mean annual permafrost
thickness, m

Figure 21-7. Correlation of changes in permafrost temperature and thickness with climatic
cycles (European Russian Northeast, Vorkuta area, borehole VK-1615)

Chronological graphs typical for the period of warming and showing the
dynamics of the most important permafrost characteristics, temperature and
thickness, are presented in Figure 21-7. Permafrost temperatures at a depth
of zero annual amplitude in directed similarly with that in air temperatures.
However, the trend in permafrost thickness, as registered in the same
borehole, is in the opposite direction. The reason is the relatively deep
position of the base of the permafrost, at a depth of about 70 m. Heat
penetration from the surface down to this depth requires a long period of
time. The trend of increasing permafrost thickness observed during the
period of warming apparently is inherited from the climatic cooling, which
occurred during the preceding quarter-century period (Figure 21-7).

It is well known that an increase in permafrost temperature leads to a
decline in ice content and an increase in liquid water content in the
permafrost. It is also well known that such a phase transition is connected
with a reduction in ground volume. The latter, in turn, includes a reduction
in pore volume under the effect of the surrounding and overlaying ground
weight. This should result in thermokarst settling of the soil surface, which,
indeed, is registered in almost all landforms (Figures 21-8 and 21-9).
Changes in mean annual height of the soil surface are relatively well
correlated with inter-annual variations in air temperature. An increase in the
air temperature was associated with thermokarst settling of the soil surface,
whereas the following cooling caused stabilization of height levels and, in
some cases, a relatively insignificant increase as a result of renewed freezing
accompanied by frost heave (Figure 21-8).

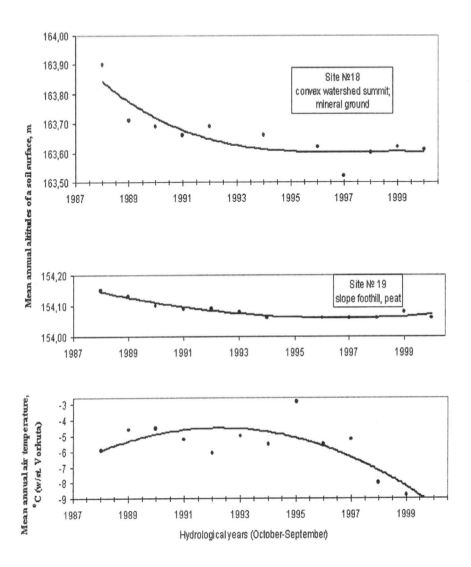

Figure 21-8. Long-term trend in the thermokarst surface settling at the end of the period of climatic warming (anchored permafrost)

Thermokarst settling is also promoted by atmospheric precipitation. However, correlation between the two is not as explicit as that between settling and air temperatures. As Figure 21-9 shows, surface settling occurred against a certain decline in atmospheric precipitation and, thus, against progressively smaller amounts of heat transferred by the infiltrating precipitation into the ground. It might be expected that thermokarst settling would weaken or completely cease in such conditions, but this is not the case. This paradox can probably be explained by a change in the length of time over which infiltrating precipitation fills up ground water stocks. An increase in the duration of this period under climatic warming, due to earlier seasonal thaw and later ground refreezing, overcompensated for the insignificant decline in annual precipitation (because of the overall excessive humidity in the area). This overcompensation is confirmed by a general long-term trend of increasing mean annual ground water levels (Figure 21-9). Hence, the thermokarst ground settling developed against a background of increasing amounts of atmospheric heat transferred into the ground by infiltrating water.

Figures 21-8 and 21-9 show that thermokarst settling values differ at different sites. For instance, they are larger at the sites formed by loamy deposits with low or medium ice content, in which thaw progress requires less heat, than in cooler and icier areas (Figure 21-8, sites 18 and 19, respectively). The reason for that is explained below. Thermokarst settling develops not only in the areas of anchored permafrost (lying directly under the active layer), but also in closed taliks (the site of melted beds, spread over permafrost). In the latter, settling results from the thaw penetrating into a frozen talik base. In the source areas of ground water supply, which include the pre-watershed part of a slope, larger amounts of atmospheric precipitation infiltrate into the ground than in transitional areas, such as lower slopes. Therefore, the warming effect of atmospheric precipitation on permafrost and on the processes developing in the permafrost in response to these effects, is significantly stronger in the pre-watershed part of a slope than on slopes. For example, the long-term mean annual surface settling in closed taliks developed in the pre-watershed part of a slope is 3.3 cm per year, whereas it is only 2.05 cm per year in closed taliks developed on lower slopes (Figure 21-9, sites 10 and 15, respectively).

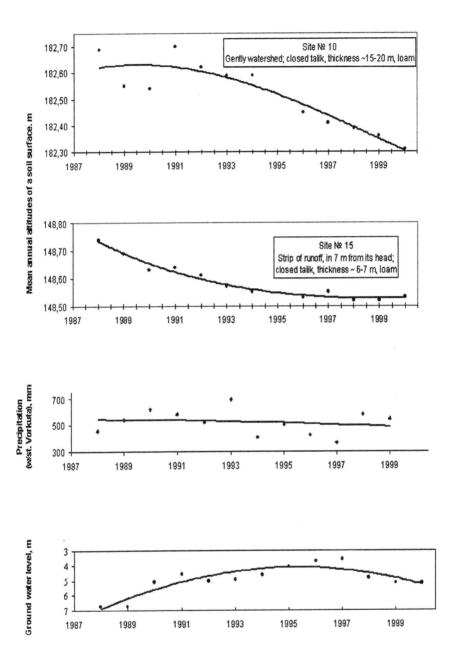

Figure 21-9. Long-term trend in the thermokarst surface settling at the end of the period of climatic warming (closed taliks)

Naturally, the time required for convective and conductive penetration of atmospheric heat down to a permafrost table depends on the table's depth. Accordingly, there are different time lags in permafrost response to changes in soil surface temperature; the response is manifested, in particular, in thermokarst settling and frost heave. For example, the climatic cooling of the late 1990s not only resulted in a slowing down of thermokarst settling in areas with anchored permafrost, but also led to the settling giving way to a weak frost heave (Figure 21-8, sites 19 and 18). Such a change in the dominant cryogenic processes has not yet taken place in shallow closed taliks affected by the same cooling (Figure 21-9, site 15). In deep closed taliks (Figure 21-9, site 10) predominant frost heave gave way to thermokarst settling at the transition from the 1980s to the 1990s, whereas, by that time, the same change already had happened at the sites with a shallower permafrost table position (sites 19, 18 and 15). Deep-table permafrost also is reacting more slowly to the climatic cooling that started at the end of the 1990s: the cooling has not yet affected the intensity of thermokarst settling at site 10, in contrast to the other sites discussed.

Based on the regression of settling against time, settling reaches 25-33 cm in 7-10 years, which corresponds to annual rates of 3.3 to 3.5 cm (Figure 21-8 and 21-9, sites 10 and 18). Such notable settling should significantly affect the stability of industrial and civil infrastructure built on permafrost. Indeed, the period during which civil buildings in Vorkuta demonstrated the highest failure rate (according to G.V.Belotserkovskaya's data) was also a period during which permafrost in the pre-watershed part of a slope and lower slopes with loamy or peaty subsurface layers, demonstrated the largest temperature rise for the whole period of observations. Increase of accident rate was marked only in buildings, constructed on frozen quaternary sediments and didn't exist in buildings, constructed on solid rocks. Similar, indirect impact of changeable cryogenic conditions on humans is also observed because of a rough increase of deformations (thermokarst settling) on the Northern Railway Road (Oberman, 2002). Direct consequence of "softening" of cryogenic conditions became fixed by N.B.Kakunov increase of the height of tundra bushes on the sites of their expansion, not far away from Vorkuta, from 0,3-0,5 till 1,5-2,0 m within last 40 years.

In conclusion, climatic changes are one of the most powerful controls over the permafrost and, indirectly, influencing by this on humans, more directly on vegetation.

PART VI:
NEWEST ECOGEOLOGICAL PROCESSES WITHIN RIVER BASINS

Chapter 22

ECOSYSTEM MONITORING UNDER DESERTIFICATION WITHIN INTERIOR SEA-LAKES AND DELTAS

NINA M. NOVIKOVA, OLGA A. ALDYAKOVA
Water Problems Institute of Russian Academy of Sciences

1. INTRODUCTION

Mapping of ecosystems and landscape dynamics is an important part of remote monitoring of the environment aimed at identifying and evaluating changes taking place in an area over a particular period of time. Maps reflecting the dynamics of ecosystems and landscapes in the Amudarya delta are urgently required to identify measures to deal with adverse phenomena and processes caused by regression of the Aral Sea and reduction of stream runoff into that sea. It is known that such changes began half a century ago. Since then problems related to natural and anthropogenic factors have been discussed in many publications. Amongst them there are several monographs devoted to the environment state at the time of our survey. These identify factors responsible for destabilization of the environment and predicting further changes (Raficov, & Tetuchin,1981; Popov, 1990; Glazovskyi,1990; Zaletaev, & Novikova,1997; Creeping..., 1999).

2.   INVESTIGATION METHODS

At present, due to the unfavorable ecological situation in the region of South Prearalye, the problem of rational environment conservation is becoming very acute. Therefore, one primarily needs objective and comprehensive cartographic information on the recent spatial structure of landscapes and their dynamics both in the area of the Amudarya delta and in the now dry bed of the Aral Sea. Given the rapidly changing conditions in the area, information from aerospace surveys undertaken at different times is a basic source for compiling the maps.

To compile the map of ecosystems and landscape dynamics at scale 1: 500 000 we used:

- digital multizonal pictures with resolution of 28 m, made by satellite "Landsat" 7 ET on July 2000;
- coloured images synthesized from 2, 3 and 4 zones of the aerospace survey;
- synthesized images with resolution of 30 m, supplied by Landsat TM in 1989, 1994, 1996, 1999;
- a picture with resolution of 5 m, made from satellite "Resurs" in 1994;
- a landscape map compiled by one of the authors (1) showing the state of the area in the 1980s, and maps of desertification, compiled by the authors in the 1980s.

When compiling the map of landscape dynamics, successive drafts at 1:500 000 and 1:200000 were prepared for visual identification of natural complexes. Literature and cartographic sources as well as original data of field surveys were generalized to obtain a more reliable interpretation of the remote sensing information. The ecological-geographical database prepared in the Laboratory of Terrestrial Ecosystem Dynamics of the Water Problems Institute of the Russian Academy of Sciences, included data from routine field surveys carried out between 1979-2001. These helped in retrospective analysis of remote sensing images.

The synthesized images at a scale of 1: 200 000 gave wide possibilities for interpreting the natural situation in detail. Color combined with the image pattern, and analysis of correlations between the subjects of the research, made it possible to observe all the changes that had taken place in ecosystems and landscapes, and to study their dynamics, during 1991-2000.

The objectives were to:

- establish trends in the development of every landscape element in the area under study for the period of our survey;
- specify the development and distribution of exogenic processes characteristic of every landscape element at the given stage of its development;

- identify sites potentially subject to ecologically dangerous and adverse processes.

In this context, a landscape map was compiled for the study area, using ArcView, to show the state of the landscape state in 2000, including hydrography, soils, ground waters, vegetation and exogenic processes.

The key feature of recent landscape development within the area of Amudarya delta is that drying out has been almost completed. All sites, where additional moistening is absent, are now at a stage of automorphic development.

Further landscape evolution proceeds according to natural laws towards the formation of climax and quasi-climax ecosystems. Because the landscapes in the delta have developed under desert zone conditions, water deficiency seems to be a major factor responsible for the dynamics of natural complexes. Therefore landscape evolution proceeds towards the formation of desert complexes.

It is evident from results of investigations undertaken over many years (Novikova, 1985; Novikova et al., 1998, 2001) that the transformation of hydromorphic landscapes into desert zonal ones has three stages, characterized by declining and changing the water sources. At the initial hydromorphic stage of landscape development the groundwater table is close to the surface and is a source for water accumulation together with precipitation. The semi-hydromorphic stage is characterized by atmospheric moistening, at the expense of deep ground waters. In the final automorphic stage the moistening of landscapes takes place only through atmospheric precipitation. Results obtained in our long-term experiments allowed us to determine threshold values of the groundwater depth for every stage of landscape development. These are as follows (Table 22-1):

- hydromorphic stage – the groundwater depth varies from 0 to 1.5 m, periodically increasing to 3 m;
- semi-hydromorphic stage – the groundwater depth is 3-5 m;
- automorphic stage – the groundwater depth reaches more than 5 m.

In the course of desertification each landscape element has a distinctive development because of patterns of changing in moistening at each stage. These determine the dynamics. Different exogenic processes affect the different surface sediments and therefore lead to distinctive plant communities (Table 22-1).

Because of the close relationships between vegetation and the other components of each landscape element, plant communities and even species could be used as indicators for each stage of landscape development (Table 22-1). Such desert plant species as *Krasheninnikova ewersmanniana, Anabasis aphylla,* and *Haloxylon aphyllum* served as indicators for the initial stage of automorphic development. At present, one can observe rapid growth of these species within the landscapes of natural levees in the delta.

The different hues of vegetation cover, in coloured synthesized images obtained by multizonal aerospace surveys, assisted in reliable interpretation of the remote information.

Since water is responsible for landscape dynamics in the region under consideration, preliminary research had the aim of studying changes in water supply for the period 1991 to 2000. This period of time was taken by the authors for the study of landscape dynamics, changes in spatial structure of landscapes, and processes that took place within them. For this purpose, fluctuations in precipitation and temperature (the mean monthly values for a year and during the vegetation period), influx of river water in the delta, and changes in water salinity in the river and other water bodies were thoroughly studied. Based upon a comprehensive analysis of diagrams showing the relationships between these indices through time, "years with much and with little water" were identified. Initial aerospace information and data from field surveys obtained in different years were used to determine changes in surface water over those these years. It was possible to identify sites, characterized by permanent or periodical flooding (from a year to the whole period of the research). Sites subject to permanent or periodical underflooding due to near-surface groundwater were also identified. In addition to open water and vegetation cover, the patterns and the colours of the aerospace images clearly showed development stages of exogenic processes including aeolian and solonchakous (water dissolved salt accumulation within the soil). These could be directly decoded. However indirect features were used to decode processes of suffusion and alluvium accumulation. These were later verified by special field surveys.

The study of changes in the depth and salinity of ground waters, using hydrogeological data, made it possible to obtain objective information on landscape dynamics in the river delta and in the dried part of the sea bed, thus verifying the results of aerospace surveys. These data alone permitted positioningof all landscape elements in the general dynamic systems of the territory under study, as illustrated in the a block diagram in Figure 22-1.

Table 22-1. Stages of the landscape dynamics, basic exogenic processes and of ecosystems

Components of the environment	Stages
	Hydromorphic
Nature of filling with water	Flood waters, ground waters, atmospheric
Ground waters	moistening
Soil-forming processes	0 - 3,5 m.; fresh-weakly saline
Phytomass reserve, t/ha	Swampy, meadow with seasonal salinization
Yield, t/ha/year	75.46 - 41.27
	10.7 - 36.87
Desertification forms:	Dominant plant communities on natural levees
1) decrease in water supplyj	Populus ariana, P. diversifolia, Elaeagnus
2) salinization	turcomanica, Halimodendron halodendron,
2) salinization	Glycyrrchiza glabra, Calamagrostis dubia
3) desalinization, takyr formation	Tamarix sp.sp.
4) sand accumulation: a) aeolian	
blowing	
b) destroying the upper compacted	
loamy layers	
5) clay formation (aeolian deposition	
into depressions between hillocks)	
Desertification forms:	Dominant plant communities in interchannel depressions
1) decline in filling with water	Typha angustifolia, Phragmites australis,
2) salinization	Limonium gmelini, Tamarix sp.
3) desalinization, takyr formation	
4) sand accumulation	

Stages	
Semihydromorphic	Automorphic
Ground waters, atmospheric water	Atmospheric water
1,5-3 (5) м.; weakly saline, saline	>5 (10-20) m.; saline
Solonchakous, takyr*-like	Zonal desert
49. 1 - 8.25	27.05 - 5.29
25.1 - 3.05	2.18 - 1.89
Dominant plant communities on natural levees	
Populus ariana, Elaeagnus turcomanica, Tamarix sp., Alhagi pseudalhagi, Aeluropus littoralis Tamarix hispida, Karelinia caspia, Climacoptera aralensis, C. Lanata Anabasis aphylla, Haloxylon aphyllum a) phytogenic hillocks of pseudosands (fine earth with salt crystals) around shrubs, b)Krascheninnikovia ceratoides (after tugai death);	a) Haloxylon aphyllum, Salsola richteri, Calligonum sp. (after solonchakous stage); b) plant communities Haloxylon persicum Ephemeric-semi-shrubby communitiesa (Artemisia sp.sp.)
Dominant plant communities in interchannel depressions	
By strong decrease in filling with water – tanatocoenoses Tamarix hispida, Halostachys belangeriana Haloxylon aphyllum, Salsola dendroides – takyr phytogenic hillocks of pseudosands (fine earth with salt crystals)	- - Anabasis salsa+ ephemers, communities Haloxylon persicum

* Takyr – type of soil, where water dissolved salts are absent only at the upper layer 10-30 cm of profile. Takyr-like soils are formed on sandy deposits.

Thus were methods developed for studying and mapping the dynamics of ecosystems and landscapes by comparing the landscape contours from aerospace pictures made in different years.

3. STRUCTURE OF THE MAP

The map of ecosystems and landscape dynamics in the area of the Amudarya delta was compiled in two versions: electronic form and paper. In the electronic version every characteristic is represented by independent layers as follows:

- change in dynamic positions of ecosystems and landscapes during 1991-2000 as changing the system of their natural development: aquatic > subaerial development: hydromorphic stage > semihydromorphic stage > automorphic stage >zonal ecosystems and landscapes;
- the nature of ecosystems and landscape development: stable and unstable areas, the latter being distinguished among those, which are striving for stabilization (Figure 22-1, N3), having a tendency to destabilization (Figure 22-1, N4, 5) or fluctuating, i.e. developing under conditions of variable moistening (Figure 22-1, NN 6-8);
- predominant exogenic processes and the resulting forms at a given stage of landscape development;
- ecologically dangerous areas where the stage of development or concentration of exogenic processes is a potential threat for human activities and health.

4. TRENDS IN ECOSYSTEMS AND LANDSCAPES DYNAMICS

Figure 22-1 illustrates basic trends in the development of landscapes within the coastal part of Amudarya delta and the dried bed of the Aral Sea (Figure 22-2). There are two groups of landscapes: zonal landscapes on plateaus; and residual-mountains. The development of these is strongly affected by zonal climatic conditions, and the aquatic landscapes of water bodies that remain stable thanks to artificial filling with water. U*nstable ecosystems and landscapes* are classified into *developing and fluctuating* ones. The first of these reveals a decrease in moistening due to increasing groundwater depth and xerophytism of soils and vegetation. The second are periodically filled with water in some years.

Unstable ecosystems and landscapes can developed towards stabilization or further destabilization. One group of such ecosystems and landscapes is

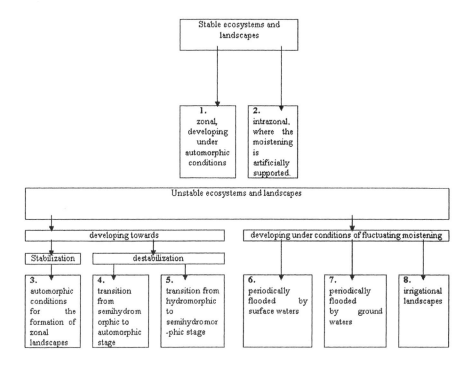

Figure 22-1. A schematic drawing (legend to the map) of the nature of, and trends in, ecosystem and landscape development within parts of the Amudarya delta and the dried bed of the Aral Sea for the period from 1991 to 2000

confined to natural levees of large distributaries. They remain at the stage of hydromorphic development for a long time. The groundwater table reaches 5-10 m below the surface and soils are represented by weakly developed takyr-like soils. In recent years desert plant communities have prevailed and, as a result, the cover of vegetation and its productivity are constantly increasing.

Landscapes tending towards destabilization are widespread within the dried sea bed. For the last 10 years the groundwater depth has increased, and surface horizons have become very dry, resulting in deflation of sand sediments and salt accumulation. These processes are responsible for environment destabilization.

Processes of suffusion and alluvium accumulation have been identified from indirect features in the landscapes, and verified by field survey.

Thus, the study and mapping of landscape dynamics was based upon the comparison of landscape contours in aerospace pictures obtained in different years and characterized by similarities in patterns of filling with water.

5.   EXOGENIC PROCESSES

Exogenic processes are reflected, as independent characteristics of recent dynamic processes in landscapes, in separate map layers. The legend for this layer is given in Table 22-2.

Table 22-2. A scheme, showing the nature, and trends in the development of, exogenic processes within the landscapes of the Amudarya delta and the dried bed of the Aral Sea for the period from 1991 to 2000.

Index at the map	Type of exogenic process	Landscape conditions, conducive to the development of processes
Processes, caused by surface waters		
П	Sheet erosion	In plateau and slopes of uplands with residual hills, derived from clays and gypsum and covered by zonal undershrub vegetation
Э	Linear erosion	In slopes of uplands with residual hills, derived from clays and gypsum under sparse vegetation
Р	River accumulation	In major branches of the river with loam-sandy and loamy sediments under shrub vegetation
Х	Hemogenic-lacustrine accumulation	Lacustrine basins with silt, loam-sandy and loamy sediments under reeds
Processes, caused by surface and ground waters		
С	Salinization of the dried sea bottom and lacustrine kettle depressions	Silt-loamy sediments of marine plains, gulfs and coastal lakes with the groundwater level at a depth of 1 m. Reed and tamarisk communities
З	Salinization of intra-channel depressions in the delta	In intra-channel depressions, derived from loam-sandy and loamy sediments with periodical under flooding and the groundwater level at a depth of 3 м. Weed-grass-halophylic-tamarisk communities.
Processes, caused by ground waters		
Ф	Solution sinkholes	In inner deltas and outer deltas, in natural levees under lost young woody tugai
Processes, caused by wind		
Д	Deflation	In dried marine gulf and coastal lakes, derived from saline sandy sediments without vegetation
В	Over-blowing	In sandy deposits of uplands with residual hills with zonal ecosystems and in sandy sediments of the bare marine plain with the groundwater level at a depth of lower than 3 m

A	Accumulation	Sand accumulation in clayey marine plain with sparse grass-shrub psammophylic communities
Processes, controlled by human activities		
H	Agro-irrigational sediments and soil salinization	Irrigated fields

Ecological danger within the area was evaluated according to the development of potentially destructive exogenic processes such as as aeolian and solonchakous ones, using the data presented in Table 22-3. The drying sea bed becomes enriched with fine earth and salts, blown by wind into the river delta in particular. The most dangerous areas, in terms of ever increasing salinity, are where groundwaters are at a depth of 0-3 m. The salts accumulate intensively on the soil surface and are an obstacle to revegetation, thus leading to blow-away during the dust storms. According to calculations made by I.P. Gerasimov in 1967 (Gerasimov et al., 1983), in the dried area of the sea bed, the salt accumulation would reach 2.4 mln/tons/year in 1990, being increased to 4.1 mln/tons/year in 1998 (Mavljanov et al., 1998). Dried areas, located between the isolines showing the coast in 1990 and 2000, proved to be ecologically dangerous due to the formation of aggressive solonchaks, that were scarcely amenable to natural amelioration. Also ecologically dangerous were areas suffering from the formation of blown sand and secondary salinization.

Table 22-3. Evaluation of ecologically dangerous developments of destructive exogenic processes in landscapes

Index at the map	Degree of danger	Remarks
0	No danger of erosion processes	In lakes, danger of pesticide accumulation from waters of collector-drainage runoff
1	Slight danger of salinization and deflation	In landscapes, periodically flooded by river waters
2	Moderate danger of salt accumulation on the soil surface and sand overblowing on solonchak and loamy surfaces, suffusion	In landscapes of the dried sea bottom and takyr-like solonchaks
3	Strong danger of Aeolian sand accumulation and formation of traveling dunes and their deflation	These landscapes are the sources for blowing away of salt and dust.

With increasing groundwater depth, capillary salt accumulation occurs in the upper soil horizons. In view of this, the landscapes, characterized by groundwater table at a depth of 3-5 m, are recognized as potentially dangerous, because the ground water salinity reaches up 47-90 g/l. Amongst potentially dangerous areas are those located in the sea bed between the coasts of 1960 and 1990, where the influence of surface waters from the delta are absent.

Processes of salinization limited development in delta sites periodically filled with water, and close to lakes. In view of the complete drying out of the delta, the groundwater depths have fallen below 5 m and local salinization does not present a serious ecological danger.

Salinization takes place in the irrigation zone and leads to the formation of solonchaks. These areas are shown in the map as ecologically dangerous. With the help of distinctive features (the absence of field boundaries a dark phototone , or locations within an oasis) these areas can be clearly identified from remote sensing images.

Aeolian processes are also widespread in the dried sea bed. These are conducive to the formation of a corrugated relief with deflation hollows in the sites with sandy and loam-sandy deposits, and by enveloping beds of sand on solonchaks or clay and loamy surfaces. In case of increasing groundwater depth, the over-blowing processes take place in the area of dried sea bed. They are active especially in warm periods and may be clearly identified from synthesized aerospace pictures. Landscapes of beach-ridges formed since the 1960s are a source of intensive sand blowing and dust storms. In the map such landscapes are also shown as ecologically dangerous. In the sea bed the intensive aeolian processes lead to sand accumulation and the formation of travelling dunes and barchan ridges of 5-10 m in height. These areas are bare, difficult to traverse and suffer from dust storms. These ecologically dangerous landscapes are reflected in the map as well. It was very difficult to identify these areas directly in aerospace pictures so additional features such as landscape lithology, groundwater depth, and cover of vegetation are used.

When preparing the paper version of the map (Figure 22-2), the first three layers are combined. Colours correspond to the nature of dynamic trends (Figure 22-1). Russian letters are used to indicate accompanying exogenic processes (Table 22-2) and an additional Arabic figures indicate ecologically dangerous processes (Table 22-3).

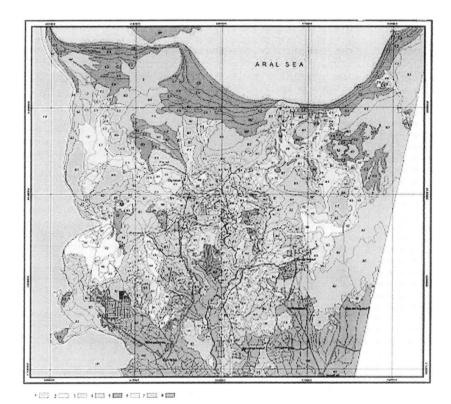

Figure 22-2. Schematic map to show the ecosystems and landscape dynamics of the Amudarya delta and the dried bed of the Aral Sea, 1: 500 000 M. The authors: N.M. Novikova, O.A. Aldyakova.

6.   CONCLUSION

A map showing landscape dynamics in the Amudarya delta and the dried sea bed for the period 1991-2000 has been compiled to provide an acceptable inventory and assessment of the nature and trends in landscape dynamics as well as the development of ecologically dangerous processes.

In preparing this map, GIS-technologies were used to reflect different processes and phenomena in the form of several interrelated layers that supplement each other. Legends provide the basis for inventory and assessment maps. The GIS facility makes it possible to monitor the development of processes and trends in landscape dynamics and to incorporate new data without additional conventional mapping.

At present , the most dangerous processes in the study area are   salt accumulation on the surface, and aeolian accumulation of sands, because these are a source of sand blowing into the oases during the dust storms. The compiled maps identify areas suffering from these processes and measures for improving them with the aim at eliminating adverse effects on adjacent territories.

Chapter 23

ECOSYSTEMS FORMING ON THE FRESH RIVER DEPOSITION

T. BALYUK[1], J.H. VAN DEN BERG[2]
[1]Water Problems Institute, Russian Academy of Sciences, [2] Faculty of Geosciences, Utrecht University, the Netherlands

1.   INTRODUCTION

The formation of islands and point bars in river systems leads to dynamic ecological changes that are very important for management of riparian areas.

There is a growing demand from society to restore natural characteristics of alluvial rivers.   However, in most large alluvial rivers in the western world, natural ecological properties have virtually disappeared, sometimes already for a long period of time. Therefore, there is a need of knowledge of natural ecosystems of comparable rivers elsewhere in order to be able to identify and restore some of the original natural conditions of morphodynamics, biodiversity and vegetation dynamics in modified rivers. The present analysis is restricted to the most active sedimentary unit of the river, the inner bend or pointbar. This implies that the vegetation succession is studied until the first stages of softwood forests have developed, and that vegetation in more remote overbank areas is not considered.

The analysis is based on a combination of spatial data from aerial photos and satellite images and field observations of morphological features and vegetation in a number of cross-sections over pointbars. As flooding is important for the development of morphology and vegetation over time on pointbars, hydrologic data is also important.

In the field, cross-sections were measured with a leveller, and located by means of GPS. Along these profiles the occurrence of plants and plant associations were determined and observations were made of bed material. Ages of trees were determined by counting tree rings in cores made at right angles to the stem. The latter information is especially valuable in the case of *Populus nigra, Salix alba, S.viminalis. Populus nigra* is important because these trees begin to grow only on freshly deposited sediment and therefore are a proxy for the age of morphodynamic events.

2.   MORPHOLOGY, MORPHODYNAMICS AND VEGETATION

All pointbars of European rivers have similar depositional and erosional morphological divisions. Accretional units are formed by scroll-bars and chute bars. Erosional units are represented by chute channels and erosional banks of outer meander bends. A general picture of these features and their dynamic change in time is given in Figure 23-1.

Outer bend erosion is followed by the formation of a scroll bar in the downstream part of the inner bend. After a number of years the morphology of a scroll bar is stabilized and a new scroll bar may start to form. During high discharge events part of the flow cross-cuts the pointbar and one, or several, chute channels may be formed, ending in chute bars. Chute bars generally form half-way across the pointbar, but sometimes one large chute channel cross-cuts the pointbar almost completely, eventually ending in a horse-shoe shaped, large chute bar that partly protrudes into the main channel.

Fundamentally, processes of outer bend erosion, inner bend scroll bar accretion and development of chute channel and chute bar morphology in European river systems are comparable.

3.   HYDRODYNAMICS AND VEGETATION

3.1   Flora

Comparative analysis of dominant families of plants for most rivers in Europe shows the same trend. The dominant families are *Asteraceae ,Poaceae, Cyperaceae, Fabaceae.* This is explained by the fact that these families contain many plants typical of the azonal characteristics of floodplain vegetation and a similarity of floral complexes.

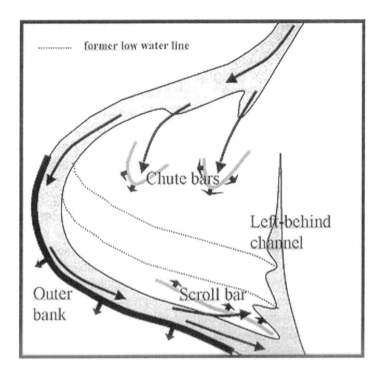

Figure 23-1. Sketch of the morphology of a pointbar (only active bars are indicated)

Communities are more diverse on the point bars of the gravel rivers, because of the larger variation of habitat, caused by greater sediment variability. Also, frequent fluctuations of water level and the absence of long periods of inundation, which are lethal for many species in major floodplains, is very important. Greater habitat variation favours diversity. A third factor that contributes to more diverse vegetation in rivers close to mountain ranges is the delivery of seeds from proximal mountainous habitats. Generally on point bars the vegetation typical of moist habitats is found, consisting of *Eleocharis palustris* and *Phalaroides arundinacea* in swamps, and aquatic vegetation including *Lemna minor, Potamogeton crispus* in ponds and standing water, together with the pioneer community *Eragrostis pilosa* + *Amaranthus alba.* The meadows dominated by *Elytrigia repens* are very similar. *Populus nigra* and *Salix alba* are generally the dominant trees in floodplain forests of most areas. Habitats of similar humidity in most areas have the same plant community or ecological analogs.

3.2   Vegetation

The vegetation on pointbars of the river plains shows many similarities. In all studied areas the communities and their dominant species show clearly the mesophytic, azonal character of the vegetation. However, there are also some clear differences related to:

- hydrodynamics (frequency and duration of flooding, flow energy),
- morphodynamics (sedimentation and erosion), and
- physical conditions (sediment characteristics and climate).

The longer duration of floods in some rivers explains the higher percentage of mesophytic plants on the point bars.

A special adaptation to the long xerophytic period following flooding is shown by *Bolboschoenus maritimus*. It forms special "tubers" on its roots, for vegetative reproduction. During the xerophytic period this plant forms "meadows" of completely dry plants.

In case of very dynamic rivers, floods change the phytocenosis of the environment very quickly. Meadows may be transformed to bare surfaces and plants may be taken away by the flow, or buried by a layer of sediment. Tree and shrub vegetation is removed during bank erosion. A sketch of the rejuvenation process in pointbars is presented in Figure 23-2.

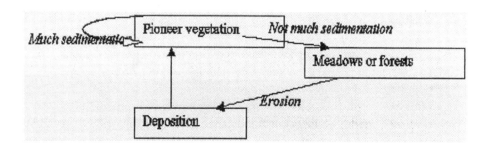

Figure 23-2. Simplified scheme of floodplain rejuvenation

Every year the flood water brings a number of seeds of Populus nigra, Salix alba, and S. purpurea to the point bar. These seeds are deposited at the water-line and their germination gives rise to rows of trees that follow the original curve of the pointbar. However the seedlings of these trees can only survive if between two high floods, enough time is available for the production of leaves and start of photosynthesis. If the time between succeeding floods is too short, no rows of soft wood forest develop and instead herbaceous pioneer vegetation will develop. A long duration of flooding may limit the growth of *Populus*: This probably explains why this tree is only found at a minimum height of 3.5 m above the low water level in case of Lower Volga point bars (Russia).

The structure of pioneer vegetation on pointbars of gravel rivers is more complex than sandy rivers due to the variable composition of the bed material. In the case of Allier, Loire rivers (France) the upstream coarse gravel part of the pointbar is generally almost without plant cover, or with some *Populus* or *Salix* seedlings. In the central part of the pointbar, with mixture of gravel and sand, communities with *Corrigiola litoralis* and *Portulaca oleracea* appear. On patches of gravel covered with a thin layer of clay the community of *Lindernia dubia* is found. On sandy areas, normally found in the downstream part of the pointbar, communities with *Eragrostis pilosa, Xanthium orientale, Datura stramonium* and *Amaranthus album* are present. The pioneer communities exist only for one summer season. During the next flood they are either covered by fresh alluvium, moved away with the flow, or develop into the next stage of succession.

The most dramatic effect of sediment transport on vegetation is rapid deposition of bed material or suspended material and erosion of river banks. Most herbs and other low plants will die if covered by a layer of several decimeters of sand or gravel. These layers prevent a continued succession of pioneer vegetation, and stop the development of perennial vegetation. Only *Salix* and *Populus* trees are able to survive, unless they become completely covered by the sediment. In a relative sense, the process of construction and erosion of pointbars in the Volga proceeds more slowly. Consequently, vegetation succession may proceed to higher levels before it is re-set by outer bank erosion. On the other hand, rapid deposition of sand from suspension on pointbars may inhibit any succession for a long period of time.

As rapid sedimentation on chute and scroll bars fronts prohibits the development of perennial vegetation, these areas form a specific habitat for one-year pioneer communities, which develop a limited group of about 20 plants that are adapted to these special conditions:

- Except for winter conditions, many of these plants are able to germinate and grow quickly directly after the passage of the river flood on the still humid substrate (*Corrigiola, Portulaca, Lindernia, Eragrostis*).
- The time for flowering always occurs after high floodwater, at the end of summer.
- Seeds have special adaptations for floating in water It is one of the major dispersal mechanisms for plants along rivers. Seeds have various means to float effectively, such as large volume, juicy or fleshy outer layers, and various appendages.

- During high water, leaves get enough light for continuing photosynthesis.
- The epidermis of the leaves is covered with a special wax, for saving water inside during a dry period.
- Pollen are transported and delivered to the receiving flowers by water.
- Due to their morphology and anatomy *Corrigiola, Portulaca, Polygonum* are able to reduce evaporation of the soil.

Pioneer trees and shrubs have resting buds on stems, and roots from it start to grow when buried by sediment.

Plants like *Polygonum scabrum* and *P. minus* have flowers on top of stems that remain above the water enabling insects to reach them.

4.   VEGETATION SUCCESSIONS

Based on the ages of trees, general remarks on changing characteristics of community cover, abundance, and number of species over time was made and succession graphs for the studied river plains were prepared.

The increase of all these characteristics commences after five years of age, at a distance of about 50 m from the low-water mark in the main channel. At this point the habitat escapes from intensive flooding. The succession enters into a stage of greater stability after 20 years, when young soils appear and forests become mature – at a distance of 200 m from the main channel and higher than 2 m above water level. In bars and channells made by chutes the succession is "reset" to pioneer stages, followed by wetland communities with plants like *Lythrum salicaria, Filipendula ulmaria, Stachys palustris, Eupatorium cannabinum, Lysimachia vulgaris, Phalaris arundinacea,* and *Thalictrum flavum.* The vegetation succession is determined by a number of factors, of which the presence of water is the most important.

5.   CONCLUSIONS

Although the river plains are located in different climate and vegetation zones, and the sediment characteristics are quite different, the riparian flora, vegetation structure and vegetation succession on the pointbars of rivers show many similarities. On similar morphological units the same communities are found. Along chutes and left-behind or abandoned channels associations of *Phalaris arundinacea + Eleocharis palustris* are found. On meadow habitats *Populus nigra + Elytrigia repens* are usually found. In both areas young alluvial forest of *Salix alba+Populus nigra* is common. Also, three similar directions of succession are present, starting from a bare surface

- Herbaceous pioneer vegetation merging into young herbaceous community and ending in *Elytrigia* or *Bromopsis* meadows;
- Development of young alluvial forest and
- Riparian vegetation of abandoned channels.

Differences of vegetation between both areas are mainly caused by the different influx of flora elements from adjacent ecosystems (steppe in case of the Volga, Don, Buzuluk rivers), or transported by the river from upstream areas (mountaneous flora elements in the case of the Allier).

PART VII:
MAIN DIRECTIONS FOR ECOGEOLOGICAL STUDIES

Chapter 24

ECOGEOLOGICAL MAPPING

G. VARTANYAN
Russian National Research Institute for Hydrogeology and Engineering Geology, Moscow region, Russia

The main objective of ecogeological investigations is to obtain an actual picture of the geoenvironmental situation in a particular region, and to assess an influence of geological factors upon the habitat conditions of the biological communities formed there.

An important element of such investigations is ecogeological mapping which needs to be undertaken at progressively greater levels of detail in a study area as knowledge increases.

Ecogeological mapping includes a complex of scientific and methodical investigations, technical and technological operations including graphic or digital modelling of the geological environment in a specific region to indicate features, state and processes (natural and man-induced) that occur underground and affect, individually or in combination, the conditions of human activity and of the ecosystem as a whole.

Ecogeological maps can be subdivided into four large groups on the basis of the amount of initial data used and the level of detail of representation:

- general purpose;
- particular;
- special;
- supporting.

General maps reflect parameters and distinctive features that characterize the geological environment of a territory under study as a whole as a natural and anthropogenically disturbed geosystem. They are based mainly on analysis and generalization of available environmental and geological data to distinguish dominating processes that predetermine the specific functioning of ecosystems in the particular geological setting (taking into account the geochemical properties of soils and shallow rocks; hydrogeological features of territory, development of exo- and endogenic processes, etc.).

Particular maps represent ecogeological models of particular geoenvironmental components showing types and intensities of anthropogenic impacts on the geological environment, scales of man-induced changes (i.e. state of geochemical and geophysical fields, underground hydrosphere, endo- and exogeodynamic situation, level of anthropogenic loading).

The characteristics of "ecological quality" of the geological environment and intensity anthropogenic impacts are a necessary component of general and particular maps.

Special maps are compiled for solving concrete, mainly practical, tasks such as:

- preparation of measures on rational use and protection of the geological environment;
- optimal placing of economic and other structures that could potentially degrade ecogeological conditions on a given territory;
- arrangements of effective ecogeological monitoring networks.

General, and sometimes also particular and special, maps require a great deal of information and may be excessively complicated if all of this is presented on a single sheet. Therefore it is often wise to compile sets of component maps that reflect selected features of the geological space and of processes having environmental significance.

Component maps belong to a group of special and supporting maps.

This group includes geological, geomorphologic, hydrogeological, and geochemical and other maps that are not ecological, but aspects of which are used in compiling ecogeological maps at different levels of detail and for various purposes.

The ecogeological situation can rapidly change as time passes, It is accepted therefore that ecogeological maps can be subdivided into:

- evaluative-information maps showing the geoenvironmental state for some period (moment) of time, coinciding, as a rule, with the time of compilation;
- predictive maps characterizing possible changes in the geological environment caused by various combinations of anthropogenic factors.

Finally, ecogeological maps can be categorized by degree of detail:

- overview (scale: over 1:1000 000);
- small-scale (1:1000 000 – 1:500 000);
- medium-scale (1:500 000 – 1:100 000);
- large-scale (1:100 000 – 1:25 000);

- detailed (below 1:25000).
  It is important to note that the overviewing ecogeological maps reflect:
- geologo-tectonic, geochemical, hydrogeological, geodynamic, engineering-geological, geophysical characteristics, parameters and processes that have a pronounced ecological impact on the habitat;
- data necessary to evaluate scales of an interaction between the geological environment and adjacent natural media (for determination of possible environmental consequences from such an interaction);
- anthropogenic factors and processes affecting the geological environment;
- results of interactions of the geological environment with anthropogenic objects and systems;
- assessment of the direction or prediction of changes in the geological environment in natural and anthropogenically disturbed conditions, etc.

An example of a successful experience of compiling such a map is the Geoecological Map of Central Eurasia, scale 1:2 500 000, prepared jointly by the geological specialists of Russia, Kazakhstan, China, Kirghizia, Uzbekistan, Azerbaijan, and Turkmenia.

This map provides a multi-factor assessment of the geological environment and its changes for a huge region. It reflects morpho-genetic complexes, landscape zoning, hydrogeological, geochemical, radio-geochemical, endogeodynamic conditions with indication of particularly hazardous (or undesirable) anomalies, anthropogenic loading within a territory under study and geoenvironmental changes under its impact. The map shows dominating factors of the geological environment, the availability (or absence) of which forms the ecological aspect of a region under consideration. In particular, it is shown that a determining role for the major part of the arid territory is the drinking groundwater resource, whereas for the mountain-folded, an influence on the conditions of ecosystems comes from endo- and exogenic geological processes. Taking these and other parameters into consideration, the territory was subdivided into the following areas:

- favourable;
- relatively favourable;
- unfavourable;
- catastrophic (such as, for example, the Aral Sea area with a high degree of aridization, groundwater mineralization, depletion of fresh groundwater reserves, etc.).

The above-named synthesis is compiled in two sheets (because of the multiplicity of factors to be taken into consideration and depicted):

- a map of natural factors determining the geoenvironmental features;

- a map of ecological assessment of the geological environment taking into consideration an anthropogenic impact.

The map highlights issues of importance for addressing specific ecogeological problems, taking into account the specific features of the region under study. Characteristics of lesser importance are not depicted on the map.

Medium scaled ecogeological maps can be used to solve concrete tasks, such as identifying locations for economic development and geological monitoring networks, preparation of environment protection programs, etc.

The following example of the legend of an ecogeological map on a scale of 1:100 000 – 1:200 000 illustrates the information "density" of such a product.

The legend includes three blocks:
- natural state of the geological environment;
- anthropogenic systems and objects;
- anthropogenically induced changes in the geological environment.

The following sections are envisaged:

Natural state of the geological environment
1. Classification of landscape systems and their geological basis
2. Geologo-tectonic conditions
3. Geochemical features of soils, unsaturated zone, bottom sediments
4. Hydrogeological conditions
5. Geodynamic (endo- and exogenic) processes
6. State of other natural media

Anthropogenic systems and objects
7. Systems and groups of systems
8. Anthropogenic structures

Anthropogenically induced changes in the geological environment
9. Changes in geochemical conditions
10. Changes in hydrogeological conditions
11. Changes in geodynamic situation
12. Changes in rock state and new formations
13. Changes in other natural media

In conclusion, it should be noted that the relatively limited experience of ecogeological mapping in the world, to date, shows considerably different approaches at both large and small scales. It is also evident that the acuteness of ecogeological problems varies between countries with high and low levels of industrial and agricultural production.

It is also obvious that processes of trans-boundary transport of contaminants and geographo-geological impacts of human activity such as aridization, or activation of exogenic or, sometimes, endogenic geological processes make it imperative to create a set of ecogeological maps for large regions of the world. Revising such maps at regular intervals would enable rates and trends of positive and negative changes in the ecogeological environment to be assessed as a basis for habitat management and protection.

This requires further work on improvements to methods for ecogeological mapping which would take into consideration the best solutions obtained by different geological schools of the world.

Chapter 25

MONITORING GEOLOGICAL PROCESSES AS PART OF GENERAL ENVIRONMENT MONITORING

MAREK GRANICZNY
Polish Geological Institute, Centre of Geological Spatial Information

1.   GROUNDWATER MONITORING

Geological boundaries usually do not coincide with the borders of states. The use of subsurface resources, pollution of groundwater and changes of landscape in the border area of one state could influence the subsurface environment of the neighbouring country. In general, groundwater is an especially sensitive element of subsurface and environment. Its resources are formed in extensive areas by recharge and could flow, crossing the administrative borders. Pollution or changes of hydrodynamics of the groundwater due to its extraction or mining of mineral resources could impact the quality and resources of groundwater over cross-border territories. It could make an impetus for hazardous geological processes such as karst or erosion. Therefore the knowledge of geological structure, potential processes and environmental risks is essential for sustainable use of cross-border areas and co-operation between neighbouring states (Satkunas, & Graniczny, 1997). Co-operation in cross-border territories is also very important for the implementation of the Principles for a European Spatial Development Policy and could contribute to reduction of environmental pollution and secure environmental capacities of European significance (Graute, 1995). Need and significance of monitoring of transboundary groundwater is stressed recently by the Economic Commission for Europe, which established the Task Force on Monitoring and Assessment of Transboundary Waters, in 1994 (Inventory, 1999).

The joint Polish-Lithuanian programme of environmental geological research "Belt of Yotvings – fragment of Green Lungs of Europe" was launched in 1992. This programme deals with collection of all significant information for assessment of geological environment, resources and possible hazards in order to ensure sustainable use of subsurface and better living conditions for the population (Atlas, 1997).

Groundwater is almost globally important for human consumption and its quality changes can have serious consequences (Tools, 1996). Therefore, the monitoring of groundwater quality is of exceptional significance and hydrochemistry of groundwater is an important direct geoindicator that provides information meaningful for environmental assessment (Berger, 1996).

1.1   Geology and hydrogeology of the monitoring area

The topography of the Lithuanian-Polish cross-border area is characteristic of the glacial moraine upland in the northeast and glaciofluvial outwash plain in the south (Atlas, 1997). The glacial advance of the Late Weichselian maximum formed the relief. The topographic altitudes vary between 130 and 298 m above sea level. The moraine upland is formed by marginal moraines, dead ice moraines, kames and kame terraces, eskers, and melt water erosion channels and other forms. During the deglaciation, glaciofluvial streams from the northern part of the upland flowed to the south and southwest forming the Augustów outwash plain, which is built of gravel and sand. The thickness of glacial and fluvoglacial deposits, accumulated during repeated glaciations of the Quaternary period, reach up to 281 m, characteristic of paleoincisions of the sub-Quaternary surface in the northern part of the project area, while the smallest thickness (about 112 m) is known to occur in the south-eastern part of the Augustów Plain. In the central part of the area the thickness of Quaternary is approximately 200 m.

The upper part pre-Quaternary section consists of Triassic, Jurassic, Cretaceous, Paleogene and Neogene sediments, exposed on the sub-Quaternary surface. Due to the SW dip, increasingly younger sediments are exposed towards the southwest. The thickness of the Triassic sediments increases northward from several to more than 150 m. In the north claystones with rare sandstone and oolithic limestone intercalations prevail. Sandy sediments are predominant in the south. Jurassic strata cover the Triassic formations, mainly terrigenous sediments - sandstones and siltstones. 45-75 m of glauconitic sand, silt and sandstone form the Lower Cretaceous. The Upper Cretaceous is composed of gaize, marl, sand and sandstone that can reach over 170 m in thickness. The Paleogene is represented by glauconitic sand, silt and marl totalling 5-25 m in thickness. The Neogene consists of sand several metres thick (Atlas, 1997).

From the hydrodynamic point of view, the monitoring area constitutes the regional groundwater recharge zone and belongs to the basins of rivers Sesupe, Juodoji Ancia and Baltoji Ancia – all are tributaries of the Nemunas river, flowing into the Baltic Sea.

Fresh groundwater occurs in Quaternary, Palaeocene and Upper Cretaceous sediments. Quaternary deposits are, as a rule, most productive and within the Quaternary sequence, 3-4 artesian aquifers occur. Tills of different thicknesses separate them from each other. The aquifers are hydraulically connected and the Quaternary deposits form one hydrogeological complex, due to presence of paleoincisions and buried valleys.

The Cretaceous aquifer in Polish territory is abstracted only from a few sites by single wells (e.g. in Augustów), while in Lithuania this aquifer is under exploitation jointly with the overlying aquifers e.g. Paleogene or Quaternary. The Paleogene aquifer overlies the Cretaceous aquifer making the same hydrogeological system in Lithuania, but its productivity is only of local significance. In the Polish territory this aquifer is absent.

1.2 Land use

To define land uses and trace changes in the Polish – Lithuanian cross-border area two Landsat MSS satellite images have been acquired as of 1979 and of 1992. They have been processed using PCI (EASI/PACE) software in the Polish Geological Institute. Land use has been analysed by supervised and unsupervised methods (Atlas, 1997). From the satellite images land uses can be classified into six classes (Table 25-1):

- coniferous forests,
- mixed forests,
- cultivated lands,
- wetlands and pastures,
- lakes,
- unclassified.

The analysed satellite images were taken in the month of September, in both 1979 and 1992. The analysis demonstrated considerably greater changes of land use in period 1979-1992 in Lithuanian, than in Polish, territory. In 1979, due to intensive land cultivation by soviet type collective farms, arable land occupied 48.5% of the Lithuanian side of the cross-border area, while cultivated land occupied 28.8% of the Polish area, at the same time. It is noteworthy that in the period 1979 - 1992 cultivated lands in Lithuania decreased nearly 20%, while in the Polish part, the cultivated land

Table 25-1. Changes in land use during 1979- 1992 according to interpreted satellite images

Class	Polish part(%)		Lithuanian part(%)	
	1979	1992	1979	1992
Coniferous forests	15.9	17.9	12.6	12.6
Mixed forests	17.0	15.1	9.6	12.8
Cultivated lands	28.8	31.3	48.5	28.5
Wetlands and pastures	26.3	26.4	14.1	35.9
Lakes	3.2	3.2	2.7	2.7
Unclassified	8.8	6.1	12.5	7.5
Total	100.0	100.0	100.0	100.0

area increased almost 3%. This dramatic change of land use in Lithuania was the result of the collapse of collective farms after the re-establishment of independence of Lithuania and corresponding economic transformation.

1.3   Groundwater monitoring systems and methods

A groundwater monitoring project has been developed jointly by the Geological Survey of Lithuania (LGT) and Polish Geological Institute (PGI) and launched in 1994. Monitoring is aimed at groundwater quality assessment of the main aquifers used in both countries, elucidation of trends and changes of the groundwater state, and forecasts of changes in the future. It is a very important initiative, because the groundwater is the only fresh water source for centralised water supply for towns, settlements and local residents in the Lithuanian – Polish cross-border area.

The monitoring system consists of 24 hydrogeological stations, 15 of them are in Lithuanian territory and 9 - on the Polish side. Quaternary aquifers are being monitored in 20 stations. In the remaining four stations - Paleogene, Cretaceous aquifers and shallow groundwater (two stations) are monitored. Quaternary monitoring wells are mainly installed into the first – Weichselian – Saalian - intertill aquifer (below the shallow - unconfined groundwater). This aquifer is particularly vulnerable to anthropogenic pollution. On the other hand this is the main fresh water supply source.

In the above mentioned wells, samples were collected in the period 1994-2000 (two times a year, spring and autumn) for determination of hydrochemical composition (Table 25-2). Hydrochemical analysis of the samples collected on the Polish and Lithuanian sides respectively were performed in laboratories of the Polish Geological Institute and Lithuanian company Grota Ltd. using standard analytical methods (Standard, 1989, Standardised, 1994). Two exchanges of test samples between laboratories in

Table 25-2. Statistical parameters of groundwater chemical composition of Quaternary aquifers. Data from Polish side in numerator, data from Lithuanian side in denominator

Statistical parameters	PH	Conductivity μS /cm	Total Hardness, Mg–eq/l	Cl⁻	SO_4^{2-}	HCO_3^-	NO_2^-	NO_3^- mg/l	Na^+
Mean	7.35/ 7.29	543/ 608	6.1/ 7.45	4.87/ 13.6	13.5/ 22.2	354/ 430	0.69/ 0.006	0.196/ 0.418	5.76/ 13.24
Median	7.31/ 7.27	548/ 636	6.2/ 7.25	4.31/ 9.15	14.7/ 13.77	360/ 427	0.01/ 0.0	0.02/ 0.0	5.7/ 9.83
Min	7/ 6.39	330/ 324	3.63/ 3.75	1.69/ 0.48	0.5/ 0.16	199/ 214	0.0/ 0.0	0.0/ 0.0	2.7/ 1.87
Max	7.84/ 8.05	701/ 734	7.81/ 11.64	12.1/ 57.07	29.4/ 92.2	488/ 663	9.84/ 0.092	4.88/ 10.762	11.8/ 37.2
Number of samples	87/ 1 02	87/ 36	87/ 102	87/ 102	87/ 102	86/ 102	87/ 102	85/ 92	86/ 102

K^+	Ca^{2+}	Mg^{2+}	NH_4^+
2.82/ 1.8	91.4/ 98.09	19.4/ 31.11	0.32/ 0.81
2.0/ 2.59	92.3/ 94	20/ 30.5	0.235/ 0.54
0.1/ 0.25	55.9/ 0.0	10.2/ 10.9	0.025/ 0.0
7.8/ 8.19	113.9/ 176	31.24/ 62.0	2.65/ 12.2
86/ 102	87/ 102	87/ 102	87/ 102

Lithuania and Poland for analysis were undertaken in order to assure the compatibility of results. The results showed insignificant differences.

Therefore, hydrochemical data could be used for comparison and assessment of groundwater quality on both sides of the border.

1.4 Results and discussion

The main focus of the present study was the Quaternary intertill aquifer, from which samples were collected and analysed in the period 1994-2000 (Table 25-2). The results show that groundwater quality of the monitored Quaternary aquifers is good and meets the legal requirements for potable water in both Poland and Lithuania (Table 25-3). However some differences of hydrochemical composition in Polish and Lithuanian parts of the cross-border area have been determined.

Statistical parameters showed that concentrations of nearly all macro-compounds in the groundwater vary much more on the Lithuanian side than on the Polish side. Groundwater on the Lithuanian side is significantly

Table 25-3. Maximum allowed concentrations of hydrochemical compounds according to potable water standards of Lithuania and Poland. In numerator: according to Lithuanian standard: HN 21:1998: Drinking water. Quality requirements and program surveillance. In denominator: * According to "Classification groundwater for monitoring purposes, Polish Inspectorate of Environment Protection 1995; ** According to Polish Norm, 937, 4.09.2000

Chemical compound	Maximum allowed concentrations depending on potable water quality class		
	Excellent	Good	Satisfactory
Conductivity, $\mu S\ cm^{-1}$	1000 300*	2000 800*	2500 1000*
Cl^- mg/l	25 60*	100 300*	250 600*
SO_4^{2+}mg/l	150 60*	250 250*	450 500*
NH_4^+mg/l	0,5 0,1*	1 1*	2 1,5*
NO_2^-mg/l		0,1 0,1**	
NO_3^-mg/l		50 50**	

harder. Even 50% of measured values vary between 6.5-8.5 mg-eq/l in the Lithuanian part. However, in the Polish part 75% of values occur below 6.5 mg-eq/l limit. Similar distributions are characteristic for the HCO_3^-, Cl^-, NH_4^+, Mg^{2+} and Na^+ ions. Higher values of all these components are probably caused by anthropogenic reasons. Similar patterns have been traced in the Lithuanian national groundwater monitoring network during over 55 years observations (Groundwater, 2000). However, climatic, land amelioration, groundwater regime and other factors could be involved too. The different types of land use that occurred on both sides of the border until 1991-1992 support the explanation of the detected differences by the influence of anthropogenic activities. Due to intensive land cultivation by soviet type collective farms, which used large amounts of mineral and organic fertilisers, arable land occupied more than 48% of the Lithuanian side of the cross-border area. At the same time, cultivated land occupied c.a. 30% of the Polish area, characterised by smaller private farms (Table 25-1). However, it is notable that values of typical agricultural pollutants – nitrates and ammonium – are very low in the whole studied area (Table 25-2).

Interesting differences of hydrochemical composition are observed also within the Lithuanian side of the cross-border area. The Lithuanian area could be subdivided into two parts according to the prevailing type of land-use: a southern part, where forest occupies a significantly larger area (there are located monitoring stations Kuciunai, Veisiejai and Kauknoriai) and a northern part, where agricultural land use prevails. Comparison of the monitoring results from these two parts showed the same differences of hydrochemical composition. It supports the assumption of possible correlation of the type of land use and the groundwater chemistry.

Monitoring results show also trends of changes of hydrochemical composition in the groundwater. Typical trends on the Lithuanian side are demonstrated in the Aukstakalnis station – increase of SO_4^{2+} and Cl^-, decrease of NO_3^- and NH_4^+. Identical trends have been determined in A. Kirsna station and are similar to those at other monitoring stations.

On the Polish side most common trends are demonstrated by the Giby station, where increase of all compounds is identified. A visible trend of increase in sulphate and chloride ions could tentatively be explained by intrusion of more mineralised waters from deeper aquifers. However, this trend is not expressed in some stations - e.g. Jenorajscie. It is important to stress that some trends – e.g. increase of NO_3^- and NH_4^+ (typical agricultural pollutants) - can cause deterioration of quality of the fresh groundwater in future.

Explanation of these trends requires a comprehensive analysis of the following factors: climatic, geological-tectonical, groundwater dynamics, technogenic impact etc. Obviously, to fully explain the reasons for these trends, monitoring must be continued and more data acquired.

1.5   Conclusion

Environmental geological data are of particular importance for sustainable use of cross-border areas and could contribute to prevention or reduction of environmental pollution and secure environmental capacities of transboundary significance. The results of monitoring of groundwater of the Quaternary intertill aquifers of the Lithuanian-Polish cross-border area, carried out in period 1994-2000, show that groundwater quality is good and meets legal requirements for potable water, valid in Poland and Lithuania. However, some differences of hydrochemical composition in Polish and Lithuanian parts of the cross-border have been determined. Higher values of most hydrochemical compounds are traced in Lithuania. The determined differences in land use support the assumption that the hydrochemical composition of the groundwater in the Lithuanian part is influenced by anthropogenic factors. This impact is most probably inherited from the period of soviet type collective farming. Monitoring results show also trends of increase of SO_4^{2+} Cl^-, NH_4^+ ions in most monitoring stations. The increase of NO_3^- and NH_4^+ is more clearly expressed on the Polish side. In order to explain reasons for these trends, monitoring must be continued and comprehensive analysis of factors performed. The data obtained demonstrate that groundwater chemistry is a direct geoindicator of the state of environment. The data are of high importance for implementation of the means to prevent groundwater pollution and for the sustainable use of water resources.

1.6   Acknowledgements

The data on groundwater monitoring have been obtained due to implementation of a joint project elaborated and funded by the Polish Geological Institute and Lithuanian Geological Survey within the common programme of environmental geological research "Belt of Yotvings – fragment of Green Lungs of Europe". The authors express their thanks to all colleagues who were involved in this project and performed field sampling, laboratory analysis and data processing.

2.   MONITORING OF HAZARDOUS EXOGENIC GEOLOGICAL PROCESSES BY MEANS OF REMOTE SENSING METHODS

Remote sensing images (and information extracted from such images), along with GPS data, have become primary data sources for modern cartography (GIS). Remotely sensed data – aerial photos and satellite images are very often used in PGI cartographic projects as updated, reliable information. The quality of the present satellite systems and its characteristics – spectral resolution and ground resolution - is higher and higher. The data from such systems as Landsat TM and ETM+, SPOT, IRS or IKONOS are very accurate and easy compatible with GIS map formats. The satellite images are very useful for evaluation of lithological boundaries, tectonic lines, land use and change detection, mining activities, geodynamic and geoenvironmental risks etc. PGI has got a collection of satellite images (mainly Landsat TM) from the polish territory and neighbouring states.

There are several ongoing thematic mapping projects using combinations of GIS, GPS, Remote Sensing and DTM (Digital Terrain Model). These projects concerns: landslide inventory in the Carpathians, geodynamic mapping of the Baltic coastal zone, pilot geo-mapping for administration units (gmina, powiat etc.). It means that maps, which will be elaborated by such means, would constitute a new product. Such product should be easily understandable not only by professionals (geologists, territorial planners, and environmentalists) but also by decision-makers and common citizens.

The modern cartographic information should be quickly and easily accessible. Therefore, PGI is developing intensively its INTERNET home page. A substantial part of this home page is and will be devoted to geological mapping. The PGI INTERNET address is as follow: http://www.pgi.waw.pl.

2.1   Landslides

There is no very sharp boundary between block falls and rock falls or between rock falls and landslides, except that rock falls only occur on bare rock walls and landslides also take place on less steep, soil covered slopes. To be defined as a landslide, four criteria must be fulfilled:

- the movement must be rapid, lasting seconds or at most a few minutes;
- the sliding surface, the shear plane, must go through the bedrock composing the largest part of the landslide;
- the sliding mass disintegrates during the movement;
- the slope area affected by the movement and the volume of the rock mass in motion must be large enough for it to be defined as a landslide by the people living in the area. A relative definition of the size of the mass seems more appropriate than one expressed by an arbitrary number of square metres, tonnes or cubic metres.

Major landslides in Poland are most characteristic of the Carpathian Mountains in the southeast part of the country.

The location of areas at risk from major landslides is controlled by two main factors: the presence of slopes with a favourable geological structure, and high levels of precipitation. Both of these conditions are fulfilled in the case of the flysch Carpathians, consisting of interbedded shales and sandstones, deeply dissected by numerous valleys. The annual precipitation here reaches 800 – 1100 mm, sometimes concentrated in rainstorms. According to investigations carried out by the Polish Geological Institute, about 672 km^2, or 4% of the area of the Carpathians within Poland, has been in the past, or is being at present, endangered by landslides or other forms of mass movement.

The temporal incidence of mass movement is strongly correlated with climate. During wet years, or soon after, an increasing number of fresh or revitalised landslides is reported. Such a situation took place after heavy rainstorms and flooding in summer 1997. Since that time numerous landslides in the Polish Carpathians were activated. Serious damages of houses and communication infrastructure were reported.

The Polish Geological Institute is developing presently a large project including registration of landslides in the Carpathians, monitoring its activity and making prognoses for the future. These prognoses are connected with necessity of changing local plans of territorial development. Such plans are prepared in smallest administration units – "gmina". All available modern mapping technologies will be applied during realisation of the project, including remote sensing, GIS and GPS measurements.

Stereo pairs of aerial photos (B&W, normal colour and IR colour) have long been used to recognise slides and slide-prone terrain. Zones of previous sliding activity are easily identified on aerial photos by characteristic crescent scarps and the hummocky topography exhibited by the debris flow. It is obviously more difficult to identify areas that have a potential for sliding or slumping, but the following characteristics may help to identify such zones. Because the key features in such cases are rather small, large-scale photos (about 1 : 10 000) have been found to be the most useful.

- Sharp Breaking Lines at the Scarp. The first sign of ground movement is the appearance of crescent-shaped, linear cracks in the surface soil. These may be difficult to see in their early stages since they may be obscured by vegetation.
- Hummocky Ground Surfaces below Cliffs. The appearance of hummocky materials indicates previous sliding activity and at least identifies materials highly susceptible to mass wasting. If this topography is graded or otherwise disturbed, sliding could start again.
- Undrained Depressions along Hillside Toe or Crest. Undrained depressions indicate water seepage and accumulation and thus indicate a potential build-up of hydrostatic pressures.
- Light Tones along Upper Edges of Hillsides or Cliffs. These tones, especially when linear, may indicate the formation of subsurface cracks; these facilitate drainage and cause the lighter tone. The appearance of these tones may precede the occurrence of actual breaks and scarps in the land surface.
- Accumulation of Debris in Valleys and Stream Channels. Accumulation of soil materials in these areas indicate previous sliding and slumping, commonly associated with stream undermining of embankments.
- Changes in Tone along Upper Areas of Cliffs or Embankments. Changes in tone near edges of embankments may indicate moisture differences in the subsoil, reflecting moisture accumulation and the development of hydrostatic water pressures.

In the regional scale, the satellite images could be useful also for landslide studies. On these images recognition of unstable terrain where slides occur could be possible. Such analysis is enriched when satellite images are applied together with DTM. Satellite images could be useful also for monitoring of land surface changes related to landslides activity.

Such studies were performed between Gorlice and Szymbark. Numerous landslides have been developed here, on the slopes of Maślana Mt. and Miejska Mt. The landslides were mapped during a geological mapping.

The contours of landslides were superimposed on the land use map, elaborated on the basis of the Landsat TM satellite image interpretation. It was found that about 50% of landslides are located at the areas covered by forests. It means that they constitute low hazard. In the next stage of analysis mapped landslides were compared with DTM. The analysis revealed that most of the mapped landslide areas correspond to the mountain slopes 8 – 10^0 .This same analysis showed that the landslide areas are located beneath slopes $11 - 14^0$. Another aspect of analysis concerned localities of landslides versus slope exposures. The above-presented digital analysis is very fast and very helpful for further terrain investigations.

During the field works, GPS measurements are performed. Pathfinder ProXL instruments enable measurements with accuracy of a half metre, which is very good for mapping in the scale 1 : 10 000. For the purpose of monitoring landslides, the other GPS instrument is used, with accuracy of 1 cm.

Finally, on the basis of remote sensing data interpretation and field works and measurements, a GIS database will be created. Maps of landslide prone areas will be elaborated, too. Besides that the landslides information system will be organised using the INTERNET. There is aspecial questionnaire on the PGI web site enabling collection of information concerning new landslides. This information will be processed and verified by PGI specialists. It will be taken into account during preparation and verification of local territorial planning policy.

2.2   Coastal dynamics

Processes and phenomena forming the present shape of the coastline are generated by many interrelated factors, such as: geologic structure, geomorphology, climatic phenomena, hydrologic and hydrodynamic conditions, biotic resources of the environment, type and way of development and utilization of the coastal zone. In the dynamic picture of the coastal zone, none of these factors has an unequivocal and long lasting priority; also none of them can be viewed, analyzed and interpreted without taking into account all other factors.

The Polish Baltic coastal zone should be considered as a region of strong conflict between economic development (urbanization, tourism, recreation, transport, industry) and the need to maintain the natural landscape and the existing geo-ecosystems. Therefore, selection of a proper method of developing of the coastal zone resulting from its natural predisposition, is a basic task in the process of economical utilization of this zone.

One of the main problems is coastal zone development in conditions of accelerated sea level rise. For the last 100 years the average sea level in the Gdansk region is about 1.5 mm/year. Beginning from the 1950th this rate increased to 5 mm/year. The frequency of dangerous storm surges increased also in the Gulf of Gdansk from 11 situations in the 1960th to 38 in the 1980th. Because of both factors, coastal processes have increased.

Traditional methods for mapping sea circulation patterns employ current meters, drift floats and direct temperature measurements. In addition to being expensive, these methods are hampered by the need for obtaining simultaneous data over a broad expanse of water. These problems are largely overcome by remote sensing systems that can provide nearly instantaneous images of circulation patterns over very large areas. Current systems are mapped by recognizing some property of water that differs from that of the surrounding water. Interpretation of Landsat MSS and Landsat TM satellite images of the Gdańsk Bay enabled recording of the following properties of water:

- colour due to suspended material such as sediment and plankton;
- radiant temperature.

On the Landsat MSS scene of 8th July 1991 water masses from the Vistula are clearly visible as they move East along the coast. The currents are clear because of the alga bloom, synonymous at this time of year with the drift of warm water masses. In this same image the Russian port of Baltijsk can be identified as a possible major source of pollution. In the Landsat TM imagery, significantly more structures of algae blooms are visible, especially in a 15 km diameter from the mouth of the Vistula river.

There are several problems, which were studied effectively with remote sensing (satellite images and aerial photos):

- the rate of erosion of cliffs (in different subenvironments: open sea, gulf and lagoon coasts)
- the dynamics (mainly strong erosion with local accumulation) of dune coast influenced by human activity like artificial beach nourishment and protection by groins and other constructions on the example of Hel Peninsula
- the dynamics of natural, non-protected dune coast on the example of Vistula Spit
- the erosion of old Vistula River outlet cones
- the rate of accretion processes of a recent outlet cone in the Vistula's mouth
- the sediment (sand) transport along the shore – its rate and range

- pathways of suspended matter and related pollutants transportation from the Vistula mouth to the Gulf of Gdansk
- assessment of the influence of harbours and coastal protection constructions on coastal processes

On the basis of field works (mapping, drilling on land and sea, seismoacoustic, sonar and microseismic sounding), laboratory analyses, collecting of archive materials, interpretation of aerial photos and studies of satellite images, a concept of the geodynamic map of the Polish coastal zone was developed.

A concept of the GIS date base structure was elaborated, too. Two pilot sheets, Władysławowo and Rewal, have been made. The MapInfo software was applied for its preparation.

Constantly occurring changes of the Polish coastline, progressing with different intensity, cause this spatially and temporally diversified accretion – erosion system becomes an extremely significant, if not the most important, element for producing a basis for integrated and sustainable development of the coastal zone.

2.3   Floods

Severe rains affect the Earth almost every year and cause floods, bringing serious damages to towns, roads, agriculture and to the environment in general, sometimes with loss of human life.

The Polish Geological Institute and BGR (Geological Survey of Germany) started common studies in the Odra valley at the beginning of the 1990th. The multitemporal Landsat TM images, satellite radar data and aerial photos were widely used for mapping purposes (Figure 25-1).

In the first half of July 1997, heavy rains falling on the border areas between Poland, the Czech Republic, Austria and Slovakia, swelled the water courses and caused floods in the southern part of this region. Within a 10-day period, over 100 people died in Poland and the Czech Republic.

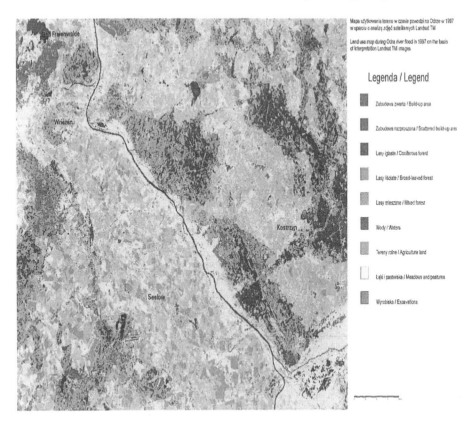

Figure 25-1. Land use map during Odra river flood in 1997 on the basis of interpretation of Landsat TM images.

On 15 July, ERS-2 SAR data revealed consistent floods near Wrocław on the Odra river and westwards, along the river course. The extent of the flooding along the Odra river was revealed by ERS-2 SAR multitemporal images on 18 July. Additional SAR data, collected from ERS-2 on 21 July, provided, up-to-date information on the event. The flooding along the Odra reached the border between Germany and Poland with a high water pressure that seriously threatened the resistance of a 160-km dike along the Odra near Frankfurt. Threatened zones of the dike could be identified at the Landsat TM data, registered 22nd July. On 23 July a 160-km dike collapsed. Two days later, the residents in the Frankfurt neighbourhood had to be evacuated. In the Czech Republic, thousands of homes were destroyed and thousands of acres of farmlands badly affected. In Poland over 149 villages were submerged and almost as many were threatened by new floods. On 26 July about 15 000 citizens had to leave the Town of Słubice on the Odra. The Polish side of the Odra was more in danger than the German side, because of the height difference between the two river banks (1-3 m lower in Poland).

On the night of 27-28 July, the water level in Frankfurt reached a record height 6.75 m. The situation improved during the second half of August. Waters retreated, thus reducing the risk of further dike cracks, with the exception of the Oderbruck region. Here, the high water level was still threatening villages and farmlands.

By using satellite information (optical and microwave), local authorities, civil protection entities and insurance and re-insurance companies are offered one more tool to monitor flood events and to assess damages. Furthermore, by combining the satellite information with topographic data (DTM), geological and hydrological data, even more end-user-oriented products can be obtained for direct utilization by entities in charge of risk management and hazard prevention.

Chapter 26

GEOLOGY AND HIGH-LEVEL NUCLEAR WASTE DISPOSAL - A BRIEF OVERVIEW

ALAN GEOFFREY MILNES
GEA Consulting, Switzerland

1. INTRODUCTION

Nuclear wastes occur in solid, liquid and gaseous forms and in a variety of isotopic compositions and radiation intensities. For the purpose of discussing long-term waste disposal, however, they can be roughly subdivided into two main categories (Milnes, 1985): low-level waste (LLW) and high-level waste (HLW). This subdivision is based on an estimation of the long-term health hazard posed by the waste, which is roughly related to its content of long-lived radioisotopes, such as plutonium, which emit α-particles and are thus highly radiotoxic if inhaled or ingested. For the purposes of this brief overview, HLW will encompass various radioactive waste categories, including spent nuclear fuel, vitrified reprocessing waste and intermediate level waste which contain long-lived radioisotopes. Emphasis is on the geological aspects of nuclear waste disposal, and these are similar for all the different categories of HLW.

To a first approximation, LLW can be regarded as material that must be isolated from the biosphere for 100 to 1000 years, after which it becomes innocuous (from the point of view of its radioactivity). HLW, by contrast, must be isolated for much longer times: 100,000-1,000.000 years is the period generally used as a guideline for developing appropriate disposal concepts. Based on our knowledge of human history, LLW management could include some degree of social commitment (i.e., disposal concepts depending on monitoring, surveillance, and if necessary, remedial action). However, HLW remains potentially hazardous for times much longer than the life span of the most stable of human societies and must be managed accordingly (i.e., disposal must depend largely on the behaviour of natural systems, although some degree of supervision is usually included in the initial stages).

2.   NUCLEAR WASTE - SOURCES AND QUANTITIES

There are over 300 nuclear power stations in operation and a large number of nuclear reactors for research and military purposes. To provide these reactors with fuel, many other nuclear facilities exist - uranium mines, ore processing plants, fuel fabrication facilities, and fuel reprocessing plants - joined in a complex, worldwide transportation network. All these facilities produce wastes that emanate ionizing radiation at appreciably higher levels than naturally occurring materials, levels that constitute a significant health hazard to present and future generations. History has forced nuclear waste, particularly HLW, into the public consciousness, and placed it at the centre of national and international controversy. Whether or not the prominence attached to it is justified, the problem of its long-term disposal has focussed considerable attention on the Earth Sciences. Demonstration of satisfactory solutions will involve a complete spectrum of geological specialists in an activity that has wide implications and that has already made major, though largely unassimilated, contributions to geological knowledge. This overview gives some background to the problem and indicates in which directions present research and development are going, with particular reference to the Earth Sciences.

The size of the nuclear waste problem is illustrated most clearly by the sequence of steps necessary to produce and use nuclear fuel. The extraction of natural uranium from mined uranium ores involves grinding them to a fine powder (milling) and then leaching out the uranium with acid or alkaline solutions (depending on ore type). Chemical separation produces U_3O_8 in the form of a yellow powder called yellow cake. To produce the 180 tons of yellow cake needed to yield enough fuel for a 1000 MWe nuclear power station for one year, more than 100,000 tons of ore are required; and more than 50,000 m^3 of mill tailings are left behind (here included in the category LLW). The yellow cake is transported to the fuel enrichment and fabrication facilities, where the production of the fuel rods (UO_2 with its content of fissile uranium-235 raised from 0.7% to 3%) results in LLW that in solid form occupies perhaps another 500 m^3. Running a 1000 MWe reactor for one year produces more LLW, in amounts of the same order. At the same time, a certain amount of the fuel is burnt up and has to be removed from the reactor core and placed in cooled and shielded water tanks. About 10 tons of spent nuclear fuel are produced per reactor per year, which, if it is treated as waste, has to be kept in storage for 10 to 50 yr, until its heat-producing capacity had died down sufficiently to allow processing and transportation, in which case it would be classified as HLW and have a volume of about 60 m^3. Reprocessing the spent fuel, i.e. separating out the recoverable uranium-235 and plutonium for reuse, in special chemical plants similar to those in which plutonium is produced for the manufacture of nuclear weapons, reduces the amount of HLW (to about 3 m^3 of vitrified waste per reactor per year), but significantly increases the amounts of LLW, which may then be contaminated with small amounts of plutonium.

The numbers given above, both the isolation times and the waste quantities, are very rough, order-of-magnitude, estimates. Also, several sources of HLW, present and future, are not covered by the brief discussion above. These include undisclosed types of military wastes, wastes derived from the decontamination and decommissioning of nuclear reactors and reprocessing plants after their expected lifetimes of 30-40 yr, and wastes associated with the running of future reactor types, such as fast-breeders (e.g. sodium-contaminated nuclear waste) and fusion reactors (e.g. the tritium problem). Everything taken into consideration, it is clear that we are dealing with a major environmental problem that is likely to increase in complexity from year to year and whose solution is increasingly urgent.

3.   DISPOSAL CONCEPTS FOR HIGH-LEVEL NUCLEAR WASTE

All HLW disposal systems are complex strategies with numerous components. Some systems are already functioning satisfactorily, others have been tried and found wanting, and still others are only in the conceptual or research and development stage. Some systems depend mainly on modern technology and engineering practice, others rely on natural retardation mechanisms to achieve isolation, and still others use the multi-barrier approach whereby the waste is enclosed behind a series of engineered barriers in a situation where natural barriers would come into play if the engineered barriers failed. Some systems aim at containment (i.e. keeping wastes within prescribed boundaries), others at dispersal (i.e. releasing wastes into the environment where rapid spreading and dilution is expected to take place), and still others at a combination of these. Which system is chosen depends strongly on the type of waste and the local conditions of geology, climate, population, topography, and so on. At the present time, purely technology-based strategies remain as more or less theoretical possibilities. These include extraterrestrial disposal (shooting HLW canisters into space), transmutation (chemical separation of all trans-uranic long-lived radioisotopes from the HLW and their conversion to stable isotopes by neutron bombardment), and permanent immobilization (conversion of HLW into a completely inert form that would remain intact no matter what the geological situation). However, most HLW disposal concepts rely wholly or partly on the geological environment for satisfactory performance in the long term, and these will be referred to as geological disposal systems.

3.1   Geological Disposal

The safe disposal of HLW presents earth scientists with a much more difficult problem than LLW because of the much longer time for which isolation should be guaranteed (10^5-10^6 yr). However, the relatively small volumes involved mean that HLW can be carefully treated and packaged and, if need be, transported to distant localities. This has resulted in an enormous upswing in research and development in three related areas: the development of industrial-scale HLW stabilization processes, the worldwide search for suitable geological environments for repository sites, and the assessment of the effects, rates, and probabilities of geological processes and events over long periods of time. In the course of this effort, some semblance of a consensus has been reached as to the most promising strategy, although the problem is by no means solved, and no HLW repository has yet been constructed and put into operation. HLW should be immobilized in an inert matrix - in the case of reprocessing waste, by incorporation into a glass or ceramic matrix, for the more voluminous spent nuclear fuel, by enclosing the fuel rods in massive iron and copper canisters - and emplaced in mined cavities at several hundred metres depth, in the most geologically stable parts of the continents. "Geologically stable" here implies the lack of tectonic or volcanic activity and the existence of a hydrogeologically stagnant environment, provable over the past 1-2 million years (Pleistocene). This concept has become known as deep geological disposal, or simply deep disposal. A second geological disposal concept which was studied extensively in the 1970s and 1980s is known as sub-seabed disposal - the disposal of HLW canisters in the sediments of the deep ocean floor - although, in spite of much research, this has since been abandoned.

Immobilization of HLW in an inert matrix is just one aspect of the production of a stable waste package. It is followed by encapsulation, in which the immobilized waste is placed in canisters composed of one or several shells of inert metal (e.g., chrome steel, lead, copper, titanium, etc.) or other materials (e.g. corundum). Also, after emplacement in the repository, it is envisaged to buffer the canisters with fill material that has special properties for retarding the migration of radionuclidcs, should the other barriers be breached (e.g. bentonite, zeolites, manganese oxides, etc.). Judging the long-term behaviour of all these materials and their efficacy as barriers to radionuclide migration has initiated a whole new field of geological research, the search for and study of natural analogues of repository-like materials and situations (Miller et al, 2000).

The disposal of HLW canisters in deep underground repositories has been the concept most consistently followed to date, and the one subjected to the most extensive research and development effort. It envisages systems of tunnels at depths of several hundred metres in the Earth's crust in which HLW canisters are emplaced individually (dimensioned and spaced to limit temperature build-up) and which can be subsequently backfilled with an appropriate buffer material and permanently sealed. Since societal commitment cannot be guaranteed for more than a few hundred years, a geological system of predictable long-term stability must be found that is not connected with potentially valuable raw materials. Since the main danger is circulating groundwater, most concepts envisage an environment in which water movement is exceedingly slow and likely to remain so under the disturbing influences of site investigation, repository construction, and waste emplacement. The potential body of rock should exhibit a high degree of homogeneity, or at least predictability, over a vertical distance hundreds of metres and over a horizontal area of several square kilometres. There should be no major fracture zones in the immediate vicinity, and the site should lie away from earthquake zones and active or dormant volcanoes. In addition, a host rock in which stable caverns can be easily mined would be an advantage. Site selection within national boundaries is thus directed by a complex evaluation of various geological factors, as well as political realities, often focussed, in the first instance, on the choice of a particular host rock on which subsequent planning is based. Three rock groups have been widely favoured as enclosing media for underground HLW repositories, namely, crystalline rocks (granite, gneiss, etc.), argillaceous rocks (clay, claystone, shale, etc.), and evaporites (rock salt, anhydrite, eetc.), but several others are or have been under consideration in particular situations (for instance, the tuffs of Yucca Mountain, in Nevada). Of these, research on deep disposal in the crystalline rocks of the Fennoscandian Shield in Sweden and Finland is most far advanced and best documented.

The problems encountered in crystalline rocks, and other so-called hard rock (i.e. basalt, tuff, etc.) involve a weighing of favourable and unfavourable attributes as far as the long-term stability of a deep HLW repository is concerned. Because of their complex structure and hydrogeology, volcanic rocks (basalt and tuff) are not regarded a priori as good enclosing media, but they are being investigated in detail in the United

States because they form the bedrock of several existing nuclear facilities (e.g. the Columbia plateau basalts under the Hanford Reservation, and the Great Basin ash-flow tuffs under the Nevada Test Site). On the other hand, granitic and other crystalline rocks of the stable Precambrian shield areas of North America and northern Europe have been or are being targeted by several countries, particularly United States, Canada, Sweden and Finland. Although crystalline rocks show several favourable geological characteristics (low porosity, high strength, widespread occurrence, and homogeneity of large rock masses), there are also disadvantages. Hard rock in general, for instance, is usually well jointed and cut through by fracture zones, which are planes of weakness for the future release of crustal stresses and pathways of rapid fluid migration. Partly out of the need to model the hydrogeology of such sites, a whole new geo-discipline has emerged: fracture geohydrology, and the need has also focussed attention on the geochemistry of fluids in such rocks (e.g. hydrothermal alteration effects, mineral deposition on joint planes), on the chemical composition of deep groundwaters (often highly saline below 600 m on the Precambrian shields), and on their stable isotope systematics. Many of these and related problems have now been studied in large-scale field experiments carried out in underground rock laboratories, such as Stripa (Sweden) and Lac du Bonnet (Canada), which are no longer operational, and Äspö (Sweden) and Grimsel (Switzerland), which are at present in operation.

4.   SELECTION, CHARACTERISATION AND MODELLING OF HLW REPOSITORY SITES

In the preceding sections we have looked at the HLW problem from the scientific and technical point of view, that is, at the various systems that have been used or proposed and at the types of geoscientific information that are relevant to long-term disposal. There is hardly a branch of the Earth Sciences that is not involved in some way in assessing the various types of enclosing media, release mechanisms, and natural analogues of repository-like situations. Conversely, the Earth Sciences have been significantly affected by the results of nuclear waste disposal research, and will continue to be. However, the attempt to ensure the development of the safest possible disposal systems involves earth scientists directly in the political decision-making process in two important areas, namely, in long-term prediction and in the selection of repository sites. On the one hand, convincing proof of the safety of nuclear waste disposal relies on long-term prediction and is an integral part of the political process that will determine the future of nuclear power in general. On the other hand, convincing procedures for the selection of repository sites are a basic necessity for overcoming the inevitable local opposition to repository construction. Unfortunately, these two requirements are almost incompatible: A convincing predictive model requires site-specific input data, but suitably specific data can be collected only from a single or very few sites (i.e. after site selection has been carried out). That nuclear waste disposal has become such a politically controversial issue arises, at least partly, from this inevitable contradiction in terms.

4.1   Predictive modelling

Long-term prediction within the framework of a complete safety analysis is based on a geological system model for a particular site. The large number of physical and chemical factors, the complexity of their interrelationships, and the long time periods involved require a systems approach and the use of computerized mathematical treatments. The geological system model is closely associated with three different geoscientific activities:

4.1.1   Predicting underground relationships

The input data to the model describe the underground geological environment of the repository as well as technical aspects such as waste inventory, waste package, and repository design. This includes such parameters as the groundwater flow pattern, fracture distribution and rock body geometry/composition in the sub-surface, and the in situ stress field. Many of these parameters have to be deduced from the integrated results of remote sensing, geophysical and geochemical surveying, borehole logging, and surface mapping, implying the development of a capacity to predict underground relationships similar to that which has become highly sophisticated in the exploration for subsurface water, energy, and mineral resources. The main difference to the latter areas of application, however, is that the data and results are extensively documented and made publically available immediately after the research is completed.

4.1.2   Predicting system evolution

Computer modeling of repository systems, using simplified analogues of the complex rock-fluid-waste interactions expected from physico-chemical theory, laboratory experimentation, and field investigations, has now reached a high level of sophistication, often far outstripping the quality of the input data. Many of the codes have been developed from those used in civil engineering, hydrogeology, and environmental studies, and are often used hypothetically, as mathematical experimentation, to explore the importance of the different parameters and to suggest system improvement (generic modelling). This is not predictive, in the sense of forecasting what will actually take place at some future time, but explorative. The aim, however, is to develop models that can be used in the final safety analysis of a specific site, i.e. to make reasoned predictions based on scientific argumentation, together with estimates of the uncertainties involved, or using conservative argumentation (pessimistic assumptions) when the input data is lacking or of poor quality.

4.1.3   Predicting disruptive events

The use of predictive modelling is limited not only by the quality of the input data but also by our ability to predict whether events that basically change the system parameters will take place in the future and what their effects will be. For the earth scientist, this means an evaluation of the probability and consequences of geological events in the next one m.y., based on expert opinion and analysis of the past geological record. The classical example of this type of prediction is the forecast that a new major continental glaciation must be assumed to take place in the next 100,000 yr (estimate of the probability of occurrence). This event will certainly change many of the system parameters (e.g., groundwater flow rates), but opinion is divided whether or not the effects of these changes will be negative from the point of view of repository integrity (estimate of the consequences of the event).

4.2   Site Selection.

The aim of site selection for a HLW repository is to identify a particular location that on the basis of a detailed safety analysis will satisfy all the requirements, not only today but also at all times in the future. This is one of the most politically explosive aspects of nuclear energy development, as many geologists know from firsthand experience. Only careful attention to the very early stages of the whole site selection process can avoid situations in which the geologist and his scientific endeavours come under intense social pressures in populated areas. This derives from the fact that a local population is interested primarily in why a site was chosen in their midst from what looks to them like a whole range of equally favourable possibilities. Thus, in the reconnaissance phase of site selection, the emphasis is not only on the development of selection criteria, but also on the development of equally valid and generally accepted exclusion criteria. The development of such criteria on geological grounds and the building up of a consensus among earth scientists is a long process, one which has rarely been achieved, in practice. However, at the time of writing, some countries are close to having selected sites which hold out promise to be demonstrably safe (once they have been investigated in sufficient detail), and which will be accepted by the local and national population. These include the Olkiluoto site in Finland (McEwen & Äikäs, 2000), one of the two sites, Forsmark and Simpevarp, in Sweden (Milnes, 2002), and the Yucca Mountain site in Nevada, USA.

Chapter 27

MONITORING ENDOGENIC GEOLOGICAL PROCESSES

G. VARTANYAN
Russian National Research Institute for Hydrogeology and Engineering Geology, Moscow region, Russia

Among the most effective and successful environmental observation systems are those operated in Japan, China, the USA, and USSR-Russia. These use a complex of mutually supporting techniques to secure information on seismically dangerous structures and territories (GPS, laser-aided telemetric measurements, high-altitude aero-geophysical surveying, repeated leveling, inclinometry, etc., carried out in combination with systematic seismic-geophysical observations).

During the closing decades of the past century (1970-2000), highly-effective methods of direct study of rock stress-strain state became available to supplement broader work on stress-strain fields over huge areas in seismically active regions.

Such monitoring provides the possibility of determining directions and rates of changes in the geoenvironmental state of a particular area from observation graphs. In combination with seismic monitoring data, this supports estimation of the likelihood and scale of a possible geodynamic catastrophe. In this case, high-technology solutions make it possible to observe the processes in real time and obtain reliable information on endogeodynamic processes.

The natural-physical pre-conditions for such monitoring are:

- Endogeodynamic processes have a global significance, and in the tectonically most stressed regions (areas) they are realized as the most destructive natural phenomenon – seismicity (including tsunami);
- An earthquake is a result of exceeding the long-term strength limit of a rock massif subjected to considerable external loads;
- "Resistance" to destructive action of external forces is determined by combined geomechanical properties of rock massifs;

- Geomechanical properties of individual rock types are varied and depend, to a considerable extent, on variations of external factors affecting the surroundings, as well as on processes occurring directly in the rock itself (metamorphization, deformation, fluid-saturation, defluidization, and so on), i.e. in addition to the initial rock characteristics it is important to establish the current state of a rock;
- Changes to the rock mass over time is manifested as a certain sequence of states, the systematic observation (monitoring) of which provides the possibility of observing the rock matter evolution and development of seismic-preparatory processes.

Accordingly, the basis for addressing the complex problem of observation over a developing endogeodynamic process and prediction of earthquake preparation lies in multi-dimensional investigation of the petrophysical medium in accordance with interconnections in the system "property – state – process – area" (Vartanyan, 1999).

A new type of geophysical field of the Earth, detected in 1982 - the HydroGeoDeformation (HGD) Field - was a turning-point in regional studies of the stress-strain state of the geological fabric and provided the possibility of instrumental investigation of the above-mentioned triad "property – state – process" over great areas in seismically active regions (Vartanyan, 1979; Vartanyan, & Kulikov, 1982).

Study of the HGD-field shows short-lived compaction and extension structures that evolve within a period of days-months.

Each short-living deformation structure forms a complicated multi-layered system having, in the cross-section, the shape of dome or cup depending on the sign of deformation.

The globally functioning HGD-field has also a layered structure, which should be taken into account during investigation and prediction of endogeodynamic phenomena.

The principles of studying the HGD-field of the Earth were used in the $1980^{th} - 1990^{th}$ to create a HGD-monitoring system – i.e. a sensitive complex for recording the stress-strain state of large rock massifs. This technology allowed continuous observation over successive changes in the geoenvironmental state for the purpose of short-term prediction of endogeodynamic catastrophes (Vartanyan, 1995).

The principal role in HGD-field investigations is played by a regional HGD-field monitoring network based on a system of observation wells where observations are carried out over groundwater level and other parameters of the subsurface space (Vartanyan, 1999; Vartanyan, et al, 1992; Vartanyan, et al, 2000).

Such HGD-monitoring networks were deployed and functioned successfully in the seismically active regions of the USSR during 1985-

1991. At the present time, some of the component networks continue to function in the Caucasian independent states (Georgia, Armenia, Azerbaijan), Ukraine and Russia.

Groundwater, being an incompressible substance, is a high-sensitiving "working body" that responds to any slight changes in the stress-strain state of water-bearing rocks and gives signals about them to a monitoring network. Specially designed hydrogeological wells can amplify the deformation signal assisting recording with the aid of accessory equipment. The measuring sensitivity is such that it enables detection of rock volumetric deformations within a range of $10^{-7} - 10^{-9}$.

Based on these measurements, a technology of regional strong earthquake regional prediction R-STEPS (Regional Short Term Earthquake Prediction System) was developed. Using such parameters as the area of short-lived deformation (compaction, extension), velocity, acceleration of deformation, deformation field gradient and other features, it is possible to make judgments about the degree of criticality of the endogeodynamic situation in a region under study.

In particular, through analyzing maps made for specified moments of time, and comparing them with data on other geophysical parameters, it becomes possible to make decisions on the character of a developing geodynamic situation and estimate the hazard of catastrophic seismic events.

Very rapid changes in the physical state of the rock mass over huge areas during seismic-preparatory processes made it necessary to develop within the R-STEPS system special techniques for processing and analysis of large quantities of monitoring data. The result was the development of a regional HGD-sounding method which provided estimation of the physical state (i.e. degree and sign of relative deformation) of particular blocks in the geological space. Current values of velocity and, especially, acceleration of deformations registered within each geological block under study proved to be particularly important.

In particular, the curves of deformation acceleration make it possible to:
- compare quantitatively the nature of development of deformation in different blocks;
- diagnose areas with the highest acceleration/inhibition of deformation process;
- observe trigger moments and critical points during rock deformation process;
- detect potentially hazardous (source) zones;
- obtain prognostic curves (of relative deformation D_e-t and seismic attack A_e-t).

Thus, for the Spitak earthquake in Armenia (07.12.1988) the A_e-curve gave a prognostic (warning) signal 6-7 days prior to the main tremor.

Similar calculations and reconstructions, undertaken retrospectively for the preparatory periods of the earthquakes in Loma Prieta (USA, October 1989), Rudbar-Tarom (Iran, June 1990), Racha-Djava (Georgia, April 1991), Kobe (Japan, January 1995), Neftegorsk (Russia, May 1995) and others, have also revealed the characteristic peaks on the A_e-t – diagrams, which could have served as prognostic signals about the coming seismic events with a warning of 3 to 15 days prior to the catastrophes.

Based on this approach, and taking account of current technology for seismic, GPS, HGD and other types of monitoring, the preconditions seem to be met for wide practical introduction of such monitoring systems, and creation of an integrated international system of endogeodynamic observations for the purpose of strong short-term earthquake prediction.

A region particularly suitable for creation of such a system could be the Pacific seismic ring involving such countries as China, Japan, USA, Indonesia, Canada, Russia and others.

The geodynamic monitoring system represents an organizational-technical structure which:

- acts in a regional or above-regional scale as a unified technological complex;
- functions according to unified scientific-methodical and engineering-technological principles;
- provides regular data (from observation networks) and generalized data on seismicity, geodynamic activity and evolutions of subsurface stress-stain state within large areas.

It follows that one of the most important components of the System is an optimal complex of methodical, instrumental and engineering solutions that obtain and processing high-accuracy information on the otherwise hidden processes of "seismic preparation".

The integration of methods would require a uniform approach to a multi-element strategy for searching for precursors and agreement on which of the available natural parameters should be used as the basis for creating a uniform prediction system. Only using such an approach would it be possible to design and construct an effective System consisting of basic monitoring technologies providing background information and additional facilities to address detailed as aspects of the developing of geodynamic situation in a region under observation.

A criterion of reliability for such a System should be its ability to provide, in cases of fast geodynamic changes, flexibility in maneuvering of the elements of the "technological arsenal" in order to reveal developing processes, and to support analysis, taking of decisions, and the supply of warning information.

Because the wide regional influence of endogenic processes of the seismic preparation is proven, the proposed System must provide:
- stationary recording of data (with a pre-specified level of precision) from a distributed network of monitoring points covering seismically active territories and adjacent "calm" areas;
- observation of "primary" geodynamic characteristics of the geological environment;
- a high sensitivity to changes in the geodynamic state;
- high technological performance;
- efficiency.

A basic element of an International Endogeodynamic Monitoring System should be a network of observations over a complex of geophysical parameters that are the derivatives of the pre-seismic state of geological space.

The most effective approach to construction of the System would be for each of the interested countries to create its own national geodynamic monitoring survey and undertake a complex of observations within an Unified International Programme.

National surveys could interact on an informational level in accordance with the rules of data exchange.

It follows from the methods and technological requirements set out above that an International Endogeodynamic Monitoring System should include, alongside seismological technologies, the R-STEPS technology in order to provide reliable observation of rapid processes of critical rock deformation.

Chapter 28

PERMAFROST MONITORING

NAUM G. OBERMAN
Mining and Geological Company "Mireko", Komi Territorial Centre for State Monitoring of Geological Environments, Russia

1.   INTRODUCTION

The objective of this section is to describe purposes, kinds and contents of permafrost monitoring and to discuss the monitoring results. Permafrost (cryolithozone) monitoring is defined as "a standardised system of observations on the condition of geological environment in the North; a system of assessment, inspection and forecast of the environment changes occurring under the effect of natural and technogenic factors" (Pavlov, 2001). Monitored variables are water and temperature regimes of grounds, regimes of cryogenic geological processes and those of other environmental components controlling the condition of permafrost, i.e., ground waters, air temperatures, snow cover, etc. All-embracing monitoring is rare. Usually only two or three variables are monitored.

This section discusses major sources of information and common methodology, as well as the main monitoring results related to permafrost temperatures, taliks, the active layer and dynamics of cryogenic processes in both natural conditions and under the most common technogenic impacts.

2.   **SOURCES OF INFORMATION, HISTORICAL REVIEW AND METHODOLOGY OF THE PERMAFROST MONITORING**

2.1   Sources of information and historical review

This section is based on the analysis of numerous publications in Russian or English devoted to various aspects of permafrost monitoring: proceedings of international (actually, world) conferences on geocryology held once in five years; proceedings of the 1st European Permafrost Conference; materials of annual international permafrost conferences in Pushchino held by the Russian Academy of Sciences, proceedings of conferences of Russian geocryologists, monographs, journal articles, etc.

The first relatively regular (annually or two yearly) observations of permafrost temperatures were, probably, conducted in 1927-1932 in Alaska, U.S.A. Since 1931, systematic observations have been conducted in Skovorodino village, Russian Far East. In the 1930s, complex studies in experimental plots were initiated by M.I. Sumgin at several North Russian field stations located near the cities of Vorkuta, Igarka, Yakutsk, and others. However, really intense permafrost studies were stimulated by the industrialisation of northern regions and started in the 1960-1970s in Russia and the U.S.A., mostly in the 1970s in Canada and Kazakhstan, in the late 1980s in West Europe, and in the 1990s in China.

2.2   Present-day monitoring organisation

In recent years, the organisation of permafrost monitoring improved as a result of a number of comprehensive national and international projects. The Circumpolar Active Layer Monitoring (CALM) program developed by the International Permafrost Association (IPA) involves more than 80 grids in 12 countries of the northern hemisphere. This long-term oriented program provides for systematic measurements of active layer thickness and for temperature monitoring of the active layer and the upper layer of permafrost. The main objective is to observe the active layer response to climatic forcing, with the further use of collected data in regional and global models (Brown, 2001). So far, over several years, scientists from several countries have been constructing a Global Terrestrial Network – Permafrost (GTN-P). This includes monitoring of the active layer and permafrost temperatures in the northern circumpolar region. More than 300 sites in 14 countries were selected as the potential sites for ground temperature measurements in boreholes; a forthcoming final selection should ensure both adequate representation of major regions and global coverage (Burgess et al., 2001). The Global Geocryological Database (GGD) is being developed: the first Circumpolar Active-Layer Permafrost System (CAPS) CD-ROM has been issued (Zhang et al., 2001), and the second one is under preparation. The Permafrost and Climate in Europe (PACE) project is aimed to arrange a monitoring network to observe the effects of climate changes on permafrost temperature regimes in European mountain systems, along a transect reaching out from Svalbard (78°11′N) to Sierra-Nevada (37°03′N). The network, composed of deep, up to 100 m boreholes, will be included in the GTN-P (Mühll, 2001). Some less general problems are being addressed by the Arctic Coast Dynamics (ACD) project and other projects.

National permafrost monitoring programs operate in a number of countries. Permafrost temperature dynamics is studied under the multidisciplinary NSF-funded ARCSS/LA II project (Romanovsky, Osterkamp, 2001). The Geological Survey of Canada supports a monitoring network, which includes more than 60 stations in the Mackenzie region alone, and many more in total (Smith et al., 2001). In Switzerland, permafrost monitoring is conducted under the PERMOS project developed by the Glaciological Commission of the Swiss Academy of Sciences. In Russia, the State System for Monitoring of Geological Environments (the Earth depths) was developed, according to a governmental decree. The system includes, in particular, geocryological monitoring. The monitoring variables are ground waters, exogenic geological processes, solid mineral deposits, hydrocarbon fields, etc. By the early 1990s, there were 25 federal-level field geocryological stations; one third of them operated year-round, and the other two thirds conducted seasonal observations (Pavlov, 1997). The observational networks of regional and object-specific levels also included a significant number of stations. Later, the number of active stations of all levels was reduced because of economic difficulties.

2.3   Monitoring methodology

Recording permafrost temperatures is the most common type of permafrost monitoring. Various equipment is used for the purpose: thermo-resistant transducers; loggers and thermocouples; exhaust resistance thermometers; mercury meteorological thermometers enclosed in heat-insulating cases; probe-thermometers; etc. Accuracy of these instruments varies from 0.002 to 0.1°C. Comparison of foreign and Russian thermometric instruments made by a joint team of Canadian, US, Japanese and Russian experts showed high accuracy and interchange ability of both (Pavlov, 1996). Depth intervals of the measurements vary from 0.5 to 25 m, and the time intervals from one min in an automatic mode to one year in the manual mode. Measurements in the active layer are usually conducted most frequently.

Differences in electric, seismic and magnetic properties, and in density between ice and water and between permafrost and thawed grounds provides the physical basis for employing geophysical methods in geocryological monitoring. Repeated vertical electric sounding, profiling with direct current, ultrasound and electric resistance logging, seismometry, and other methods are used to trace permafrost - talik borders where the latter are strongly affected by large industrial objects. Intervals between such measurements vary from one to several years.

Several methods allow changes in landscape and permafrost conditions to be monitored, including the development of cryogenic geological processes. These methods include medium- and large-scale geocryological mapping repeated after 3-15 years, sometimes longer; analyses of chronoseries of aerial photos; comparison of national bathymetric and topographic maps, periodically revised, based on repeated surveys; in the Mackay study (as cited in Are, 1980) and other publications. These methods are useful in natural environments, but are especially effective in environments strongly affected by the technogenesis.

Seasonal and long-term subsidence, thermoerosion and cryogenic heave development, stone fields and rock glaciers are monitored with the instrumental levelling of a soil surface repeated usually twice a year or less. In Western Europe a set of methods is used to monitor the cryogenic processes, which includes radar interferometry, laser scanning, inclinometry to measure deformations within a rock glacier, etc. (Kaufmann, Ladstaedter, 2001; Paar et al., 2001; and other publications). Observations on thermo-abrasion are conducted along sections perpendicular to a shore line, usually at a frequency of once a year or less.

Permafrost monitoring aimed at testing various engineer decisions related to construction is conducted at experimental-industrial polygons. The objective of this kind of monitoring is selection of the optimum construction methods from ecological and economic points of view (Esch, 1973).

3.   MAJOR RESULTS OF PERMAFROST MONITORING IN UNDISTURBED NATURAL ENVIRONMENTS

The objective of the monitoring is evaluation of the effects of climatic and regional landscape factors on the active layer, permafrost and cryogenic geological processes.

3.1 Active layer

Active layer dynamics is controlled by the corresponding dynamics of several factors: climatic variables (air temperature, precipitation, and snow thickness and snow density), water content in the active layer and fluctuations of the surface ground water levels. Ground temperatures and the intensity of ground freezing correlate with 10-year running averages of mean annual and mean summer air temperatures; seasonal thaw is controlled by mean winter temperatures. Depths of seasonal thaw in warm and cool summers differ 1.3-1.4 times in sites with a relatively thick soil organic layer, and 1.1-1.2 times in sites lacking the organic layer (Pavlov, 1997). The warming effect of snow can exceed 6°C on an annual basis, according to the same author. Therefore, differences in thaw depths between the CALM sites results partly from variations in snow thickness (Hinkel, Nelson, 2001). Ranges of variations in mean annual ground temperature at the bottom of an active layer and ranges of active layer depths is 2.6-2.8 times greater in regions with a maritime climate, compared to those with a continental climate. Water content in the active layer is the most efficient control on seasonal freezing and thaw dynamics in moderately maritime and moderately continental climates (Parmuzin, Shatalova, 2001). The latter authors summarised the results of many studies and reported that, in many regions, possible annual deviations of thaw depth from the long-term mean thaw are quite comparable with this mean itself. Also, the thawing depth can change 2-5 times in 10 to 15 years of observations.

Anomalous active layer thickness, up to nine meters, is characteristic for shallow closed taliks developed in the solid rocks of mountain areas, which are almost completely dry in winter. A seasonally frozen layer up to 6-8 m thick is observed in the aufeis - bearing segments of mountain rivers. In this case, the anomalous thickness is due to the deep freezing of coarse alluvium, which has very low water (ice) content in winter. In summer time the alluvium is characterised by a deep thaw resultant from the powerful warming effect of the flowing water absorbed by the alluvium (Oberman, 1989 a).

A large number of factors controlling an active layer depth results in a rather complex relationships. Even the effects of such an obvious control as air temperature are complicated. For example, it has been shown that the summer air temperature does not explain all the inter-annual variability in thaw depths; in the Mackay study (as cited in Pavlov, 1997). In general, responses of the active layer depth to recent climatic variations are differential (Brown, 2001).

The climatic warming observed in the 20th century in many regions of the globe changed permafrost conditions in these regions. Dominant trends are common to all continents in the Northern hemisphere and for the Antarctic, as well as for both lowlands and mountains. They are, in particular, an increase in active layer temperatures and in thaw depths, and a decrease in the depths of seasonal freezing: in the Lachenbruch, Marshall study (as cited in Romanovsky, Osterkamp, 2001); in the Haeberli et al. study (as cited in Gorbunov, 1998); Pavlov, 1994; Wang, French, 1994; Osterkamp et al., 1994; Mühll et al., 1998; Gorbunov et al., 1999; Guglielmin, Dramis, 2001; Oberman, Mazhitova, 2001; Pavlov et al., 2002; and other publications). Many regions share common climatic trends, however, the climatic changes are not completely identical. For example, a trend to a later offset of climatic warming is observed in the west-east direction in European Russia and West Siberia, whereas in East Siberia the warming started earlier than in those regions (Parmuzin and Shatalova, 2001). Temperature increments during warming are highly spatially variable, and the same is true of the degree of discrepancy between precipitation and snow depth, which are commonly negatively correlated (Pavlov, 1997). Naturally, therefore, offsets in the timing of trends demonstrated by active layer characteristics, as well as the magnitude of changes, differ between regions. Moreover, within a particular region, the trends fairly often differ not only in increments, but are negatively correlated: examples have been reported from Alaska (Osterkamp, Lachenbruch, 1990), Central Yakutia (Pavlov et al., 2002), Tien-Shan (Gorbunov et al., 1999), and other regions.

Under the climatic warming of 1970-1995, depths of seasonal thaw in the lowlands of the European Northeast increased in average by 0.3-1.0 cm year^{-1} and the depths of seasonal freezing decreased by 0.4-1.6 cm year^{-1};. Ranges are explained by variability in lithological composition of deposits (Oberman, Mazhitova, 2001). In the Tien-Shan Mountains, the warming of 1974-1999 characterised by the same increase in air temperature as that registered in the European Northeast, caused increases in seasonal thawing of an average, 4.0 cm year^{-1} and decreases in seasonal freezing (observation period 1974-1996) of 2.0 cm year^{-1} (Gorbunov et al., 1999). Similar long-term increments of the depth of seasonal freezing were observed in the Ural Mountains. More rapid development of the ground warming in mountain regions, as compared to lowlands, is, probably, due to greater snow accumulation and higher heat conductivity of rocks in the former regions, as well as more effective convective deliverance of atmospheric heat by ground waters.

3.2   Permafrost

The main monitoring result is the registering of increases in permafrost temperature during the 20th century in Russia, U.S.A., Canada, Kazakhstan,

China and Switzerland. In Alaska, the increased temperature at the permafrost surface was 2-4°C during the 20th century, and even higher, 4.0-4.5°C, in the Prudhoe Bay area in 1987-1998 (Romanovsky, Osterkamp, 2001). It is interesting that permafrost temperature increased at a depth of three meters in the European Northeast during the period 1970-1995 and in Central Yakutia during 1966-1996. The former region demonstrated an increase of 1.6°C and the latter of 0.6°C, whereas the increases in air temperature were 0.2-0.8°C and 2.05°C, respectively (Pavlov et al., 2002; Pavlov, 1997).

Comparison of the same two regions with regard to changes in permafrost temperatures at the depth of zero annual amplitude (geocryological term, meaning the depth of occurrence of base of layer with annual temperature fluctuations) shows similar relationships. In the European Northeast, the temperature increase was 0.7-1.2°C, if areas of new permafrost formation are not considered (Oberman, Mazhitova, 2001). In Central Yakutia, the changes ranged from a 0.1-0.5°C decrease[10] to a 0.2°C increase (Pavlov et al., 2002). The permafrost temperature anomaly in Central Yakutia is explained in the latter publication by an "apparent tendency to a decrease in snow thickness in the region during the last" 10-20 years. Data from other regions can be interpreted in a similar way. One example is the Marmot River basin, where ground temperatures depend more on snow thickness and on a snow period length, than on air temperatures (Harris, 1999). In the European Northeast, opposite changes in heat and moisture characteristic for Brikner cycles have been observed, similar to those in Central Yakutia. However, permafrost warming is well manifested in this region. We suppose that the main reason for these differences between the two regions is their fundamentally different hydro-geological characteristics. In Central Yakutia, the precipitation norm is half that in the European Northeast, therefore, supra-permafrost waters in the cohesive grounds composing an active layer "practically do not occur", and the discharges of supra-permafrost/inter-permafrost water springs decreased in the period 1970-1995 (Shepelev et al., 2002). It is known, that inter-permafrost waters, along with water contents in an active layer, exert a "prevailing influence on the ground temperature regime in the terrain type characterised by occurrence of sand ridges" (Skriabin et al., 1996). Therefore, one can expect that a heating effect of ground waters on permafrost, which is currently moderately strong, will weaken under climatic warming. A different situation exists in the European Northeast, where excessive moistening and ground water abundance are characteristic. In this region, mean annual ground water levels increased during the period

1970-1995 from two to 15 meters, according to our data (Shepelev et al., 2002). In our opinion, it is the heating effect of ground waters which explains the fact that the increase in permafrost temperatures was larger than that in air temperatures, in spite of decreased precipitation.

Annual minimum and maximum temperatures do not always coincide in the air and in the ground, because of the powerful effect of snow cover (Pavlov, 1997), to which the effects of ground waters and, possibly, those of some other landscape components could be added. This author reported also, that the amount of long-term warming of the high-temperature permafrost in West Siberia is less than that of low-temperature permafrost occurring within the same small areas. A similar phenomenon is observed in the European Northeast (Oberman, Mazhitova, 2001). This tendency results in smoothing of the temperature heterogeneity in the upper layers of permafrost under the climatic warming (Pavlov et al., 2002).

The longest records available show that ground temperatures in the Northern Tien-Shan and in the Polar Urals increased by 0.2-0.5°C during a 22-25 year long period in the 1970-1990s; in the Qinghai-Xizang plateau, south-western China they increased by 0.1-0.3°C during the period 1974-1989; and in the Alaska Range, U.S.A. the increase during the period 1960-1995 was 0.3°C (Gorbunov et al., 1999; Yongjian, 1998; Oberman, 1999; Romanovsky, Osterkamp, 2001). These values are smaller than those from lowlands, in spite of a usually larger snow thickness in the mountain areas. One of possible reason for the weaker warming of the mountain permafrost may be well drained rocks and, thus, weaker warming effects exerted on these by ground water.

According to the last two publications cited, changes in permafrost temperatures are observed at present down to depths of 50 meters and even more. Unpublished data of N.G. Oberman and N.B. Kakunov, and a personal communication to V.E. Romanovsky, show that both in the piedmont of the Urals and in Alaska, ground temperatures increased between 1971 and 1995-1996 at depths down to 35 meters, whereas they decreased during the same period in the deeper layers.

The long periods of climatic warming observed in the 20th century caused shallow closed talik formation in the anchored (lying directly under active layer) permafrost sites of the European Russian North, West Siberia, Chinese mountain systems and the Tien-Shan (Akimov, Bratsev, 1959; Osterkamp, Lachenbruch, 1990; Pavlov, 1997; Yongjian, 1998). Water-bearing inter-permafrost taliks located at relatively shallow depths transformed into supra-permafrost taliks (Oberman, 1999). Downward retreat of the permafrost table ranged from 0.1 m year^{-1} in Alaska (Romanovsky, Osterkamp, 2001) to 0.1-0.3 m year^{-1} in Chinese mountain systems. At the same time, the thin permafrost partly thawed out. In the

Tien-Shan, its thickness decreased by 20-25 m in 25 years (Gorbunov et al., 1999). A northward shift of the southern border of permafrost of 200 km in the Lake Baikal area occurred between the 1920-1930s and the 1950s-1960s. In the Bolshezemelskaya Tundra of the European Russian North it was 35-40 km in 25 years during approximately the same period (Maksimova, 2001; Akimov, Bratsev, 1959). In the Chinese mountains, the shift occurring during recent decades varies from 8-10 to 1500-3000 m year^{-1} (Yongjian, 1998).

3.3   Cryogenic geological processes

Frost heave mounds grow rapidly in the early stages of formation, but later the growth slows down. The lower segment of a 48-meter high mound grew 0.2 cm per year, and its upper segment grew 2.8 cm per year; in the Mackay study (as cited in Sukhodrovsky, 1979). Seasonal hydrolaccolith defined formation is closely correlated with the dynamics of active layer freezing: the growth speed is 7-8 cm day^{-1} at the start of freezing becoming 4 cm day^{-1} or even less after 1.5-2.0 months, and zero after another two months, i.e., in spring. By the spring, the hydrolaccoliths reach 2.0-3.5 m in height, however, they are completely destroyed during the first half of summer. Migration of seasonal hydrolaccoliths is 30-50 m year^{-1} (Oberman, 1989 b). Rather rapid progress of long-term thermokarst surface subsidence is proved by occurrence of numerous small lakes about 1 m deep, which preserve half-flooded, yet still living, trees. Analyses of N.B. Kakunov's observations of seasonal ground subsidence and heave, conducted on practically all major landforms and patterned ground components of the glacial-marine plain in the Vorkuta area, show that, during the observation period 1988-2000, long-term ground subsidence occurred almost everywhere. Its maximum average values of 2.7 and 2.0 cm year^{-1} were both observed in watersheds, the first value in a closed talik and the second in a site with anchored permafrost. Rapid subsidence in the talik was due to equally rapid permafrost thawing because of the warming effect of infiltrating atmospheric precipitation and relatively high temperatures of upper layers of permafrost in closed taliks. Such heat losses decrease during thawing. Subsidence rates in peat plateaus were 1 cm year^{-1}, however, with the offsetting of climatic cooling, subsidence gave way to heave.

Solifluction monitoring in Europe, Asia and North America revealed major peculiarities of this process which involve an upper ground layer 30 to 50 cm thick. Speeds of ground sloughing vary from 1 to 300 cm year^{-1}, reaching, in the case of "rapid solifluction" 6-15 m year^{-1}. The speeds increase with increased slope steepness, summer precipitation, fine particle content of soils, and decreased slope stabilization by vegetation (in Rapp,

Furer, French studies, as cited in Sukhodrovsky (1979); Sukhodrovsky, 1979; and other publications). The longest record available from the Baikal area shows 25 to 42 m ground sloughing during 1966-1987 (Leschikov, 1996). Climate warming caused an increase in the seasonal thaw of grounds on fixed slopes in the Northern Tien-Shan (2900-3000 m a.s.l.), accompanied with decreased water content and plasticity of the ground. As a result, solifluction rates dropped from 97 mm year^{-1} in 1978-1984 to 45 mm year^{-1} in 1985-1993 (Gorbunov et al., 1996).

Rates of stone field drift in the French Alps varied during 16 years of observations from 2 to 41 mm year^{-1} on gentle slopes, and reached 150 mm year^{-1} on 17° slopes; in Pissart's study (as cited in Sukhodrovsky, 1979). Comparable values, 20-120 mm year^{-1}, were registered in the Angara area in Siberia, where maximum values were observed in the sites with a permanent "groundwater flow"; in Voiloshnikov's study (as cited in Sukhodrovsky, 1979). Stone field dynamics is controlled, in particular, by cryo-lithological conditions of an area, seasonal characteristics and climate fluctuations (Turin et al., 1982).

Among rock glaciers, both active and inert forms occur. The latter are known, for example, from Japan (Fukui et al., 2001). Where a rock glacier is active, only the upper layer containing fine grained particles moves, rather than the whole mass (Haeberli et al., 1998). Rock glaciers in the Swiss Alps demonstrate movements of 0.08-0.19 mm day^{-1}, where the temperature at the bottom of the moving layer of a glacier is from -0.5 to -1.5°C. Rates increase abruptly to 1.2 mm day^{-1} where temperatures are close to 0°C (Arenson et al., 2001). This difference is readily explicable: higher content of unfrozen water provides additional hydrostatic pressure "lubricant" while a rock glacier drifts. Also, at higher glacier temperatures heat-bearing ground waters more easily penetrate to the glacier body. It has been stated that "pure heat conduction is, however, not able to fully explain the observed variability" in glacier drift rates (Kääb and Frauenfelder, 2001). This statement indirectly confirms that factors other than ground water affect drift rates of rock glaciers. Undoubtedly, the main control is fluctuating climate.

The latter conclusion was based on the results of monitoring conducted since 1938 (Schneider, 2001). Air temperatures exceeding the long-term mean usually cause an immediate down-slope acceleration of a rock glacier by as much as 2.0-6.6 m year^{-1}. Conversely, temperatures lower than the long-term mean cause an immediate decrease in glacier activity. However, rock glacier dynamics during the period 1970-1990 "cannot be explained in this way". It is supposed in that particular case that speeds of glacier drift were controlled by the glacier bed topography. A firmer conclusion was made by A.P. Gorbunov et al (1996) with regard to the 1923-1994 monitoring record from the Gorodetsky rock glacier in the Northern Tien-

Shan: "a speed of the rock glacier drift follows changes in mean annual air temperature with a 5-to-7 year lag". Data presented in the latter publication allowed us to subdivide the observation period into three sub-periods, 1923-1946, 1947-1977 and 1978-1994, roughly coinciding with quarter-century climatic phases: the first and the last sub-periods with warming phases, and the period in the middle with a cooling phase. During each of the three sub-periods, a drift of the most active central segment of the glacier's frontal bluff proceeded with the speeds averaging 90, 80 and 100 cm year^{-1}, respectively.

Speeds of thermoerosional destruction of shore bluffs in the Lena, the Yana and the Indigirka River deltas vary from 10 to 15 m year^{-1}. These bluffs are composed of frozen sand, peat and aleurite; in Brice's study (as cited in Sukhodrovsky, 1979). An average of the maximum speeds of lateral thermoerosion determined for 14 Alaskan rivers by the comparison of the 1950 and 1969 aerial photos, is 3 m year $^{-1}$. Speeds of the thermoerosion expressed in gully formation reach sometimes 10-20 m year^{-1} in areas of ice-rich ground distribution in the Indigirka River basin; in Tolstov's study (as cited in Sukhodrovsky, 1979); and other publications.

F.E. Are (1980) developed many of the modern concepts related to sea coast thermo-abrasion in the Russian and North American Arctic. He based his concepts on critical evaluation and analysis of scarce data in the literature and on his own long-term observations. Some of his statements are cited below. The speed of thermo-abrasion during an ice-free period exceeds common speeds of the abrasion developing in comparable conditions in non-permafrost regions by a factor of 3 to 4. Thermo-abrasion is most intense in specific years at the coastlines of small islands surrounded by open water or capes jutting out far to sea. In these cases, meteorological and hydrological conditions, as well as rapid removal of released materials, favor thermo-abrasion development. The highest recorded speed of recent thermo-abrasion is 55 m year^{-1}. Common speeds at the coastlines of continents and large islands are 2-6 m year^{-1}. During "the last 30 years" a "slowdown in thermo-abrasion development" was observed. Allowing for when F.E. Are's monograph was published, the 30 years mentioned corresponded to a quarter-century period of climatic cooling. The highest recorded speed, and values close to this, were all registered in the mid-1940s, i.e., at the very end of the warming phase, which preceded the cooling. The maximum coastal retreat during the last 48 years, 650 m and 800 m, are known from the Muostakh Island, Laptev Sea and from the Cape Halkett, Beaufort Sea, respectively (Rachold et al., 2001; Are, 2001).

4.   MAJOR RESULTS OF PERMAFROST
        MONITORING IN TECHNOGENICALLY-
        DISTURBED ENVIRONMENTS

4.1   Mining industries

Responses of geocryological conditions to major human activities have characteristic features. However, mining is obviously the kind of activity which exerts the most powerful effect of these conditions. Therefore, it is monitoring of mining impacts that we have chosen to discuss.

Development of hydrocarbon fields leads to various changes in geocryological conditions. Anchored permafrost may form within shallow closed taliks, whereas a permafrost table may fall to a lower level in taliks that are more than 5-6 meters deep. Also the mean temperature of anchored permafrost increases. Sites with non-anchored permafrost become less active are more inert under technogenic influences (Riazanov, 2001). It should be noted, that permafrost aggradation is almost always followed by ground heave. Construction of such structures as drilling waste storage in the areas with wedge ice distribution causes abrupt activation of thermokarst and thermoerosion, with up to five m^3 of ground-ice carried away during one summer season from each meter of gully length (Glavatskikh and Chistotinov, 1996). Increased snow accumulation around industrial constructions causes degradation of the underlying permafrost and talik formation. The process is especially intense around the heat-emitting constructions. Thawing rates of the permafrost under a boiler house, where initial temperatures were not lower than $-0.6°C$, were 4.0-4.3 m in two years and irregular continuous subsidence of the soil surface proceeded with rates of 0.10-0.12 m year^{-1} (Popkov, 1996).

The commonest impact of an oil pipeline on extensive discontinuous and the sporadic discontinuous permafrost zones is soil surface subsidence due to thawing. The subsidence is greatest along the main axis of the pipeline. Thermokarst micro-topographical patterns developed in an area with "ice-rich permafrost" within two years of installation of the Norman Wells pipeline (Harry, 1990). Observations made in 1994-1997 at a 100 meter long segment of the same pipeline showed seasonal upward and downward movements with an amplitude of 22 cm, and that the movements were oppositely directed in different portions of the segment (Burgess et al., 1998). A similar situation was identified in 1989-1992 in a 40 km long experimental-industrial segment of the Khar-Yakha – Usinsk pipeline[11]. The

[1] Original data were kindly provided by N.S. Kirikova; the data summarizing and analysis were conducted by the author and upported by the SPICE project ICA-CT-2000-10018.

pipeline is located in the same permafrost zones, as the Norman Wells, but in the European Russian North. Seventy four per cent of 182 bearings settled down during this period. The settling values ranged, in general, from 0.6 to 20.0 cm (seasonal subsidence subtracted). Five bearings demonstrated an average settling equal to 77.9 cm: this value should be an underestimate, because floating of the pipeline in boggy sites was not taken into account. Heave was registered for 21% of bearings with its range generally the same as that of the above mentioned settling; only two bearings demonstrated heave averaging 53.2 cm. The gas pipelines of the Yamburg gas field demonstrated heave that exceeded settling. The heave predominance increased in 1996-1998, compared to 1995 (Emelianova et al., 2001). We suppose this tendency, observed since 1996, to be related to offsetting effects of a climatic cooling trend.

Ground temperatures at a depth of 7.5 m increased from –0.7°C in 1987 to 0.0°C in 1992 at 0.5 m distance from the axis of the Khar-Yakha – Usinsk pipeline. A similar trend, i.e., a progressive increase in ground temperatures as compared to background values, was registered in 1970-1995 along the Nadym-Punga gas pipeline in the northern taiga (Moskalenko, 1996). As a result, seasonal freezing decreased, the permafrost table dropped down to five m below ground surface in peat plateaus, and basins filled with standing water appeared, up to one m deep and up to 170 m^2 in area. According to A.N. Kozlov et al. (2001), activation of cryogenic processes is strongest along the pipelines with a mean annual temperature of transported gas above 0°C. Thermoerosional gullies appear and grow with a speed of 20 m year^{-1} and more, and winter freezing in the areas of pipeline-induced permafrost thaw lead to intense heave.

In areas underlain by firm mineral deposits, for example, in the Norilsk area, the following processes were registered (Grebenets et al., 2001; Grebenets, 2001): activation of solifluction; activation of the cryogenic weathering of solid rocks induced by heat pollution and repeated water spills from water-bearing communications; an overall trend to permafrost degradation; and transformation of waste rock dumps into technogenic rock glaciers. One of these glaciers not only had the region's largest, but the world's largest, displacements with an average speed of 40 mm day^{-1}, but the highest speeds recorded were 800-1000 mm day^{-1}. About 40 years of relative immobility of the glacier was followed in 1992-2000 by a 180-360 m drift.

Remarkable features are: firstly, the speed of the movements of this technogenic body, which are an order of magnitude higher than the maximum speed known for its natural analogue (see above); and secondly that the offset of the glacier "animation" coincided with the closing years of the 25-to-30 year long climatic warming cycle, i.e., with the peak of this warming.

The ground temperature at a depth of 40 m to the side of an open-cut mine in Yakutia dropped from –5.1°C to –6.2°C during the period 1982-1986, due to lateral cooling. In the same region, an 80 m high waste rock dump has a complex temperature field, which is vertically dynamic. The "warmest" layers of the dump cooled and froze in 1990-1994 with a rate of 0.1-0.6°C year^{-1}, whereas the "colder" (from –4.1 to –4.7°C) layers, including the dump bottom, warmed by 0.4-0.5°C per year (Gotovtsev and Klimovsky, 1996). In the same open-cut mine, burial of more than nine mln m^3 of drainage brines in the permafrost, at depths of 40-260 m resulted in the formation in 15 years of a technogenic aquifer up to 120 m thick and of 6 to 8 km in horizontal extent (Alekseev et al., 2001).

The base of the permafrost moved 33 m upward in 47 years in one Mongolian coal fields under the effect of sub-permafrost mines. The maximum mean annual ground temperature increased by 2.2°C during 20 years of the exploitation of an open-cut mine in another coal field (Sharhuu, 1998). The permafrost, with temperatures from –1 to -2°C and 50-60 m thick, thawed out completely above spontaneously combusting waste rock dumps in the Vorkuta area, Russia. Even after two years after the complete dismantling of such a dump, the underlying rock temperature at a depth of 20 m reached 7°C. A drainage shaft allowed infiltration of warm river water. As a result, the mean annual ground temperature at the coastal zone of the river at depths of 30-40 m increased by 2-3°C leading to permafrost degradation.

5.   ACKNOWLEDGEMENTS

The author is grateful to N.B. Kakunov and N.S. Kirikova for the opportunity to use their original data in this paper.

Chapter 29

ECOLOGICAL EDUCATION OF GEOLOGY STUDENTS

VIKTOR T. TROFIMOV[1], VLADIMIR M. SHVETS[2]
[1]*Department of Engineering and Ecological Geology, Faculty of Geology, Moscow State University,* [2] *Faculty of Hydrogeology, Moscow State Geology Academ, Russia*

The ecological orientation of geological education in Russia was started about 30 years ago. This was realized in the form of "Geological Environment Protection" specialization in the cadre of "Hydrogeology and Engineering Geology" speciality. Later in some institutes of our country such an approach was realized in the cadre of specialities "Geochemistry" and "Geophysics".

However, the defectiveness and narrowness of such an approach became clear as soon as geologists formulated the theoretical and methodological principles of Ecological Geology. It became clear that successful solution of ecological - geological problems could be realized only by specialists sufficiently educated not only in geology but also in ecology. In the classic geological curriculum the latter aspect was absent, which is why it was important to correct the situation.

It is evident that the social and professional institutions are in need of qualified ecologist – geologists. Firstly, there are the requirements of federal and municipal services for environmental protection. Secondly, the requirements of the Mining, Gas and Oil industries, Energetics, Civil Engineering and other branches of industry, that exploit the subsurface, i.e. lithosphere space, are also in great need of qualified experts. Thirdly, there are the needs of research and scientific institutes, universities and colleges.

Since the 1990th years of the last century the realization of ecological – geological education in Russian geological universities and institutes has been carried out in three alternative ways (Trofimov, & Shvetc, 2000).

The first way represents the organization of ecologically oriented specialization in the cadre of existing geological specialities ("Applied Geology" option).

Table 29-1 General professional and special disciplines of the educational program of the "Ecological geology" speciality

№	Discipline	Semester	Auditory hours Total	Lectures
1	Landscape science	1	36	36
2	Mineralogy with the principles of crystallography	2	84	36
3	Geodesy and the beginnings of airborne cosmophotography	2	36	24
4	Meteorology and climatology	3	54	54
5	Hydrology	3	54	36
6	Petrography	3	90	36
7	Paleontology	3	36	12
8	Structural geology and mapping	3,4	90	18
9	Historical geology	4	72	36
10	Geomorphology	4	48	24
11	Lithology	4	72	49
12	Pedology	4	36	36
13	The fundamentals of geoecology	4	48	48
14	Chemistry and toxicology of the environment	4	36	24
15	Life safety	4	24	64
16	Hydrogeology, p. 1	5	96	48
17	Engineering geology, p. 1. Soilscience	5	96	32
18	Instrumental methods for substance analysis	5	112	64
19	Cryology	5,6	96	82
20	Geochemistry	5,6	96	24
21	The fundamentals of physical geochemistry	6	36	36
22	Hydrogeology, p. 2	6	60	24
23	Engineering geology, p. 2. Engineering and ecological geodynamic	6	36	48
24	The fundamentals of geophysics	6	48	48
25	Geology of mineral deposits	6	48	28
26	Engineering geology, p. 3. Methods of engineering geological survey	7	56	42
27	Geochemistry of natural water	7	56	28
28	Hydrogeoecology	7	42	42
29	Ecological geochemistry	7	56	56
30	Geology of Russia	7	84	52
31	Economics of nature use	7,8	52	36
32	Tectonics	8	36	24
33	Geology and facial analysis of quaternary deposits	8	36	48
34	Ecological geology	8	48	48
35	Industrial ecology	8	48	24

36	Law regularities, economics and organization of the geological exploration	8	36	24
37	Geology and geochemistry of combustible minerals	8	24	42
38	History and methodology of the geological sciences	9	42	28
39	Ecological expertise	9	56	28
40	Human ecology (medical aspects)	9	28	64
41	Special disciplines	7,8	102	110
		9	294	

The second way is the formation of geological specialization in the cadre of the speciality called "Geoecology" ("Ecology and Rational Use of Natural Resources" option).

The third way consists of creation of a new speciality: "Ecological Geology".

According to educational standards the first way ("Applied geology" option) is based on 70 auditory hours course, including the following subjects: "Biosphere and humanity", "Structure of the Biosphere", "Ecosystems", "Relationships between organism and environment", "Ecology and human health", "Global problems of the environment", "Ecological principles of rational use of nature resources and environmental protection", "Fundamentals of nature resources economics", "Ecoprotective technique and technology".

This list shows that the student geologist can acquire only a general understanding of Natural resources ecology and nothing of Ecogeological investigations.

The student training in an ecologically oriented specialization in the cadre of existing diverse geological specialities permits creating of specialists, capable of solving applied ecological and ecological - geological problems by means of their own special methods (i.e. geophysical, geochemical, hydrogeological, engineering geological etc.). The volume of ecological disciplines in such specialities is about 500-600 hours. As a rule it includes new disciplines, closely connected with a basic speciality, for example "Ecological radioecology", which was introduced in the curriculum of "Ecological geology" specialization. "Ecological hydrogeology and engineering geology" curriculum includes such disciplines as "Natural and technical hydrogeological systems", "Hydrogeodynamics and Hydrogeochemistry" of such systems as well as their "Monitoring and management". The education of a specialist in "Ecological geochemistry" includes: "Antropogenic geochemical anomalies", "Technogenic geochemistry", "Methods of ecological geochemical data processing", etc.

The "Ecological geology" program is absolutely necessary for all ecogeological specialization. The main purpose of it is the student's acquaintance with structure and contents of Ecological geology as a new

Table 29-2 Special disciplines of the "Ecological geology" master program

№	Discipline	Semester	Auditory hours	
			Total	Lectures
1	Engineering constructions	7	42	28
2	Ecological geochemistry, p.2	8	24	12
3	Decontamination of soils from the pollutants	8	36	36
4	Ecological geological cartography	9	42	28
5	Improvement of ground	9	56	28
6	Technogenic soils	9	28	28
7	Landscape geochemistry	8	56	42
8	Geological substantiation of territories, constructions and population protection	9	28	28
9	Ecological functions of the lithosphere	9	28	28
10	Lithotechnical systems	9	28	28
11	Mathematical methods for the solution of ecological geological problems	10	60	24
12	Ecological aspects of the engineering geology	10	36	36
13	Monitoring of the geological environment	11	42	28
14	Pollution and protection of underground water	11	42	28
15	Dynamics of the lithosphere and insurance problems	11	28	28
16	Selected disciplines	9, 10	76	76

scientific branch of Geology, as well as with Methods of ecological – geological investigations.

The second way of geologist ecological training is formation of a geological specialization in the cadre of comparatively new interdisciplinary speciality - "Geoecology". The educational plan of this speciality includes geological disciplines (geology, geophysics, geochemistry etc.). However the number of auditory hours for these disciplines does not exceed 400 hours (i.e. less than 9 % of auditory time). In comparison with geologists (whose geological training is more than 30 % of auditory time) their geological knowledge is not sufficient. They permit one only to obtain a general idea of problems connected with evaluation and forecast of ecological lithospheric functions variation due to anthropogenic factor influence.

The analysis of programs and experience of student training according to the first and second ways of geological education ("ecologization") shows that quality of such education is insufficient. The first way is not sufficient to give a specialist the serious and competent ecological knowledge. The second one does not give to the specialist - geoecologist the fundamental geological ground for his activity in such important spheres as the Rational

Table 29-3 Special disciplines of the "Ecological geochemistry" master program

№	Discipline	Semester	Auditory hours	
			Total	Lectures
1	Engineering constructions	7	42	28
2	Ecological geochemistry	8	24	12
3	Colloid geochemistry	8	36	24
4	Landscape geochemistry	9	56	42
5	Biogeochemical cycles of elements	9	56	42
6	Mathematical methods for data processing (computer graphics)	9	84	14
7	Actual problems of ecological geochemistry	9	28	28
8	Thermodynamics of natural processes	10	72	24
9	Geochemical activity of microorganisms	10	48	48
10	Ecological geological cartography	11	42	28
11	Pollution and protection of underground water	11	42	28
12	Additional chapters of ecological geochemistry	10,11	52	52
13	Selected disciplines	9	70	28

Usage of Earth Resources, Evaluation and Forecast of Natural and Man-caused Catastrophes, monitoring of lithosphere space, groundwater pollution, construction and mining security, disposal of waste etc.

The most effective is the third way, which gives specialists the possibility to solve a multitude of complex problems in different spheres of human activity. This way consists of creation of a new speciality – "Ecological geology". The Moscow Lomonosov State University, the Universities of St. Petersburg and Voronezh have enjoyed the positive experience of such training. It is useful to examine the details of its contents.

The Ecological Geology program investigates ecological functions of the lithosphere, their behaviour under the influence of natural and human factors connected with the life and activity of the biota and especially with human beings. On the one hand, ecological geology is a new scientific branch of geology; on the other hand it is a constituent part of geoecology (Trofimov & Ziling, 2000, 2002).

The Educational and Methodical Board of Universities included "Ecological geology" in the number of disciplines representing constituent part of the four years bachelor curriculum of "Geology". The Russian Ministry of Education instituted the master's 2-years curriculum of

"Ecological   Geology",   "Ecological   geochemistry",   "Ecological
Geophysics", "Hydrogeoecology", "Ecological gas and oil geology".

Students involved in the bachelor curriculum of "Ecological Geology"
get a complete geological education. Moreover, they study Biology,
Fundamentals   of   Ecology,   Landscapes   investigation,   Pedology,
Meteorology,   Hydrology,   Fundamentals   of   Geoecology,   Industrial
economics and Human ecology. As special disciplines they study such
principally   new   subjects,   such   as   Ecological   Geology,   Ecological
Geodynamics, Ecological Geochemistry, Ecological Functions of the
Lithosphere, Ecological Hydrogeology, Methods of Ecological - Geological
Investigations, Ecological - Geological Mapping, Ecological Audit, Law
Regulations in environment use and conservation.

Table 29-1 presents an example of the General professional and special
educational disciplines list studied by "Ecological - geology" students in
Moscow State University of Lomonosov. Tables 29-2 and 29-3 present
principal special disciplines of "Ecological geology" and "Ecological
geochemistry" of master's curriculum studied in the same university.

In addition, students of the "Ecological geology" speciality like
everybody else have the geological training stage (I academic year, 4
weeks), a stage on geological survey and prospecting methods (II, 9 weeks),
ecological - geological stage (III, 5 weeks) and ecological - geological
professional training after IV and V academic years.

So the structure of the "Ecological geology" curriculum presumes a good
balance of time devoted to different disciplines: (20 %) for natural
scientific, (20 %) for professional geological and (25 %) for special
ecological disciplines. This approach permits the ecological geologist to be
able to solve any complex ecological problems in such spheres of human
activity as Monitoring of natural subsurface resources exploitation,
Ecological analysis and Audit of design solutions, concerning the resources,
geodynamic, geochemical and geophysical ecological functions of the
lithosphere, evaluation and prognostic of ecological risk related to man-
caused influence on the lithosphere and subsurface hydrosphere, solution of
complex ecogeological and geoecological (environmental) problems
requiring fundamental knowledge of geology, ecology and geoecology.
Ecogeologists will be able to work successfully as researchers and experts in
scientific institutes, production organizations and companies, carrying out
the geological, ecogeological, mining and construction works as well as
designing and monitoring.

Three alternative ways of geologist's formation, specializing in
ecological - geological researches, assume essentially different contents of
general professional and special education. Experience of Russian
universities shows that the first and the third ways are preferable due to their

fundamental geological and ecological content. This gives specialists more opportunity to find jobs because of theirs higher competence and educational level.

In conclusion it should be pointed out that ecological - geological education has a long way to go. Only the first steps have been made and the first monographs and textbooks been published (Bogoslovsky et al, 2000; Trofimov & Ziling, 2000, 2002; Trofimov et al, 2002).

We expect that in 20th century these paramount beginnings of geological education will receive the further development.

REFERENCES

Abrahams, P. (2005). Geophagy and the involuntary ingestion of soil. In: Selinus, O., Alloway, B., Centeno, J. A., Finkelman, R. B., Fuge, R., Lindh, U., and Smedley, P. (Editors), Essentials of Medical Geology. Amsterdam: Elsevier

Adams, G.F., & Wyckoff, J. (1971). Landforms, Golden Press, New York.

Ahnert, F. (1996). Introduction to Geomorphology, London: Arnold

Akimov, A.T., Bratsev, L.A. (1959). On permafrost degradation in the Bolshezemelskaya Tundra. Journal of Trudy Komi filiala Vsesoyuznogo geograficheskogo obshchestva, 5, 53-66.

Alekseev, S.V., Alekseeva, L.P., Borisov, V.N. (2001). Natural and technogenic processes in the cryolithozone of the Yakutskaya diamond province. Proceedings of the 2nd Conference of Geocryologists of Russia, Moscow, 3-8 June 2001, 2 (3-9). Moscow: Moscow State University.

Alexander, D. (2002). Principles of emergency planning and management. Terra publishing (Harpenden).

Alexander, E. B. (2002). Serpentine geoecology based on soil survey in the Rattlesnake Creek terrane, Klamath Mountains, California. Geological Society of America (Boulder, Colorado), 2002 Denver Annual Meeting, October 27-30, Abstract of oral presentation, session 240.

Alker, S et al (2002). Integrating environmental information into a decision support tool for urban planning - an environmental information system for planners (EISP) 23rd Urban data Management Symposium Conference (Prague)

Allen, R.H., Gottlieb, M., Clute, E., Pongsiri, M.J., Sherman, J., & Obrams, G.I. (1997). Breast cancer and pesticides in Hawaii: the need for further study. Environmental Health Perspectives, 105 (3), 679-683.

Alley, R. B., Marotzke, J., Nordhaus, W. D., Overpeck, J. T., Peteet, D. M., Pielke, R. A., et al. (2003). Abrupt climatic change. Science, 299(5615), 2005-2010.

Anthony, E. J. (1996). Evolution of estuarine shoreline systems in Sierra Leone. In K.F. Nordstrom & C.T. Roman (editors), Estuarine Shores: Evolution, Environments and Human Alterations (pp. 39-61). Chichester: John Wiley & Sons Ltd.

Appleton, D. (2005). Radon in air and water. In Selinus, O., Alloway, B., Centeno, J. A., Finkelman, R. B., Fuge, R., Lindh, U., and Smedley, P. (Editors), Essentials of Medical Geology. Amsterdam: Elsevier

Appleton, J D and Ball, T K (1995). Radon and background radiation from natural sources: characteristics, extent and relevance to planning and development in Great Britain. BGS Technical Report WP/95/2. British Geological Survey (Keyworth) 93pp

Are, F.E. (1980). Thermo-abrasion of sea coasts. Moscow: Nauka.

Are, F.E. (2001). Land-ocean interaction in polar regions as a laboratory of the subaqueous cryolithozone dynamics. Abstracts of the International Conference "Conservation and Transformation of Matter and Energy in the Earth Cryosphere", Pushchino, 1-5 June, 2001 (pp. 184-185). Pushchino: Russian Academy of Sciences.

Arenson, L., Hoelzle, M., Springman, S. (2001). Borehole deformation measurements in Alpine rock glaciers. Abstracts of the 1st European Permafrost Conference, Rome, 26-28th March 2001 (pp.48-49). Rome.

Arkley, R. J., and Brown, H. C. (1954). The origin of Mima mound (hogwallow) relief in the far western states, Proceedings, Soil Science Society of America, 18(2), 195-199.

Atlas "Geology for environmental protection and territorial planning in the Polish-Lithuanian cross-border area" (1997). Scale 1:500 000. Warsaw.

Atwater, T. C. (1989). Plate tectonic history of the northeast Pacific and western North America. In Winterer, E., Hussong, D. and Decker, R., The eastern Pacific Ocean and Hawaii. Geological Society of America Decade of North American Geology. N, 21-72.

Avakyan, A.B., Sanin, M.V., & Elpiner, L. I. (1987). Water desalination in nature and national economy. Moscow: Nauka.

Bailey, E. H., et al. (1964). Franciscan and related rocks, and their significance in the geology of western California. California Division of Mines and Geology Bulletin 183, Sacramento.

Barbieri, M. (1997). – Monitoring the Summer 1997 floods in North-Eastern Europe, through "Earth watching", Earth Observartion Quarterly, December

Barker, G M A. (1996). Earth science sites in urban areas: the lessons from wildlife conservation. In: Bennett, M R; Doyle, P; Larwood, J G and Prosser, C D Geology on your doorstep: role of urban geology in earth heritage conservation. Geological Society (London) 194-200

Bartlein, P. J. (1997). Past environmental changes: characteristic features of Quaternary climatic variations. In Huntley, B., Cramer, W., Morgan, A. V., Prentice, H. C., and Allen, J.R.M. (Eds.), Past and future environmental changes: the spatial and evolutionary responses of terrestrial biota: NATO ASI Series, 147, (pp. 11-29). Berlin: Springer-Verlag.

Bastian, O., Steinhardt, U., and Naveh, Z. (2002). Development and perspectives of landscape ecology. 527 pp. Dordrecht, The Netherlands: Kluwer Academic Publishers,

Bear, J., Tsang, C-F & de Marsily, G. (1993). Flow and Contaminant Transport in Fractured Rock. Academic Press, Inc.

Be'er, S.A. (1996). Causal and effective relationship of different pollutions and parasitosis human morbidity. Regional problems and management of population health of Russia. Moscow, 116-123.

Belkin, H. E., Kroll, D., Zhou, D.-X., Finkelman, R. B., & Zheng, B. (2003). Field test kit to identify arsenic-rich coals hazardous to human health. Abstract in Natural Science and Public Health – Prescription for a Better Environment. U.S. Geological Survey Open-file Report 03-097. Unpaginated.

Belkin, H.E., Zheng, B., Zhou, D., and Finkelman, R.B. (1997). Preliminary results on the Geochemistry and Mineralogy of Arsenic in Mineralized Coals from Endemic Arsenosis in Guizhou Province, P.R. China: Proceedings of the Fourteenth Annual International Pittsburgh Coal Conference and Workshop. CD-ROM p. 1-20.

Belousova A. P. (2003). Methodology of the groundwater pollution risk assessment. Natural risk assessment and management. All Russian Conference "Risk – 2003", 124 – 128.

Berg, A. W. (1990a). Formation of Mima mounds: a seismic hypothesis. Geology, 18(3), 281-284.

Berg, A. W. (1990b). Reply in Comment and Reply on Formation of Mima mounds: a seismic hypothesis. Geology, 18(12), 1260-1261.

Berger, A.R. (1996). The geoindicators concept and its application: an introduction. Geoindicators: assessing rapid environmental changes in earth system. Eds. A. Berger, W. Iams, Balkema, 1-14.

Bezdnina, S.Ya. Irrigational water quality. Principles and assessment techniques. (1997). Moscow: Roma.

Biennial Report 1997-1998 Polish Geological Institute, 1999, Warszawa

Bilsniuk, P. M., Hacker, B. R., Glodny, J., Ratschbacher, L., Slwn, B., Wu, Z. et al. (2001). Normal faulting in central Tibet since at least 13.5 Myr ago. Nature, 412, 628-632.

Bogoslovsky, V.A., Zhigalin, A.D., & Khmelevskoi V.K. (2000). Ecological geophysics. Moscow: MSU.

Bolton, K. & Evans, L J. (1997). Contaminant Geochemistry of Urban Sediments & Soils In: Eyles N [Editor] 373-382

Bourgoin, B. P., Risk, M. J., Evans, R. D., & Cornett, R. J. (1991). Relationships between the partitioning of lead in sediments and its accumulation in the marine mussel, *Mytilus edulis* near a lead smelter. Water, Air and Soil Pollution, 57-58, 377-386.

Bredehoeft J.D. (1967). Response of well-aquifer systems to earth tides. Journal of Geophysical Research, (72), 3057-3087.

Brook, D. & Marker, B R. (1987). Thematic Geological Mapping as an Essential Tool in Land Use Planning In: Culshaw M G et al 211-4 [separately referenced]

Brook, D. & Marker, B R. (1998). Geomorphological information needed for environmental policy formulation In: Hooke, J M [Ed] Geomorphology in environmental policy J Wiley (Chichester) 247-262

Bro-Rasmussen, F. (1996). Contamination by persistent chemicals in food chain and human health. Science of the Total Environment, 188 (1), 45-60.

Brown, J. Long-term observations of soil thaw in the Arctic, Sub-arctic and Alpine: Initial results based on the CALM network. Abstracts of the 1st European Permafrost Conference, Rome, 26-28th March 2001 (pp.13). Rome.

Brown, S. L. (1986). Faeces of intertidal benthic invertebrates: influence of particle selection in feeding on trace element concentration. Marine Ecology Progress Series, 28, 219-231.

Brozovic, N., Burbank, D. W., and Meigs. A. J. (1997). Climatic limits on landscape development in the northwestern Himalaya. Science, 276, 571-574.

Bruner, M.A., Rao, M., Dumont, J.N., Hull, M., Jones, T., & Bantle, J.A. (1998). Ground and surface water developmental toxicity at a municipal landfill: description and weather-related variation. Ecotoxicology and Environmental Safety, 39(3), 215-226.

Bryan, G. W. & Langston, W. J. (1992). Bioavailability, accumulation and effects of heavy metals in sediments with special reference to United Kingdom estuaries: a review. Environmental Pollution, 76, 89-131.

Bryan, G. W., Gibbs, P. E., Hummerstone, L. G. & Burt, G. R. (1987). Copper, zinc and organotin as long-term factors governing the distribution of organisms in the Fal estuary Southwest England. Estuaries, 10, 208-219.

Bryan, G. W., Gibbs, P. E., Hummerstone, L. G. & Burt, G. R., 1986. The decline of the gastropod *Nucella lapillus* around south-west England: evidence for the effect of tributyltin from antifouling paints. Journal of the Marine Biological Association U.K, 66, 611–640.

Bukantis, A., Gulbinas, Z., Kazakevicius, S., et al. (2001). The influence of climatic variations on physical geographical processes in Lithuania. Geografijos institutas ir Vilniaus universitetas, Vilnius

Bukantis, A., Rimkute, L., & Kazakevicius, S. (1998). Atmospheric precipitation. The variability of climatic elements in the Lithuanian territory. P. 19–34.

Bullen, M. E., Burbank, D. W., Abdrakhmatov, K. Y., and Garver, J.(2002). Late Cenozoic tectonic evolution of the northwestern Tien Shan: Constraints from magnetostratigraphy, detrital fission track, and basis analysis. Geological Society of America Bulletin, 113(12), 1544-1559.

Burgess, M., Smith, S., Romanovsky, V., Brown, J. (2001). The Global Terrestrial Network for Permafrost (GTN-P): Borehole Thermal Monitoring. Abstracts of the 1st European Permafrost Conference, Rome, 26-28th March 2001 (pp.11). Rome.

Burgess, M., Tarnocai, C., Nixon, M., Wright, F. (1999). Active layer depth and soil, and ground temperature monitoring in permafrost areas of Canada. Abstracts of the International Conference "Monitoring of Cryosphere", Pushchino, 20-23 April, 1999 (pp.92-93). Pushchino: Russian Academy of Sciences.

Burgess, M.M., Nixon, J.F., Lawrence, D.E. (1998). Seasonal Pipe Movement in Permafrost Terrain, KP2 Study Site, Norman Wells Pipeline. PERMAFROST – Seventh International Conference (Proceedings), Yellowknife (Canada). Journal of Collection Nordicana, 55, 95-100.

Campbell, P.G.C. (1995). Interactions between trace metals and aquatic organisms: a critique of the free-ion activity model. In A.Tessier & D.R. Turner (editors), Metal speciation and bioavailability in aquatic systems (pp 45-102). Chichester: John Wiley & Sons Ltd.

Cantor, KR (1997) Drinking water and cancer. Cancer Causes and Control, 8(3), 292-308.

Cendrero, A., Panizza, M., & Marchett, M .[Eds] (2001). Geomorphology and EIA Rotterdam: Balkema

Centeno, J.A., Mullick, F.G., Martinez, L., Gibb, H., Longfellow, D., Thompson, C. (2002b). Chronic Arsenic Toxicity: An Introduction and Overview. Histopathology. 41(2), 324-326.

Centeno, J.A., Mullick, F.G., Martinez, L., Page, N. P., Gibb, H., Longfellow, D., Thompson, D., Ladich, E.R. (2002a). Pathology Related to Chronic Arsenic Exposure. Environmental Health Perspectives. 110 (5), 883-886.

Chang-Jo F. Chung, & Fabbri, A.G. (1998). Probabilistic prediction models for landslide hazard monitoring, Deposit and Geoenvironmental Models for Resource Exploitation and Environmental Security, September 6-18 1998, Hungary: Matrahaza

Cherkassky, B.L. (1981). Transformation of the nature and human health. Moscow: Medgiz.

Chian-Min Wu, (1992). Groundwater development and management in Taiwan. J. Geol.Soc.China, 35(3), 293-311.

Cloetingh, S., Horvath, F., Dinu, C., Stephenson, R. A., Bertotti, G., Bada, G. et al., and the TECTOP Working Group. (2003). Probing tectonic topography in the aftermath of continental convergence in central Europe. EOS, 84(10), 89-93.

Combs, Gerald T. (2005). Geological impacts on nutrition. In Selinus, O., Alloway, B., Centeno, J. A., Finkelman, R. B., Fuge, R., Lindh, U., and Smedley, P. (Editors), Essentials of Medical Geology. Amsterdam: Elsevier

Corti, S., Colteni, I F., Palmer, T. N. (1999). Signature of recent climate change in the frequencies of natural atmospheric circulation regimes. Nature, 398, 799–802.

Cox, A. (Ed.). (1973). Plate tectonics and geomagnetic reversals. 702 pp., San Francisco: W. H. Freeman.

Cox, G. W. (1984). The distribution and origin of Mima mound grasslands in San Diego County, California. Ecology, 65(5), 1397-1405.

Cox, G. W. (1990a). Comment and Reply on Formation of Mima mounds: A seismic hypothesis. Geology, 18(12), 1259-1260.

Cox, G. W. (1990b). Soil mining by pocket gophers along topographic gradients in a Mima moundfield. Ecology, 71(3), 837-843.

Creeping environmental Problems and Sustainable Development in the Aral Sea Basin. (1999). Ed. by M.Glantz. Cambridge:Univ. Press.

Culshaw, MG., Bell, F G., Cripps, J C. & O'Hara, M. [Eds] Planning and engineering geology. Engineering Geology Special Publication 4 Geological Society (London)

Cushing, C. E., Cummins, K. W., and Minshall, G. W. (Eds.), (1995). River and stream ecosystems. Ecosystems of the world, no. 22. Amsterdam: Elsevier.

Cutter, S.L. (1993). Living with risk: The geography of technological hazards. Edward Arnold.

Danilov-Daniliyan, V.I., Losev, K.S., & Zalikhanov, M.Ch. (2001). Ecological Safety. General principles and Russian aspect. Moscow: UEPS Publishers.

Decho, A. W. & Luoma, S. N. (1991). Time - courses in the retention of food material in the bivalves *Potamocorbula amurensis* and *Macoma balthica*: significance to the absorption of carbon and chromium. Marine Ecology Progress Series, 78, 303-314.

Decho, A. W. & Luoma, S. N. (1996). Flexible digestion strategies and trace metal assimilation. Limnology and Oceanography, 41(3), 568-572.

Decho, A.W. & Luoma, S. N. (1994). Humic and fulvic acids: sink or source in the availability of metals to the marine bivalves *Macoma balthica* and *Potamocorbula amurensis*. Marine Ecology Progress Series, 108, 133-145.

Denton, G. H., and Hughes, T. (Eds.). (1981). The last great ice sheet. (484 pp). New York: Wiley Interscience.

Department of Food and Rural Affairs, (2003). Interim shoreline management plan guidance. www.defra.gov.uk /corporate/ consult/ smpguidance.htm

Department of the Environment, Transport and the Regions (DETR) (2000). EIA – a Guide to Procedures. Thomas Telford (London)

Department of Transport, Local Government and the Regions (DTLR) (2002). Planning Policy Guidance Note 14. Development on Unstable Land. Annex 2. Subsidence and planning. The Stationery Office (London)

Derbyshire, E. (2005). Natural aerosolic mineral dusts and human health. In: Selinus, O., Alloway, B., Centeno, J. A., Finkelman, R. B., Fuge, R., Lindh, U., and Smedley, P. (Editors), Essentials of Medical Geology. Amsterdam: Elsevier

Desk Study on the Environment in Iraq. (2003). UNEP, ISBN: 9211586283.

Despande, B.G. (1987). Earthquakes, Animals and Man, Pune, India: The Maharashtra Association for the Cultivation of Science..

Di Toro, D. M., Mahony, J. D., Hansen, D. J., Scott, K. J., Carlson, A. R. & Ankley, G. T. (1992). Acid volatile sulfide predicts the acute toxicity of cadmium and nickel in sediments. Environmental Science and Technology, 26, 96-101.

Dissanayake, C.B., Chandrajith, R. (1999). Medical geochemistry of tropical environments. Earth Science Reviews 47, 219-258.

Dobson, D. P., Meredith, P. G., and Boon, S. A. (2002). Simulation of subduction zone seismicity by dehydration of serpentine. Science, 298, 1407-1410.

Dogdeim, S.M., Mohamed, el-Z., Gad-Alla, S.A., el-Saied S., Emel, S.Y., Mohsen, A.M., et. al. (1996). Monitoring of pesticide residues in human milk, soil, water, and food samples collected from Kafr El-Zayat Governorate. The Journal of AOAC International, 79(1), 111-116.

Dourison, M.C. & Felter, S.P. (1997). Route-to-route extrapolation of the toxic potency of MTBE. Risk Analysis, 17(6), 717-725.

Drake, J. (1980). The effect of soil activity on the chemistry of carbon groundwater. Water Resor Res 16: 381–386.

Dunn, S M., & Ferrier, R C. (2003). Catchment management: the scientific challenges Macauly Land Research Institute www.mluri.sari.ac.uk /biogeochem / cmreport.html

Duursma, E.K. & Gross, M.G. (1971). Radioactivity in the marine environment (pp 147-160). Washington, D.C.: National Academy of Science.

Dyer, K. R. (1997). Estuaries: a physical introduction (2nd edition). Chichester: John Wiley & Sons Ltd.

Dzektser E. S. (1992). Geological hazard and risk. Russian journal of Engineering geology, 6, 3 – 10.

Earnst, W. G. (Ed.). (1981). The geotectonic development of California. Rubey (1), 706 pp. Englewood, New Jersey: Prentice-Hall.

Earthwise (2001). Geology and health. British Geological Survey. Issue 17.

Edmunds , M., & Smedley, P. (2005). Fluoride in natural waters. In Selinus, O., Alloway, B., Centeno, J. A., Finkelman, R. B., Fuge, R., Lindh, U., and Smedley, P. (Editors), Essentials of Medical Geology. Amsterdam: Elsevier.

Elpiner, L.I. (1975). Water supply of sea vessels. Moscow, Transport.

Elpiner, L.I. (1986). Medical ecological criteria of ecosystems' assessment. Abstracts of SEV countries symposium: Complex methods of environmental quality control, 62-69.

Elpiner, L.I. (1995). On the Influence of the Water Factor on the Health of Russia's Population. Water Resources, 22(4), 418-425.

Elpiner, L.I. (2003). Scenario of possible influence of hydrological sutiation changes on medical ecological situation (on the problem of global hydroclimate changes). Water resources, 30(4), 473-484.

Elpiner, L.I., & Bezdnina, S.Ya. (1986). On standardization of water quality for agricultural purposes. Water resources, 4, 102-110.

Elpiner, L.I., & Vasiliev, V.S. (1983). Problems of Drinking Water Supply in USA. Moscow: Nauka.

Elpiner, L.I., Shapovalov, A.E., & Zeegofer, Y.O. (1998). Subteranean water under conditions of intensive technogenes: hydroecological and medical aspects. Melioratsya i vodnoe khozaistvo, 3, 66-67.

Embleton, C&C. (1997). Geomorphological hazards of Europe, Developments in Earth Surface Processes 5, Amsterdam: Elsevier

Emelianova, L.V., Kaurkin, V.D., Kozlov, A.N. (2001). Oppositely directed deformations of pile-supported substructures at the objects of the Yamburg gas-condense field. Proceedings of the 2nd Conference of Geocryologists of Russia, Moscow, 3-8 June 2001, 4 (pp.97-100). Moscow: Moscow State University.

Energy and mineral potential of the Central American-Caribean Region, San Jose, Costa Rica, 6-9 March (1989). Episodes. (13), 1.

Engel, D. W., Sunda, W. G. & Fowler, B. A. (1981). Factors affecting trace metal uptake and toxicity to estuarine organisms. In J. Vernberg, A. Calabrese, F.P. Thurberg & W.B. Vernberg, (editors), Biological monitoring of marine pollutants, I. Environmental parameters (pp.127-145). New York, Academic Press.

Eriksen, C., and Belk, D. (1999). Fairy shrimps of California's puddles, pools, and playas. 196 pp. Eureka, California: Mad River Press.

Esch, D.C. (1973). Control of permafrost degradation beneath a roadway by subgrade insulation. Permafrost: North American contribution [to the]. Second International Conference, Yakutsk, USSR, 13-28 July 1973 (pp.608-622). Washington, D.C.: National Academy of Sciences.

Feder, G.L., Radovanovic, Z., & Finkelman, R.B. (1991). Relationship between weathered coal deposits and the etiology of Balkan endemic nephropathy. Kidney International, 40(34), s-9 – s-11.

Fell, R. (1994). Landslide risk assessment and acceptable risk. Canadian Geotech Jl 31, 261-272

Ferren, W., Jr., and Fiedler, P. (1993). Rare and threatened wetlands of central and southern California. In Keeley, J. E. (Ed.). Interface between ecology and land development in southern California (pp. 119-131). Los Angeles: Southern California Academy of Sciences.

Ferren, W., Jr., Fiedler, P., and Leidy, R. (1995). Wetlands of the central and southern California coast and coastal watersheds: A methodology for their classification and description. San Francisco: United States Environmental Protection Agency, Region IX.

Ford, D.C. & Williams, P.W. (1989). Karst geomorphology and hydrology. London: Unwin Hyman.

Fordyce, F. (2005). Selenium deficiency and toxicity in the environment. In Selinus, O., Alloway, B., Centeno, J. A., Finkelman, R. B., Fuge, R., Lindh, U., and Smedley, P. (Editors), Essentials of Medical Geology. Amsterdam: Elsevier.

Foster, A. (1995). Active ground fissures in Xian, China. Quart. J. Eng.geol., 28(1), 1-4.

Freeze, R. A., & Cherry, J. A. (1979). Groundwater. Prentice-Hall, Inc., New Jersey.

Fried, J.J., (1975). Groundwater Pollution: Theory, Methodology, Modeling and Practical Rules. Oxford, N.Y.

Fuge, R. (2005). Soils and iodine deficiency. In: Selinus, O., Alloway, B., Centeno, J. A., Finkelman, R. B., Fuge, R., Lindh, U., and Smedley, P. (Editors), Essentials of Medical Geology. Amsterdam: Elsevier.

Fukui, K., Iwata, S., Ikeda, A., Matsuoka, N. (2001). Rock glaciers and a protalus rampart in relation to mountain permafrost in the northern Japanese Alps. Abstracts of the 1st European Permafrost Conference, Rome, 26-28th March 2001 (pp.53). Rome.

Fule, P. Z., Covington, W. W., Moore, M. M., and Heinlein, T. A. (2002). Natural variability in forests of Grand Canyon, USA. Journal of Biography, 29, 31-47.

Garrett, R.G. (2000). Natural sources of Metals in the Environment. Human and Ecological Risk Assessment, (6), 945-963

Geoindicator Checklist (1996). Tools for Assessing Rapid Environmental Changes, ITC Publication 46, Enschede

Geoindicators (1999). Abstracts of Workshops in Vilnius 11-16 October 1999, IUGS, Geological Survey of Lithuania, Vilnius

Geoindicators (2000). Symposium and Field Meeting Poland, September 2000, IUGS, Polish Geological Institute, Gdańsk

George, S. G. & Coombs, T. L. (1977). The effects of chelating agents on the uptake and accumulation of cadmium by *Mytilus edulis*. Marine Biology, 39, 261-268.

Gerasimov, I.P., Kuznetzov, N.T., Kes, A.S., & Gorodetskaja, M.E. (1983). Problem of Aral Sea and antropogenous desertification. Problems of desert development, (6).

Gimeno Ortiz, A., Jimenez Romano, R., Blanco Aretio, M., & Castillo Moreno, A. (1990). (Relationship of several physico-chemical components in drinking water, hypertension and cardiovascular disease mortality). Revista de Sanidad E Higiene Publica (Madr), , 64(7-8), 377-385.

Glavatskykh, V.V., Chitotinov, L.V. (1996). Effects of technogenic disturbances on thermoerosion development. Proceedings of the 1st Conference of Geocryologists of Russia, Moscow, 3-5 June 1996, 1 (pp.456-465). Moscow: Moscow State University.

Glazovskyi, N.F. (1990). Aral Sea crisis. Causes of appearance and ways of decision. M.:Nauka, 136 p.

Goimerac, T., Kartal, B., Bilandzic, N., Roic, B., & Rajcovic-Janje, R. (1996). Seasonal atrazine contamination of drinking water in pig-breeding farm surroundings in agricultural and industrial areas of Croatia. Bulletin of Environment Contamination and Toxicology, 56(2), 225-230.

Goldberg, V.M. (1987). Interrelation between Groundwater Pollution and Environment. Hydrometeoezdat, Leningrad.

Gorbunov, A.P. (1998). Alpine permafrost in West Europe. In K.A.Kondratieva, V.V.Baulin, E.D.Ershov (Ed.). Regional and Historical Geocryology of the world (pp.237-249). Moscow: Moscow State University.

Gorbunov, A.P., Marchenko, S.S., Seversky, E.V. (1999). Permafrost and seasonally frozen ground response on climate change in the northern Tien-Shan. Abstracts of the International Conference "Monitoring of Cryosphere", Pushchino, 20-23 April, 1999 (pp.98). Pushchino: Russian Academy of Sciences.

Gorbunov, A.P., Seversky, E.V., Titkov, S.N. (1996). The Tien-Shan: a tendency to changes in geocryological situation. Proceedings of the 1st Conference of Geocryologists of Russia, Moscow, 3-5 June 1996, 2 (pp.329-335). Moscow: Moscow State University.

Gordon, R. G., and Stein, S. (1992). Global tectonics and space geodesy, Science, 256, 333-342.

Gotovtsev, S.P., Klimovsky, I.V. (1996). Ground temperatures in the objects of the open-cut mine field of the Udachnaya diatreme. Proceedings of the 1st Conference of Geocryologists of Russia, Moscow, 3-5 June 1996, 3 (pp.363-370). Moscow: Moscow State University.

Graniczny, M. (1992). Wykorzystanie teledetekcji do monitoringu środowiska przyrodniczego oraz konstruowania map zagrożeń geodynamicznych, Przegląd Geologiczny, nr 1, 8 – 12.

Graniczny, M. (1994). Remote Sensing Techniques for Monitoring the Gulf of Gdańsk, GIS in Ecological Studies & Environmental Management, Warszawa, Poland 26 - 28 September 1994, GRID Warsaw

Graniczny, M. (1998). Geoenvironmental study of the Vistula Bay and its surrounding using modern cartographic techniques, Przegląd Geologiczny, (12)

Graniczny, M. (1998). Preventing Floods, Przegląd Techniczny, (18)

Graniczny, M. (1998). Satellite Remote Sensing Systems in the end of XX Century – Present Possibilities and Perspectives, Przegląd Geologiczny, (2)

Graniczny, M. (2001). Computer Analysis of Spatial Data in Aspect of Landslides Risk, Przegląd Geologiczny, (1)

Graniczny, M.(1998). Monitoring of the Gulf of Gdansk, ER Mapper and ER Storage software and documentation – Applications, Earth Resources Mapping Pty Ltd, West Perth

Graute, U. (ed.) (1995). Newsletter 5. Network of Spatial Research Institutes in Central and Eastern Europe. Dresden.

Grebenets, V.I. (2001). Formation of peculiar natural-technogenic complexes in the Norilsk industrial district. Proceedings of the 2nd Conference of Geocryologists of Russia. Moscow, 3-8 June 2001, 4 (pp.59-65). Moscow: Moscow State University.

Grebenets, V.I., Titkov, S.N. and Malorossiyanov, V.A. (2001). Technogenic rock glaciers: genesis, structure, mechanism of movement, forecast of development. Abstracts of the 1st European Permafrost Conference, Rome, 26-28th March 2001 (pp.53-54). Rome.

Green, J. (1968). The Biology of Estuarine Animals. London: Sidgwick and Jackson.

Grim, N. B., Chacon, A., Dahm, C. N., Hostetler, S. W., Lind, O. T., Starkweather, P. L. et al. (1997). Sensitivity of aquatic ecosystems to climatic and anthropogenic changes. The Basin and Range, American Southwest and Mexico, In Cushing, C. E. (Ed.). Freshwater ecosystems and climatic change in North America; a regional assessment (pp. 205-223). New York: John Wiley and Sons.

Groundwater monitoring in Lithuania (2000). Bulletin, Geological Survey of Lithuania, Vilnius.

Gruza, G.V. & Rankova, E.Y. (1980). Structure and variability of the observed climate. Air temperature in the northern hemisphere. Leningrad: Gidrometeoizdat.

Guglielmin, M. and Dramis, F. Permafrost monitoring Antarctic Network in Northern Victoria Land (Antarctica): preliminary results. Abstracts of the 1st European Permafrost Conference, Rome, 26-28th March 2001 (pp.100-101). Rome.

Guidelines for drinking-water quality. (2nd ed.). (1993). Geneva , WHO.

Gvozdeckij, N.A. (1981). Karst. Moscow, p.214

Haeberli, W., Hoelzle, M., Kääb, A., Keller, F., Mühll, D.V., Wagner, S. (1998). Ten years after drilling through the permafrost of the active rock glacier Murtel, Eastern Swiss Alps: answered questions and new perspectives. PERMAFROST – Seventh International Conference (Proceedings), Yellowknife (Canada), 1998. Journal of Collection Nordicana, 55, 403-410.

Harmancioglu, N.B. (1997). Integrated Approach to Environmental Data Management Systems, Kluwer Academic Publishers, Dodrecht, published in cooperation with NATO Scientific Affairs Division

Harris, S.A. (1999). Twenty Years of Data on Climate-Permafrost-Active Layer Variations at the Lower Limit of Alpine Permafrost, Marmot Basin, Jasper National Park, Canada. Abstracts of the International Conference "Monitoring of Cryosphere", Pushchino, 20-23 April 1999 (pp.100). Pushchino: Russian Academy of Sciences.

Harry, D.G. (1990). Terrain Performance, Norman Wells oil pipeline, Northern Canada. International Symposium "On Geocryological studies in Arctic Regions", Yamburg. USSR, August 1989. II (pp.20-22). Tumen: Russian Academy of Sciences.

Harvey, R. W., & Luoma, S. N. (1985). Separation of solute and particulate vectors of heavy-metal uptake in controlled suspension-feeding experiments with *Macoma balthica*. Hydrobiologia, 121, 97-102.

Haryomo, (1995). Relation between groundwater withdrawal and land subsidence in Kalantan, Malaysia. IAHS Publ., 234, 31-33.

Haupert, T.A., Wiersma, J.H., & Goldring, J.M. (1996). Health effects of ingesting arsenic-contaminated groundwater. Wisconsin Medical Journal, 95(2), 100-104.

Haven, D. & Morales-Alamo, R. (1972). Biodeposition as a factor in sedimentation of fine suspended solids in estuaries. In B. Nelson (editor), Environmental framework of coastal plain estuaries. Geological Society of America Memoirs, 133, 121-130.

Healy, T. R., Cole, R. & De Lange, W. (1996). Geomorphology and ecology of New Zealand shallow estuaries and shorelines. In K.F. Nordstrom & C.T. Roman (editors), Estuarine Shores: Evolution, Environments and Human Alterations (pp. 115-154). Chichester: John Wiley & Sons Ltd.

Hermanns, R. L., Strecker, M. R., Niedermann, S., Villanueva-Garcia, A., and Sosa-Gomez, J. (2001). Neotectonics and catastrophic failure of mountain fronts in the southern intra-Andean Puna Plateau, Argentina. Geology, 29(7), 619-622.

Hickman, J. C. (Ed.). (1993). The Jepson Manual: Higher plants of California. 1400 pp. Berkeley: University of California Press.

Hiller, J.R. (1993). Conf. "Aqulfera at Risk Tpwards Nat.Groundwater Qual.Perapest." Canbera 13-15 February, 1993. AGSO J. Austral.Geol. and Geophys., 14(2-3), 213-217.

Hilmer, M. & Jung, T. (2000). Evidence for recent change in the link between the North Atlantic Oscillation and Arctic Sea ice export. Geophys. Res. Lett., 27, 989–992.

Hilmer, M., Hander, M., Lemke, P. (1998). Sea ice transport a highly variable link between Arctic and North Atlantic. Geophys. Res. Lett., 25, 3359–3362.

Hinkel, K.M. and Nelson, F.E. (2001). Controls on the spatial and temporal patterns of thaw at CALM sites on the North Slope of Alaska: 1995-2000. Abstracts of the 1st European Permafrost Conference, Rome, 26-28th March 2001 (pp.16). Rome.

Hiscock, K. (editor). (1996). Marine Nature Conservation Review: rationale and methods. Coasts and seas of the United Kingdom MNCR series. Peterborough: Joint Nature Conservation Committee.

Hopenhayn-Rich, C., Biggs, M.L., Fuchs, A., Bergoglio, R., Tello, E.E., Nicolli, H., et al. (1996). Bladder cancer mortality associated with arsenic in drinking water in Argentina. Epidemiology, 7(2), 117-124.

Hughes, T., Borns, H. W., Jr., Fastook, J. L., Kite, J. S., Hyland, M. R., and Lowell, T. V. (1985). Models of glacial reconstruction and deglaciation applied to Maritime Canada and New England. In Borns, H. W., Jr., LaSalle, P., and Thompson, W. B. (Eds.). Late Pleistocene history of northeastern New England and adjacent Quebec. Geological Society of America (Boulder, Colorado) Special Paper 197, 139-150.

Hurn J. (1993). Differential GPS. Explained, Trimble Navigation

Hurn, J. (1989). GPS A Guide to the Next Utility, Trimble Navigation

Hurrell, J. W., H. van Loon. (1997). Decadal variations in climate associated with the North Atlantic oscillation. Clim. Change, 36, 301–326.

Hydrogeological Base of Groundwater Protection. (1984). UNESCO-UNEP, Moscow.

Inventory of transboundary groundwaters (1999).. UN/ECE Task Force on Monitoring and Assessment. (1). Lelystad.

IRPTS. (1996). Human risk assessment. UNEP chemicals.

Jenny, H. (1989). The soil resource, origin and behavior. Ecological Studies 37, 337 pp. New York: Springer-Verlag.

Jepson Flora Project (2005). Jepson Interchange: Index to California plant names: Current status categories. Retrieved March 1, 2005 from http://ucjeps.berkeley.edu/interchange/I-indexes.html. Berkeley: University of California.

Kääb, A. and Frauenfelder, R. (2001). Temporal variations of mountain permafrost creep. Abstracts of the 1st European Permafrost Conference, Rome, 26-28th March 2001 (pp.56). Rome.

Kabata-Pendias, A. (2001). Trace elements in soils and plants. 3rd ed. CRC press.

Kakunov, N.B. & Pavlov, A.V. (1997). Evaluation and prediction of the cryogenic soil thermal regime in the Russia's North in connection with the anticipated climate. In I.V.Zaboeva (Ed.), Cryopedology'97: Abstract International Conference. Russia, Syktyvkar, 5-8 August, 1997 (pp.43-44). Syktyvkar: Institute of Biology.

Kalytyte, D., Zvikas, A. Paskauskas, R. (2002). Spatial and temporal changes of microplankton structure in North Lithuanian karst lakes (in Lithuanian). Botanica Lithuanica, 8(4): 309–323.

Kaufmann, V. and Ladstaedter, R. (2001). Glaciers (Oetztal Alps, Austria) by means of digital photogrammetric methods. Abstracts of the 1st European Permafrost Conference, Rome, 26-28th March 2001 (pp.57). Rome.

Kearey, P., and Vine, F. J. (1990). Global tectonics. 302 pp. Oxford: Blackwell Scientific Publications.

Keeler-Wolf, T., Elam, D. R., and Flint, S. A. (1995). California vernal pool assessment, preliminary report, State of California, The Resource Agency, Department of Fish and Game. Sacramento, California.

Keeley, J. E., and Zedler, P. H. (1998), Characterization and global distribution of vernal pools. In Witham, C. W., Bauder, E. T., Belk, D., Ferren, W. R., Jr., and Ornduff, R. (Eds.). Ecology, conservation and management of vernal pool ecosystems (pp. 1-14).

Proceedings from a 1996 Conference. Sacramento, California. California Native Plant Society.

Kerrick, D. (2002). Serpentinite seduction. Science, 298, 1344-1345.

Khantush, M. (1964). New of flow theory. Questions of hydrogeological calculations. Moscow: Mir, . 43-61.

Kinniburgh, D. G., & Smedley, P. L. (eds.) (2001). Arsenic concentrations of groundwater in Bangladesh. British Geological Survey Technical Report WC/00/19, (1).

Kling, G.W., & Kusakabe M. (1990).Conclusions from Lake Nyos disaster. Nature, (348), 6398.

Koshkina, N.A., Guerasimov, A.N., & Belyakov, V.D. (1996). Distribution of acute intestinal morbidity in the Russian Federation territory. Regional problems and Management of population health of Russia. Moscow, 201-209.

Kouzmina, J. V., & Treshkin, S. E. (2001). The Modern Status Flora and Vegetation of the "Baday-Tugay" reservation area. Botanical journal, 86, 1, 73-84.

Kouzmina, J. V. , & Treshkin, S. Y. (2003). Evalution of the effect exerted by South-Karakalpakian main collector on Baday-Tugai Nature reserve. Arid ecosystems, 9, 19-20, 93-105.

Kovalevsky, V.S. (1994). Effect of changes in hydrogeological conditions on the environment. Moscow: Nauka.

Kozlov, A.N., Parmuzin, S.Y., Pustovoit, G.P., Khrenov, N.N. (2001). Heat-related gas pipeline - permafrost interactions at the Yamburg gas-condense field. Proceedings of the 2nd Conference of Geocryologists of Russia, Moscow, 3-8 June 2001, 4 (pp.114-120). Moscow: Moscow State University.

Kozlovsky, E., (ed) (1988). Landslides and mudflows. Moscow.

Krassovsky, G. N., Avaliani, S.P, & Zholdakova, Z.I. (1992). System of criteria for complex chemical risk assessment. Gigiena i sanitaria, 9-10, 15-17.

Krassovsky, G. N., Elpiner, L.I., & Beim, A.M. (1982). Principles of ecological hygienic standardization of water bodies' quality. Water resources, 4, 3-19.

Kroop R.H. & Nocido M.E. (1988). Water Supply Critical areas. Water World Dev.: Proc. 6th IWRA World Congr. Water Resour., (pp. 34-44). Ottawa.

Kruckeberg, A. R. (1984). California serpentines: flora, vegetation, geology, soils and management problems. 180 pp. Berkeley: University of California Press.

Kruckeberg, A. R. (1992). Plant life of western North American ultramafics. In Roberts, B. A., and Proctor, J. (Eds.). The ecology of areas with serpentinized rocks: A world view, (pp. 31-73). Dordrecht: Kluwer Academic Publishers.

Kruckeberg, A. R. (2002). Geology and plant life: the effects of landforms and rock types on plants. 362 pp. Seattle: University of Washington Press.

Kuehn, F. (2000) Remote Sensing for Site Characterization, Springer-Verlag, Berlin

Kuhn, G. G., Johnson, J. A., and Shlemon, R. J. (1995.). Paleo-liquefaction evidence for pre-historic earthquakes in north-central San Diego County, California. American Geophysical Union, Abstracts with Programs, (7), 362, Session S12C, San Francisco, California.

Kukkula, M., Arstila, P., Klossner, M.L., Maunula, L., Bonsdorff, C.H., & Jaatinen, P. (1997). Waterborn outbreak of viral gastroenteritis. Scandinavian Journal of Infectious Diseases, 29(4), 415-418.

Kuz'mina, Zh., V., & Treshkin, S., E. (1997). Soil Salinization and Dynamics of Tugai Vegetation in the Southeastern Caspian Sea Region and in the Aral Sea Coastal Region. Eurasian Soil Science, 30, 6, 642-649.

Kuz'mina, Zh., V., & Treshkin, S., E. (2004). Ecological consequences of construction of South-Karakalpakian main collector. Problems of desert development, 1, 13-16.

Kyuntsel, V.V. (1980). The regularities of the landslide process and its regional prediction on the European territory of the USSR. Moscow: Nedra.

Lacey, R.F., & Shaper, A.G. (1984). Changes in water hardness and cardiovascular death rates. International journal of epidemiology, 13(1), 18-24.

Låg, J. (1990). Geomedicine. CRC Press.

Lampe, R. (1996). Shoreline changes in the bodden coast of northeastern Germany. In K.F. Nordstrom & C.T. Roman (editors), Estuarine Shores: Evolution, Environments and Human Alterations (pp. 63-88). Chichester: John Wiley & Sons Ltd.

Langston W.J., Bebianno, M.J. & Burt, G.R. (1998). Metal handling strategies in molluscs. In W.J. Langston & M.J. Bebianno (editors), Metal metabolism in Aquatic Environments (pp. 219-283). Norwell, MA: Kluwer Academic Publishers.

Langston, W. J. & Spence, S. K. (1995). Biological factors involved in metal concentrations observed in aquatic organisms. In A. Tessier & D.R. Turner, (editors), Metal Speciation and Bioavailability in Aquatic Systems, (pp. 407-478). Chichester: John Wiley & Sons Ltd.

Langston, W. J. (1990). Toxic effects of metals and the incidence of metal pollution in marine ecosystems. In R.W. Furness & P.S. Rainbow (editors), Heavy metals in the marine environment (pp. 101-122). Boca Raton: CRC Press.

Langston, W. J., Bryan, G. W., Burt, G. R. & Gibbs, P. E. (1990). Assessing the impact of tin and TBT in estuaries and coastal regions. Functional Ecology, 4, 433-443.

Langston, W.J. & Pope N.D. (1995). Determinants of TBT adsorption and desorption in estuarine sediments. Marine Pollution Bulletin, 31, 32-43.

Langston, W.J., Chesman, B.S., Burt, G.R., Hawkins, S.J., Readman J., & Worsfold P. (2003a). Characterisation of the South West European Marine Sites: The Severn Estuary pSAC, SPA. Marine Biological Association of the UK, Occasional Publication No.13 (pp 206). Plymouth.

Langston, W.J., Chesman, B.S., Burt, G.R., Hawkins, S.J., Readman J., & Worsfold P. (2003b). Characterisation of the South West European Marine Sites: The Fal and Helford cSAC. Marine Biological Association of the UK, Occasional Publication No. 8 (pp. 160). Plymouth.

Lee, E M., Doornkamp, J C., Brunsden, D. & Noton, N H. (1991). Ground Movement in Ventnor, Isle of Wight. Geomorphological Services Ltd (Newport Pagnell)

Lee, H. & Swartz, R.C. (1980). Biological processes affecting the distribution of pollutants in marine sediments. In R.A. Baker (editor), Contaminants and Sediments, Part II. Biodeposition and bioturbation (pp. 555-606). Michigan: Ann Arbor.

Legett, R F. (1973). Cities and Geology. New York: McGraw Hill

Leshchikov, F.N. (1996). Peculiar features of cryogenic processes development in the natural-technogenic systems of the Baikal area. Proceedings of the 1st Conference of Geocryologists of Russia, Moscow, 3-5 June 1996, 1 (pp.389-398). Moscow: Moscow State University.

Lillesand, T.M., & Kiefer, R.W. (2000) Remote Sensing and Image Interpretation, New York: John Wiley&Sons Inc.

Lizuma, L. (2000). An analysis of a long-term meteorological data series in Riga. Living with diversity in Latvia. Folia Geographica, vol. VIII, Riga, 53–60.

Luoma, S. N. & Bryan, G. W. (1982). A statistical study of environmental factors controlling concentrations of heavy metals in the burrowing bivalve *Scrobicularia plana* and the polychaete *Nereis diversicolor*. Estuarine, Coastal and Shelf Science, 15, 95-108.

Luoma, S. N. & Jenne, E. A. (1976). Factors affecting the availability of sediment-bound cadmium to the estuarine deposit-feeding clam, *Macoma balthica*. In: C. E. Cuming, Jr.

(editor), Radioecology and energy resources, Proceedings of the 4th National Symposium on radioecology, Oregon State University, May 12-14 1976. The Ecological Society of America, Special Publication No. 1. (pp238 – 290). Cornwallis, Oregon, ,

Luoma, S. N. & Jenne, E. A. (1977). The availability of sediment-bound cobalt, silver and zinc to a deposit-feeding clam. Paper presented at the 15th Life Sciences Symposium, Biological Implications of Metals in the Environment, Sept. 29 - Oct 1, 1977. (30 pp.) Hanford, Washington.

Luoma, S. N., Johns, C., Fisher, N. S., Steinberg, N. A., Oremland, R. S. & Reinfelder, J. R. (1992). Determination of selenium bioavailability to a benthic bivalve from particulate and solute pathways. Environmental Science and Technology, 26 , 485-491.

Luoma, S.N., Bryan, G.W. & Langston, W.J. (1982). Scavenging of heavy metals from particulates by brown seaweed. Marine Pollution Bulletin, 13, 394-396.

Maksimova, L.N. (2001). Regional features of the influence of short-term climate fluctuations on the dynamics of geocryological conditions. In L.S.Garagulya and E.D.Ershov (Ed.), Dynamic Geocryology (pp.362-372). Moscow: Moscow State University.

Mapping Systems (1994) General Reference, Trimble Navigation

Marchenko, N.G. (2001). Results of the active layer monitoring in the Northern Tien-Shan Mountains. Abstracts of the International Conference "Conservation and Transformation of Matter and Energy in the Earth Cryosphere", Pushchino, 1-5 June, 2001 (pp.125-126). Pushchino: Russian Academy of Sciences.

Marcinkevicius, V., & Buceviciute, S. (1986). Geological and hydrogeological conditions of sulfate karst development in North Lithuania (in Russian). Proc. of Lithuanian higher schools. Geologija 7: 104–119.

Marcinkevicius, V., (1998). Sulphate karst of Devonian Suoas formation in North Lithuania (in Lithuanian). Proc. of Lithuanian higher schools. Geologija 24: 65–72.

Marker, B R. & McCall, (1990). Applied Earth Science Mapping: the Planners' Requirement. Engineering Geology 29, 403-11

Marker, B R. (1996). Urban development – Identifying Opportunities and Dealing with Problems. In: McCall et al [Eds] 181-214

Marker, B R. (1998). Incorporating Information on Geohazards into the Planning Process. In: Maund J·C & Eddleston M [Eds] Geohazards in Engineering Geology. Geological Society Lond. Special Pub. 15, 385-9

Marker, B R; Pereira, J J. & de Mulder, E F J. (2003). Integrating geological information into urban planning and management: approaches for the 21st century. In: Heiken, G; Fakundiny, R & Sutter, J Earth science in the city: a reader. American Geophysical Union (Washington DC) 379-411

Masironi, R., Pisa, Z., & Clayton, D. (1980). Myocardial infarction and water hardness in European towns. Journal of Environmental Pathology and Toxicology, 4(2-3), 77-87.

Mather, P.M. (1999) Computer Processing of Remotely-Sensed Images, New York: John Wiley&Sons

Mavljanov, T.E., Pinchasov, B.I.., Oteev, R., & Kurbanijazov, A.K. (1998).Sources of salt-dust transport at the dry Aral Sea bottom // Problems of desert development. № 3-4.

McCall, G J H. & Marker, B R. [Eds] (1989). Earth Science Mapping for Planning, Development and Conservation. London: Graham and Trotman

McCall, G J H., de Mulder, E F J. & Marker, B R. (1996). Urban Geoscience. AGID Special Publication Series 20. A A Rotterdam: Balkema

McEwen, T., & Äikäs, T. (2000). The site selection process for a spent fuel repository in Finland - summary report. Posiva Oy, report POSIVA 2000-15.

McLachlan, D.R., Bergeron, C., Smith, J.E., Boomer, D., & Rifat, S.L. (1996). Risk for neuropathologically confirmed Alzheimer's disease and residual aluminum in municipal drinking water employing weighted residental histories. Neurology, 46(2), 401-405.

Melloul A., & Collin M. (1994). Inst. J.Earth Sei., 43(2), 105-116.

Miller, W., Alexander, R., Chapman, N., McKinley, I., & Smellie, J. (2000). Geological Disposal of Radioactive Wastes and Natural Analogues. Amsterdam, etc.: Elsevier (Pergamon).

Millward, R.D. & Grant, A. (2000). Pollution induced tolerance to copper of nematode communities in the severely contaminated Restronguet Creek and adjacent estuaries, Cornwall, United Kingdom. Environmental Toxicology and Chemistry, 19, 454-461.

Milnes, A.G. (1985). Geology and Radwaste. London: Academic Press.

Milnes, A.G. (2002). Swedish deep repository siting programme. Guide to the documentation of 25 years of geoscientific research (1976-2000). Swedish Nuclear Fuel and Waste Management Co., Technical Report SKB TR-02-18.

Minchin, D., Oehlmann, J., Duggan, C. B., Stroben, E., & Keatinge, M. (1995). Marine TBT antifouling contamination in Ireland, following legislation in 1987. Marine Pollution Bulletin, 30, 633-639.

Ministry for Public Health of RSFSR. (1989). Assessment of hygienic effectiveness of water protective measures. Methodical guidelines. Moscow.

Mishra S.K. & Singh R.P. (1993). Prediction of subsidence in the Indo-Gangetic basin carried by groundwater withdrawal. Eng. Geol., 33(3), 227-239.

Mitch, W. J., and Gosselink, J. G. (2000).Wetlands. 920 pp. New York: John Wiley & Sons, Inc.

Mogi, K. (1985). Earthquake Prediction. Tokyo: Academic Press.

Moore, A.C., Herwaldt, B.L., Craun, G.F., Calderon, R.L., Highsmith, A.K., & Juranek, D.D. (1993). Surveillance for waterborne disease outbreaks – United States, 1991-1992. MMWR CDC Surveillance Summaries, 42(5), 1-22.

Moore, J., Smith, J., Northen, K. & Little, M. (1998). Inlets in the Bristol Channel and approaches: area summaries. Marine Nature Conservation Review Sector 9. 137pp. Peterborough: Joint Nature Conservation Committee

Moore, J.J., Smith, J. & Northern, K.O. (1999). Marine Nature Conservation Review. Sector 8. Inlets in the western English Channel. Area summaries. 171pp. Peterborough : Joint Nature Conservation Committee

Morner, N. A. (1979). The Fennoscandian uplift and late Cenozoic geodynamics: geological evidence. GeoJournal, 3(3) 287-318.

Moseley, H. (1998). Archaeology and Development. In: Corfield M, Hinton P, Nixon T & Pollard M [Eds] (1998) Preserving Archaeological Remains in Situ. Proc. Conference 1-3 April 1996. Museum of London Archaeological Service (London) 47-50

Moskalenko, N.G. (1996). Effects of gas pipeline laying on geosystems of the cryolithozone of West Siberia. Proceedings of the 1st Conference of Geocryologists of Russia, Moscow, 3-5 June 1996, 2 (pp.399-407). Moscow: Moscow State University.

Mühll, D.V. (2001). Permafrost Monitoring in the High Mountains of Europe: the PACE and PERMOS Project in its Global Context. Abstracts of the 1st European Permafrost Conference, Rome, 26-28th March 2001 (pp.11-12). Rome.

Mühll, D.V., Stucki, T., Haeberli, W. (1998). Borehole temperatures in Alpine Permafrost: a Ten Year Series. PERMAFROST – Seventh International Conference (Proceedings), Yellowknife (Canada), 1998. Journal Collection Nordicana, 55, 1089-1095.

Mumford, L. (1961). The City in History. New Jersey: Harcourt Brace

Murphy, D. D., and Weiss, S. B. (1988). Ecological studies and the conservation of the Bay Checkerspot Butterfly, *Euphydryas esitha bayensis.* Biological Conservation 46, 183-200.

Murphy, D. D., Freas, K. E., and Weiss, S. B. (1990). An environmental-metapopulation approach to population viability analysis for a threatened invertebrate. Conservation Biology, 4, 41-51.

Najem, G.R., Strunck, T., & Feuerman, M. (1994). Health effects of a Superfund hazardous chemical waste disposal site. American Journal of Preventive Medicine, 10(3), 151-155.

Nathanail, P. & Symonds, A. [Eds] (2001). Geographical Information Systems. In: Griffiths, J S [Ed] Land Surface Evaluation for Engineering Practice. Geol Soc Engineer Gp Spec Pubn 18, 57-58

Novikova N.M., Kuzmina, J.V., Dikareva, T.V., & Trofimova, G.Yu. (2001). Preservation of the tugai bio-complex diversity within the Amudarya and Syrdarya river deltas in conditions of aridization. In D. Keyser (Ed.) Ecological research and monitoring of the Aral sea deltas (pp. 155-188). UNESCO. Aral sea project 1997-2000. Final scientific reports. Paris: UNESCO.

Novikova, N. M., Kust, G. S., Kuzmina, J. V., Trofimova, T. U., Dikariova, T. V., Avetian, S. A., et al. (1998). Contemporary plant and soil cover changes in the Amu-Dar'ya and Syr-Dar'ya river deltas. In S. Bruk, D. Keyser, J. Kutscher & V. Moustafaev (Ed.), Ecological research and monitoring of the Aral sea deltas (pp. 55-80). UNESCO: Paris.

Novikova, N. M., Kuz'mina, J. V., Dikareva, T. V., & Trofimova, T. U. (2001). Preservation of the tugai biocomplex diversity within the Amu-Darya and Syr-Darya river deltas in aridization conditions In S. Bruk, D. Keyser, J. Kutscher & V. Moustafaev (Ed.), Ecological research and monitoring of the Aral sea deltas, Boock 2 (pp. 155-188). UNESCO: Paris.

Novikova, N.M. (1985). Dynamics of vegetation of arid deltas under antropogenic river runoff transformation. In D.D. Vyshivkin (Ed.) Biogeographic aspects of desertification (pp. 36-40). M.:MFGO.

Novikova, N.M. (1999). Priaralye and creeping environmental changes in the Aral Sea. Creeping environmental Problems and sustainable development in the Aral Sea basin. Cambridge University Press, 100-127.

Novikova,, N.M., Kust, G.S., Kuzmina, J.V., Dikareva, T.V., & Trofimova, G.Yu. (1998). Contemporary plant and soil cover changes in the Amudarya and Syrdarya deltas. In D. Keyser (Ed.) Ecological research and monitoring of the Aral sea deltas (pp. 55-80). UNESCO Aral sea project 1992-1996. Final scientific reports. Paris: UNESCO.

Ob'edkova, G. Yu. (1983). Study of the influence of highly mineralized water on reproductive function of female body and substantiation of measures aimed at prevention of its adverse impact. Abstract of MD degree work, Saratov.

Oberman, N.G. & Kakunov, N.B. (2002). Ground water monitoring in the cryolithozone of Northeast European Russia and the Ural region. In V.T.Balobaev, V.V.Shepelev (Ed.), Monitoring in the Cryolithozone. (pp.18-43). Yakutsk: Institute for Geocryology.

Oberman, N.G. & Mazhitova, G.G. (2001). Permafrost dynamics in the north-east of European Russia at the end of the 20th century. Norwegian Journal of Geography, 55(4), 241-244.

Oberman, N.G. & Yudina, E.A. (2000). Features of manifestation of intra-age rhythm of the characteristics of an active layer and the layer of annual temperature fluctuations in main periglacial landscapes of the European Northeast of Russia. Abstracts of the International Conference "Rhythms of natural processes in the earth cryosphere". Pushchino, 12-15 May 2000, (pp.244-245). Pushchino: Russian Academy of Sciences.

Oberman, N.G. (1989a). Some aspects of the interrelation between ground and river waters in the aufeis belt of the Urals. In B.M.Piguzova (Ed.), Permafrost-Hydrogeological Studies of the Free Water-Exchange Zone (pp.64-73). Moscow: Nauka.

Oberman, N.G. (1989b). Processes and phenomena accompanying aufeis formation in the Northern Urals. In E.V.Seversky, I.V.Klimovsky (Ed.). Geocryological Studies in the USSR Mountains (pp.67-75). Yakutsk, Permafrost Institute SD USSR AS.

Oberman, N.G. (1996). Geocryological specificity and current trends in natural and anthropogenic dynamics of the cryolithozone of the East-European Subarctic. Proceedings of the 1st Conference of Geocryologists of Russia. Moscow, 3-5 June 1996, 2 (pp.408-417). Moscow: Moscow State University.

Oberman, N.G. (1999). Some Line Seasonal and Perennial Dynamic of Permafrost of Petchora-Urals Region and Its Consequences. Abstracts of the International Conference "Monitoring of Cryosphere", Pushchino, 20-23 April 1999 (pp.108-109). Pushchino: Russian Academy of Sciences.

Oberman, N.G. (2002). Activization of Cryogenic Geological Processes in Zone of Influence of the Northern Railway Road. In O.N.Gryaznov (Ed.), Man-caused Transformation of Geological Environment.- Materials of International Scientific - Practical Conference. Russia, Ekaterinburg, December 17-19, 2002 (pp.182-184). Ekaterinburg: AMB Publishing House.

Oberman, N.G. and Mazhitova, G.G. (2001). Permafrost dynamics in the north-east of European Russia at the end of the 20th century. Norwegian Journal of Geography, 55, 241-244.

On the State of Water Supply to the Population in Russia and Measures to Improve the Quality of Drinking Water. (1996). Proceedings of Interdepartmental Commission on Ecological Safety (Sept. 1994 - Oct. 1995). Russia's Ecological Safety, 9, 178-190.

Onischenko, G.G., Novikov, S.M., Rakhmanin, Yu.A., Avaliani, S.P., & Bushtueva, K.A. (2002). Basics of human health risk assessment under the impact of hazardous chemical substances. Moscow: NII ECH I GOS.

Onishchenko, G.G. (2002). Assessment of the risk of the influence of environmental factors on health in the sociohygienic monitoring system. Gigiena & Sanitariya, 6, 3-5.

Orem, W. H., Feder, G. L., & Finkelman, R. B., (1999). A possible link between Balkan endemic nephropathy and the leaching of toxic organic compounds from Pliocene lignite by groundwater: preliminary investigation. Int. Jour. of Coal Geol., 40 (2-3), 237-252.

Osipov, V.I. (ed) (1999). Hazardous Exogenic Processes. Moscow: GEOS.

Osterkamp, T.E. and Lachenbruch, A.H. (1990). Current Thermal regime of permafrost in Alaska and the predicted global warming. International Symposium on Geocryological studies in Arctic Regions, Yamburg, USSR, August 1989 (pp.40-45). Tumen: Russian Academy of Sciences.

Osterkamp, T.E., Zhang, T., Romanovsky, V.E. (1994). Evidence for a cyclic variation of permafrost temperatures in Northern Alaska. Journal of Permafrost and Periglacial Processes, 5, 137-144.

Oxley, J. (1998). Planning and the Conservation of Archaeological Deposits. In: Corfield M, Hinton P, Nixon T & Pollard M [Eds] 1998 Preserving Archaeological Remains in Situ. Proc. Conference 1-3 April 1996. Museum of London Archaeological Service (London) 51-54

Paar, G., Bauer, A. and Kaufmann, V. (2001). Rock Glacier Monitoring using Terrestrial Laser Scanning. Abstracts of the 1st European Permafrost Conference, Rome, 26-28th March 2001 (pp.58-59). Rome.

Pankova, E. I., Kouzmina, J. W., & Treshkin, S. Y. (1994). The Influence of flooding area on soil-vegetation cover of the South Gobi oasis. Water resources, 21, 3, 358-364.

Parmuzin, S.Yu and Shatalova, T.Yu. (2001). Dynamics of seasonally thawed and seasonally frozen ground layers in response to short-term climate fluctuations. In L.S.Garagulya and E.D.Ershov (Ed.). Dynamic Geocryology (pp.284-302). Moscow: Moscow State University.

Parulekar, A. M., Ansari, Z. A. & Ingole, B. S. (1986). Effect of mining activities on the clam fisheries and bottom fauna of Goa estuaries. Proceedings of the Indian Academy of Sciences - Animal Sciences, 95, 325-339.

Paskauskas, R., Antanyniene, A., Budriene, S., Krevs, A., Kucinskiene, A., Slapkauskaite, G., Sulijiene, R. (1998). The diversity and peculiarity of biological processes in the karst Kirkilu lake (in Lithuanian). Annual Proc. of Geography 31: 62–70.

Paukstys, B., Cooper, A.H., & Arustiene, J. (1997). Planning for gypsum geohazards in Lithuania and England. The Engineering Geology and Hydrogeology of Karst Terranes, Beck and Stephenson (eds), Balkema, Rotterdam, 127–135.

Paul, T., Chow, F. & Kjeksted, O. (2002). Hidden Aspects of Urban Planning – Surface and Underground Development. London: Thomas Telford

Pavlov, A., Anan'eva, G., Drozdov, D., Moskalenko, N., Dubrovin, V., Kakunov, N. et al. (2002). Monitoring of active layer and the temperature of frozen grounds in the North of Russia. Journal of Earth Cryosphere, VI, 4, 30-39.

Pavlov, A.V. (1994). Current changes of climate and permafrost in the Arctic and Sub-Arctic of Russia. Journal of Permafrost and Periglacial Processes, 5, 101-110.

Pavlov, A.V. (1996). Current condition of geocryological monitoring in Russia and some problems of its development. Proceedings of the 1st Conference of Geocryologists of Russia, Moscow, 3-5 June 1996, 3 (pp.327-336). Moscow: Moscow State University.

Pavlov, A.V. (1997). Permafrost-climate monitoring in Russia: methodology, observation results, forecast. Journal of Earth Cryosphere, 1, #1, 47-58.

Pavlov, A.V. (2001). Major points of the concept of a geocryological information system. Proceedings of the 2nd Conference of Geocryologists of Russia, Moscow, 3-8 June 2001, 3 (pp.198-204). Moscow: Moscow State University.

Pavlovsky, Ye. N. (1964). Natural centers of inoculable diseases with regard to landscape epidemiology of zooanthroposes. Moscow-Leningrad.

Pederson, J. L., Mackley, R. D., and Eddleman, J. L. (2002), Colorado Plateau uplift and erosion evaluated using GIS. GSA Today, 4-10, August 2002, Boulder, Colorado.

Perkins, E J. (1974). The Biology of Estuaries and Coastal Waters. London: Academic Press.

PFINDER (1992). Software User's Guide, Trimble Navigation

Pickett, S. T. A., and Cadenasso, M. L. (2002). The ecosystem as a multimensional concept: Meaning, model, and metaphor. Ecosystems, 5, 1-10.

Pletnev, A.A. (1972). Mathematical analysis methods in studying the groundwater regime and determination of aquifer parameters. Moscow: The Dissertation Abstract.

Plitman, S.I. (1988). Evaluation of Hygienic Efficiency of Water Protection Measures. Methodological Recommendations.

Plumlee, G., S., & Ziegler, T., L. (2004). The Medical Geochemistry of Dusts, Soils and other Earth Materials. Treatise on Geochemistry, (9), Elsevier.

Popkov, O.N. (1996). Development of the natural-industrial system of an oil field in the cryolithozone. Proceedings of the 1st Conference of Geocryologists of Russia, Moscow, 3-5 June 1996, 3 (pp.269-278). Moscow: Moscow State University.

Popov, V.A. (1990). Problem of Aral Sea and landscapes of delta of Amudarya river. Tashkent: FAN.

380

Porfiriev, B.N. (1990). Ecological examination and risk of technologies. Results of a science and techniques, 27, Moscow: VINITI.

Post-Conflict Environmental Assessment – Afghanistan. (2003). UNEP, ISBN: 928072309X.

Pratt, M. (1993). Remedial Processes for Contaminated Land Institute of Chemical Engineers (Rugby)

Profiles of Disasters in the World Summary of Statistics by Continent. (1994). Brussels: CRED Bulletin.

Prognosis and counteraction of the landslide results, (2000) Conference Materials, Kraków 7th September 2000, PGI Warszawa

Punsar, S., & Karvonen, M.J. (1979). Drinking water quality and sudden death: observations from West and East Finland. Cardiology; 64(1), 24-34.

Pyrkin, V.I. (1977). Temporal methodical recommendations on a computerized groundwater data processing system. Moscow: VSEGINGEO.

Rachold, V., Are, F.E., Grigoriev, M.N., Hubberten, H.W., Rasimov, S. and Schneider, W. Coastal erosion of ice-rich, permafrost-dominated coastlines in the Laptev Sea Region. Abstracts of the 1st European Permafrost Conference, Rome, 26-28th March 2001 (pp.111-112). Rome.

Raficov, A.A., Tetuchin, G.F. (1981). Change of the Aral Sea level and transformation of environment within downflow of Amudarya. Tashkent: FAN.

Ragozin A. L. (2001). Natural risk assessment and management. Russian journal of Geoecology, 2, 183-187.

Rakhmanin, Yu.A., Novikov, S.M., & Rumyantsev, G.I. (2002). Assessment of the human health risk upon acute and chronic exposures to environment-polluting chemicals. Gigiena & Sanitariya, 6, 5-7.

Rasomavicius, V. (red.) (2001). Europinės svarbos buveinės Lietuvoje. Lietuvoje aptinkamų Europos Sąjungai svarbių buveinių tipų aiškinamasis vadovas. Vilnius.

Renfro, W. (1973). Transfer of ^{65}Zn from sediments by marine polychaete worms. Marine Biology, 21, 305-316.

Revis, N.W., Major, T.C., & Norton, C.V. (1980). The effect of calcium, magnesium, lead, or cadmium on lipoprotein metabolism and atherosclerosis. Journal of Environmental Pathology and Toxicology, 4 (2-3), 293-304.

Rhoads, D. C. (1974). Organism-sediment relations on the muddy sea floor. In H. Barnes (editor), Oceanography and Marine Biology. Annual Review. 12, 263-300.

Riazanov, A.V. (2001). Peculiar features of the geocryological conditions at the Trans-Polar GNKM under its development and the possible global climatic warming. Proceedings of the 2nd Conference of Geocryologists of Russia, Moscow, 3-8 June 2001, 2 (pp.269-274). Moscow: Moscow State University.

Ridgway, J. & Shimmield, G. (2002). Estuaries as repositories of historical contamination and their impact on shelf seas. Estuarine, Coastal and Shelf Science, 55, 903-928.

Ridgway, J., Breward, N., Langston, W.J., Lister, R. Rees, J. G. & Rowlatt, S. M. (2003). Distinguishing between natural and anthropogenic sources of metals entering the Irish Sea. Applied Geochemistry, 18, 283-309.

Riefner, R. E., Jr., and Pryor, D. R. (1996). New locations and interpretation of vernal pools in southern California. Phytologia, 80(4), 296-327.

Roberts, P. & Sykes, H. (2000). Urban Regeneration: a Handbook. London: Sage Publications

Robertson, G D. & Speiker, A M. (1978). Nature to be commanded: earth-science maps applied to land and water management. USGS Professional Paper 950 (Washington DC)

Rodriguez-Iturbe, I. (2000), Ecohydrology: A hydrologic perspective of climate- soil-vegetation dynamics. Water Resources Research, 6, 3-9.

Roman, C. T. & Nordstrom, K. F. (1996). Environments, processes and interactions of estuarine shores. In: K.F. Nordstrom & C.T. Roman. (editors), Estuarine Shores: Evolution, Environments and Human Alterations (pp. 1-12). Chichester: John Wiley & Sons Ltd.

Romanovsky, V.E., Osterkamp, T.E. (1999). Permafrost monitoring system in Alaska: structure and results. Abstracts of the International Conference "Monitoring of Cryosphere", Pushchino, 20-23 April 1999 (pp.112-113). Pushchino: Russian Academy of Sciences.

Romanovsky, V.E., Osterkamp, T.E. (2001). The system of permafrost monitoring in Alaska (structure and results). Journal of Earth Cryosphere, V, 4, 59-68.

Rose, J.B. (1997). Environmental ecology of Criptosporidium and public health implications. Annual Review of Public Health, 18, 135-161.

Rostron, D. (1985). Surveys of harbours, rias and estuaries in southern Britain:. Falmouth. Volume 1. A report to the NCC by the Field Studies Council Oil Pollution Research Unit, Orielton Field Centre, Pembroke, Dyfed. FSC/OPRU/49/85, 109pp. Peterborough: Nature Conservancy Council.

Rowden, A. A. & Jones, M. B. (1995). The burrow structure of the mud shrimp *Callianassa subterranea* (Decapoda: Thalassinidea) from the North Sea. Journal of Natural History, 29(5), 1155-1165.

Rowley, D. B. (2002). Rate of plate creation and destruction: 180 MA to present. Geological Society of America Bulletin, 114(8), 927-933.

Rubenowitz, E., Axelsson, G., & Rylander, R. (1996). Magnesium in drinking water and death from acute myocardial infarction. American Journal of Epidemiology, 143(5), 456-62.

Rygg, B. (1985). Effect of sediment copper on benthic fauna. Marine Ecology Progress Series, 25, 83-89.

Sabins, F.Jr. (1986) Remote Sensing, Principles and Interpretation, New York: W.H. Freeman and Company

Safford, H. D. (2002). Geoecology: historical roots and contemporary practice. Geological Society of America, 2002 Annual Meeting, October 27-30, Abstract of oral presentation, session 240. Boulder, Colorado.

Sahagian, D., Proussevitch, A., and Carlson, W. (2002). Timing of Colorado Plateau uplift: Initial constraints from vesicular basalt-derived paleoelevations. Geology, 30(9), 807-810.

Sakamoto, N., Shimizu, M., Wakabayashi, I., & Sakamoto, K. (1997). Relationship between mortality rate of stomach cancer and cerebrovascular disease and concentrations of magnesium and calcium in well water in Hyogo prefecture. Magnesium Research, 10(3), 215-223.

Sashourin, A.D. (1996). Formation of centres of technogeneous catastrophes in areas of intense mineral mining. Mining in the Arctic. Proceedings international symposium. Swalbord. Norway: Trondheim, 201-206.

Satkunas, J., & Graniczny, M. (1997). Lithuanian Polish cross-border geoenvironmental mapping and its relevancy to spatial development. Engineering Geology and Environment (Marinos, Koukis, Tsiambaos and Stoumaras eds.), Balkema, 1483- 1486

Schneider, B. (2001). Fluctuations of air temperature as a reason for short-term velocity changes at the rock glacier Äuberes Hochebenkar (Ötztal Alps, Tyrol) Abstracts of the 1st European Permafrost Conference, Rome, 26-28th March 2001 (pp.62-63). Rome.

Schoenherr, A. A. (1992). A natural history of California. California Natural History Guide 56, 772 pp. Berkeley and Los Angeles, University of California Press.

Scholz, C. H., & Contreras, J. C. (1998). Mechanics of continental rift architecture. Geology, 26(11), 967-970.

Schumm, S. A. (1986). Alluvial river response to active tectonics. Chapter 5, 80-94. Washington, DC: International Academies Press.

Schumm, S. A., Dumont, J. F., and Holbrook, J. M. (2002). Active tectonics and alluvial rivers. 276 pp. New York: Cambridge University Press.

Scientific council on the problems of biosphere. (1990). Guidelines for forecasting medical biological consequences of hydrotechnical construction. Ed. by Elpiner, L.I., Be'er, S.A. Moscow: the USSR Academy of sciences.

Selinus, O., Alloway, B., Centeno, J. A., Finkelman, R. B., Fuge, R., Lindh, U., & Smedley, P. (Editors) (2005). Essentials of Medical Geology. Amsterdam: Elsevier

Selinus, O., Frank, A., (1999). Medical geology. In Möller, L., (Editor): Environmental medicine. Stockholm: Joint Industrial Safety Council, 164-183.

Shapovalov, A.Ye. (2002). Medical geographical approach in the methodology of polluted drinking water risk assessment. (Smolensk oblast as an example). Melioratsia I selskoe khozyaystvo, 2, 18-20.

Sharhuu, N. (1998). The Dynamics of Permafrost Development in the Nalaikh and Baganur coal fields (Mongolia). Abstracts of the Conference "Problems of Earth cryology", Pushchino, 20-24 April 1998 (pp.94-95). Pushchino: Russian Academy of Sciences.

Sheko A.I., & Krupoderov V.S. (1994). Assessment of hazard and risk of exogenic geological processes. Moscow: Journ. Geoecology, 3, 11-21.

Sheko, A.I. (1980). The regularities of mudflow formation and prediction. Moscow: Nedra.

Sheko, A.I., & Grechishchev, S.E. (1988). The technique of studying and prediction of exogenic geological processes. Moscow: Nedra.

Sheko, A.I., & Krupuderov, V.S. (1984). The methods of long-term regional predictions of exogenic geological processes. Moscow: Nedra.

Shepelev, V.V., Boitsov, A.V., Oberman, N.G., Petshenko, M.F., Anisimova, N.P., Kakunov, N.B. et al. (2002). Monitoring of groundwater in cryolithozone. Yakutsk: Permafrost Institute Press.

Shlemon, R. J., Kuhn, G. G., Boka, B., and Riefner, R. E., Jr. (1997). Origin of a Mima-mound field, San Clemente State Park, Orange County, California: A test of the seismic hypothesis. Association of Engineering Geologists, Abstracts with Programs, 40th Annual Meeting, pp. 148-149, Portland, Oregon.

Sholtz, C. (1997). Whatever happened to Earthquake Prediction? Geotimes, (3).

Sigurdsson M., et al. (1987). J.Volcan.Geotherm. (31),1.

Simonds, J O. (1978). Earthscape - a Manual of Environmental Planning and Design. New York: Van Nostrand Reinhold

Simpson, T G. (1996). Urban Soils. In: McCall, G J H et al [Editors] 35-60

Site Investigation Steering Group (1993). Site investigation in construction: 1: Without site investigation, ground is a hazard. London: Thomas Telford

Skryabin, P.N., Varlamov, S.P., Skachkov, Y.B., Shender, N.I., Tetelbaum, A.S., Murzen, Y.A. (1996). Effects of technogenic disturbances on the ground temperature regime in Central Yakutia. Proceedings of the 1st Conference of Geocryologists of Russia, Moscow, 3-5 June 1996, 1 (362-371). Moscow: Moscow State University.

Smedley, P., & Kinniburgh, D.G. (2005). In: Selinus, O., Alloway, B., Centeno, J. A., Finkelman, R. B., Fuge, R., Lindh, U., and Smedley, P. (Editors), Essentials of Medical Geology. Amsterdam: Elsevier.

Smith, A. & Ellison, R A. (1999). Applied Geology Maps for Planning and Development - a Review of Examples from England and Wales 1983-1996. Quarterly Journal of Engineering Geology 32 (supplement) S1-S44

Smith, S.L., Burgess, M.M., Nixon, F.M., Taylor, A.E. (2001). Canadian High Arctic and Western Arctic permafrost observatories. Abstracts of the 1st European Permafrost Conference, Rome, 26-28th March 2001 (pp.19-20). Rome.

Society and Culture of Southeast Asia. Continuities and Changes. Int. Academy of Indian Culture and Aditya Prakashan, (2000). New Delhi.

Soil Survey Staff. (1999). Soil taxonomy: U.S. Department of Agriculture, Natural Resources Conservation Service, Agricultural Handbook 436, Second Edition, 869 pp. Washington, DC: Government Printing Office.

Sonneborn, M., & Mandelkow, J. (1981). German studies on health effects of inorganic drinking water constituents. Science of the Total Environment, 18, 47-60.

Spink A.E.F., & Welson E.E.M. (1990). Groundwater resource management in coastal aquifers. IAHS Publ., 173, 475-484.

Standard methods for the examination of water and wastewater (1989). 17th edition, USA.

Standardised methods for the examination of quality of surficial water and wastewater (1994). Vilnius

Stein, S., and Freymueller, J. T. (Eds.). (2002). Plate boundary zones. American Geophysical Union Geodynamic Series 30, 425 pp. Washington, DC.

Stern, B.R., & Tardiff, R.G (1997). Risk characterization of methyl tertiary butyl ether (MTBE) in tap water. Risk Analysis, 17(6), 727-43.

Sukhodrovsky, V.L. (1979). The exogenic relief-formation in the cryolithozone. Moscow: Science.

Taminskas, J., & Marcinkevicius, V. (2002). Karst geoindicators of environmental change: The case of Lithuania. Environmental geology, 42(7), 757–766.

Taminskas, (1994). Hydrological regime and its impact on the development of karst processess (in Lithuanian). Annual Proc. of Geography 28: 152–177.

Tatu, C. A., Orem, W. H., Finkelman, R. B., & Feder, G. L. (1998). The etiology of Balkan Endemic Nephropathy: still more questions than answers. Environmental Health Perspectives. 106(11), 689-700.

Tchounwou P.B., Patlolla A.K., & Centeno J.A. (2003). Carcinogenic and systemic health effects associated with arsenic exposure – a critical review. Toxicologic Pathology, 31, 575-588.

Tessier, A., Couillard, Y., Campbell, P. G. C. & Auclair, J. C. (1993). Modelling Cd partitioning in oxic lake sediments and Cd concentrations in the freshwater bivalve Anodonta grandis. Limnology and Oceanography, 38(1), 1-17.

Testa Stephen M. (1991). Elevation changes associated with groundwater withdrawal and reinjection in the Wilmington area, Los Angeles coastal plain, California. IAHS Publ., 200, 485-502.

Theis Ch. (1935). The relation between rivers and lowering of the piezometric groundwater surface. Tran.Amer.Geophys.Union.

Thomas, K. V., McHugh, M, Hilton, M, & Waldock, M. (2003). Increased persistence of antifouling paint biocides when associated with paint particles. Environmental Pollution, 123, 153-161.

Thompson, A,, Hine, P D., Greig, J R. & Peach, D W. (1996). Assessment of gypsum subsidence arising from dissolution (with special reference to Ripon, North Yorkshire) Symonds Travers Morgan (East Grinstead)

384

Thompson, A., Hine, P D., Poole, J. & Greig, J R. (1998). Environmental Geology in Land Use Planning - a Guide to Good Practices Symonds Group (East Grinstead)

Thompson, G.R., & Turk, J. (1998). Introduction to physical geology, Saunders College Publishing, Fort Worth, Philadelphia.

Thornton, I. (1993). Environmental geochemistry and health in the 1990s: a global perspective. Applied Geochem., Suppl. Issue, (2), 203-210.

Thornton, I. (Editor) (1983). Applied environmental geochemistry. Academic Press.

Tohyama, E. (1996). Relationship between fluoride concentration in drinking water and mortality rate from uterine cancer in Okinawa prefecture, Japan. Journal of Epidemiology and Community Health, 6(4), 184-191.

Tools for assessing rapid environmental changes. The 1995 geoindicator checklist (1996). International Institute for Aerospace Survey and Earth Sciences (ITC), Publication Number 46, Enschede.

Treshkin, S. E. (2001). Transformation of tugai ecosystems in the floodplain of the lower reaches and delta of the amu-dar'ya and their protection. In S. Bruk, D. Keyser, J. Kutscher & V. Moustafaev (Ed.), Ecological research and monitoring of the Aral sea deltas, Boock 2 (pp. 189-201). UNESCO: Paris.

Tripe, J.P., Doerfliger, N., Zwahlen, F., Delporte, C. (2000). Vulnerability mapping in karst areas and its uses in Switzerland. Acta carsologica, 29/1: 163–171.

Trofimov, V.T., & Ziling, D.G. (2000). Theoretical and methodological basis of ecological geology. St. Petersburg: St. Petersburg university.

Trofimov, V.T., & Ziling, D.G. (2002). Ecological geology. Moscow: Geoinformmark.

Trofimov, V.T., Shvetc, V.M. (2000). Improvement of the system of ecological education. In: "Ecology of Russia", (1) European part. Moscow: Geoinformmark.

Trofimov, V.T., Ziling, D.G., Baraboshkina, T.A., & Kharkina, M.A. (2002). Ecological geological maps. St. Petersburg: St. Petersburg university.

Tucker, P., Ferguson, C. & Tzilivakis, J. (1998). Integration of Environmental Assessment Indicators into Site Assessment Procedures In: Lerner D N and Walton N R G Contaminated Land and Groundwater - Future Directions Geol. Soc. Lond. Eng Group Spec Pub 14, 127-133

Turin, A.I., Romanovsky, N.N., Poltev, N.F. (1982). Permafrost-Facial Analysis of Stone Fields. Moscow: Nauka.

UNESCO (1997) Geology for Sustainable Development Bulletin II Urban Geology Royal Museum for Central Africa. Belgium: Tervuren

Unrug, R. (1997). Rodinia to Gondwana: the geodynamic map of Gondwana supercontinent. GSA Today, 7, 1-6.

Vartanyan, G.S. (1979). A technique of earthquake prediction. – A Russian Patent.

Vartanyan, G.S. (1979). A technique of earthquake prediction. – A Russian Patent.

Vartanyan, G.S. (1990). Geodynamic catastrophes and socio-behavial syndrome. Moscow: Otechestvennaya Geologia (National Geology) Journal, (3), 90.

Vartanyan, G.S. (1993). Geodynamic catastrophes and sociobehavial syndrome. Moscow: Otechestvennaya Geologia, 6, 90-92.

Vartanyan, G.S. (1994). Hydrogeodeformation Field of the Earth and some Problems of Ecogeology. Journal of Russian Mineral Resources.. Economy and Management, 6,16-20.

Vartanyan, G.S. (1995). Hydrogeodeformation field in studying the mechanisms of geodynamics. Otechestvennaya Geologia, (4), 29-37.

Vartanyan, G.S. (1997). Geodynamic catastrophes and relationships to changes in societal behavior. COGEOENVIRONMENTAL News. Sciences, (11), 15-18.

Vartanyan, G.S. (1997). Geodynamic catastrophes and relationships to changes in societal behaviour. Cogeoenvironment News, 11.

Vartanyan, G.S. (1997). Geodynamic catastrophes and relationships to changes in societal behaviour. Cogeoenvironment News, 11.

Vartanyan, G.S. (1999). Regional geodynamic monitoring system in the problem of sustainable development of the countries in the earthquake-prone provinces of the world. Otechestvennaya Geologia, (2), 37- 45.

Vartanyan, G.S. (1999). Regional geodynamic monitoring system in the problem of sustainable development of the countries in the earthquake-prone provinces of the world. Otechestvennaya Geologia, (2), 37- 45.

Vartanyan, G.S., Bredehoeft, J.D., & Roelloffs, E. (1992). Hydrogeological methods for studying tectonic stresses. Sovetskaya Geologia, (9), 3-12.

Vartanyan, G.S., & Kulikov, G.V. (1982). The Hydrogeodeformation Field of the Earth. Reports of the Academy of Sciences of the USSR, (2), 310-314.

Vartanyan, G.S., Goncharov, V.S., Krivosheev, V.P., et al. (2000). Methodical guidelines on hydrogeodeformation monitoring aimed at seismic prediction. Moscow: Geoinformmark.

Vitinsky, Yu.I. (1973).Cyclicity and prediction of solar activity. Leningrad: Nauka.

Walsh, A. R. & O'Halloran, J. (1997). The accumulation of chromium by mussels *Mytilus edulis* (L.) as a function of valency, solubility and ligation. Marine Environmental Research, 43, 41-54.

Walton, G. & Lee, M K. (2001). Geology for our Diverse Economy. Keyworth: British Geological Survey

Wang, B., French, H.M. (1994). Climate controls and high-altitude permafrost, Qinghai-Xizang (Tibet) Plateau, China. Journal of Permafrost and Periglacial Processes, 5, 87-100.

Wang, Binbin, Finkelman, Robert B., Belkin, Harvey E., & Palmer, Curtis A. (2004). A possible health benefit of coal combustion. Abstracts of the 21st Annual Meeting of the Society for Organic Petrology, 21, 196-198.

Ward, T. J., Warren, L. J. & Tiller, K. G. (1984). The distribution and effects of metals in the marine environment near a lead-zinc smelter, South Australia. In: J. O. Nriagu (editor), Environmental Impacts of Smelters (pp. 1-73). New York: John Wiley & Sons.

Warwick, R. M., George, C. L., Pope, N.D. & Rowden, A.A (1989). The prediction of post-barrage densities of shorebirds. Volume 3. Invertebrates (112pp). Harwell, UK: Department of Energy.

Warwick, R.M., Langston, W.J, Somerfield, P.J., Harris, J.R.W., Pope, N.D., Burt, G.R., & Chesman, B.S. (1998). Wheal Jane Minewater Project, Final Biological Assessment Report for the Environment Agency, 169pp. Bristol, UK: Environment Agency

Waters, C N., Northmore, K., Prince, G. & Marker, B R. (1996). Geological background to planning and development in the City of Bradford Metropolitan District. BGS Technical Report WA/96/1 Vol. 2 A technical guide to ground conditions. Keyworth: British Geological Survey

Way, S.W. (1978) Terrain Analysis, A Guide to Site Selection Using Aerial Photographic Interpretation, New York: McGraw-Hill Book Company

Weinstein, P,, & Cook, A. (2005). Volcanic emissions and health. In: Selinus, O., Alloway, B., Centeno, J. A., Finkelman, R. B., Fuge, R., Lindh, U., and Smedley, P. (Editors), Essentials of Medical Geology. Amsterdam: Elsevier.

White, R. (2003). Early hominids—diversity or distortion?. Science, 299(5615), 1994-1997.

Woldai, T., & Fabbri A. (1998) The impact of mining on the environment, Deposit and Geoenvironmental Models for Resource Exploitation and Environmental Security, September 6-18 1998, Hungary: Matrahaza

Wolf, A. (2001). Conservation of endemic plants in serpentine landscapes. Biological Conservation, 100, 35-44.

Woodruffe, C. D. (1996). Late Quaternary infill of macrotidal estuaries in northern Australia. In: K.F. Nordstrom & C.T. Roman. (editors), Estuarine Shores: Evolution, Environments and Human Alterations (pp. 89-114). Chichester: John Wiley & Sons Ltd.

World Conference on Natural Disasters Mitigation, Iokogama (1994).

World Health Organization. (1993). Guidelines for drinking-water quality. (2nd ed.). Geneva.

Wright, D. A. & Zamuda, C. D. (1987). Copper accumulation by two bivalves: salinity effect is independent of cupric ion activity. Marine Environmental Research, 23, 1-14.

Yakovlev, S.V., ,.Elpiner, L.I., Nechaev, A.P. Mazaev, V.T., Myasnikova, Ye.V, Shlepnina, T.G., Kocharyan, A.G., et al. (1998). Ecological engineering problems of drinking water supply. Engineering ecology, 1, 2-18.

Yakubovskaya, E.L. (1998). Semipalatinsk nuclear polygon, 50 years. Novosibirsk.

Yamamoto Soki, (1986). Groundwater resources in Japan with special reference to its use and conservation. IAHS Publ., 151, 381-389.

Yang, C., & Hung, C. (1998). Colon cancer mortality and total hardness levels in Taiwan's drinking water. Archives of Environment Contamination and Toxicology, 35(1), 148-51.

Yang, C.Y., Cheng, M.F., Tsai, S.S., & Hsieh, Y.L. (1998). Calcium, magnesium, and nitrate in drinking water and gastric cancer mortality. Japanese Journal of Cancer Research, 89(2), 124-130.

Yang, C.Y., Chiu, J.F., Chiu, H.F., Wang, T.N., Lee, C.H., & Ko, Y.C. (1996). Relationship between water hardness and coronary mortality in Taiwan. Journal of Toxicology & Environmental Health, 49(1), 1-9.

Yeats, R. S., Sieh, K., and Allen, C. R. (1997). The geology of earthquakes. 568 pp. New York: Oxford University Press.

Ying, W., Batley, G. E. & Ashanullah, M. (1992). The ability of sediment extractants to measure the bioavailability of metals to three marine invertebrates. The Science of the Total Environment, 125, 67-84.

Yongjian, D. (1998). Recent degradation of permafrost in China and the response to climating warming. PERMAFROST – Seventh International Conference (Proceedings), Yellowknife (Canada), 1998. Journal Collection Nordicana, 55, 225-230.

Zaletaev, V.S., Novikova, N.M. (1997). Phytoecological mapping for dynamics of desertification within arid regions (At the sample of South Priaralye). In Z. Karamysheva (Ed.) Geobotanical mapping – 1996 (43-52). St.Pb.: Nauka.

Zedler, P. H. (1987). The ecology of southern California vernal pools: A community profile. U.S. Fish and Wildlife Service Biology Report 85(7.11), Washington, DC.

Zektser I.S., & Rogachevskaya, L.M. (2002). Groundwater Resources and Regional Environmental Radionuclide Contamination. Proceedings of International Nuclear Atlantic Conference Proceedings, Rio de Janeiro.

Zektser, I. S. (2000). Groundwater and the environment. Boca Raton: Lewis Publishers.

Zemla, B. (1980). (Geography of the incidence of stomach cancer in relation to hardness of drinking water and water supply). Wiadomosci Lekarskie, 33(13), 1027-1031.

Zeyegofer, Yu., Vilkovich, R. V., & Daniluchev, N. V. (1998). Federal ecological program, Volga River in Moscow region. Moscow.

Zeyen, H., Volker, F., Wehrle, V., Fuchs, K, Sobolev, S. V., Altherr, R. et al. (1997). Styles of continental rifting; crust-mantle detachment and mantle plumes. Structure and dynamic processes in the lithosphere of the Afro-Arabian Rift system. Tectonophysics, 278(1-4), 329-352.

Zhang, T., Barry, R.G., Brown, J. (2001). Global geocryological database: a long-term goal for the International Permafrost Association. Abstracts of the International Conference "Conservation and Transformation of Matter and Energy in the Earth Cryosphere", Pushchino, 1-5 June, 2001 (pp.15-16). Pushchino: Russian Academy of Sciences.

Zhorov, A.A. (1995). Groundwater and environment. Moscow.

Zielhuis, R.L., & Haring, B.J. (1981). Water hardness and mortality in Netherlands. Science of the Total Environment, 18, 35-45.

Zonnenshain, L. P., Kuzmin, M. I., and Natapov. L. M. (1990). Geology of the USSR: A plate-tectonic synthesis. American Geophysical Union Geodynamics Series 21, 240 pp. Washington, DC.

Zschau J. (1996) SEISMOLAR   Ein Schritt in Richtung   Erdbeben vohesage. Geowissenschaften 14..

Zvikas, A., Paskauskas, R., Kucinskiene, A., (2002). Static evaluation of structure and functioning of hydromicrobes in karst lakes. Ekologija, 1: 40–47.

INDEX